Stress Ribbon and Cable-supported Pedestrian Bridges
Second edition

Stress Ribbon and Cable-supported Pedestrian Bridges

Second edition

Jiri Strasky

Published by ICE Publishing, 40 Marsh Wall, London E14 9TP.

Distributors for ICE Publishing books are
USA: Publishers Storage and Shipping Corp., 46 Development Road, Fitchburg, MA 01420

First published 2005
Second edition published 2011

Also available from ICE Publishing

Bridge design, construction and maintenance
Institution of Civil Engineers. ISBN 978-0-7277-3593-5
Current and Future Trends in Bridge Design, Construction and Maintenance
Institution of Civil Engineers. ISBN 978-0-7277-3475-4
ICE Manual of Bridge Engineering, Second edition
Edited by G. Parke and N. Hewson. ISBN 978-0-7277-3452-5
Concrete bridge strengthening and repair
I. Kennedy-Reid. ISBN 978-0-7277-3603-1

www.icevirtuallibrary.com

A catalogue record for this book is available from the British Library

ISBN 978-0-7277-4146-2

© Thomas Telford Limited 2011

ICE Publishing is a division of Thomas Telford Ltd, a wholly-owned subsidiary of the Institution of Civil Engineers (ICE).

All rights, including translation, reserved. Except as permitted by the Copyright, Designs and Patents Act 1988, no part of this publication may be reproduced, stored in a retrieval system or transmitted in any form or by any means, electronic, mechanical, photocopying or otherwise, without the prior written permission of the Publisher, ICE Publishing, 40 Marsh Wall, London E14 9TP.

This book is published on the understanding that the author is solely responsible for the statements made and opinions expressed in it and that its publication does not necessarily imply that such statements and/or opinions are or reflect the views or opinions of the publishers. Whilst every effort has been made to ensure that the statements made and the opinions expressed in this publication provide a safe and accurate guide, no liability or responsibility can be accepted in this respect by the author or publishers.

Typeset by Academic + Technical, Bristol
Printed and bound in Great Britain by CPI Antony Rowe Limited, Chippenham and Eastbourne

Contents

Foreword vii

01	**Introduction**	**1**
	1.1. Eurocode	11
02	**Structural systems and members**	**13**
	2.1. Structural systems	13
	2.2. Structural members	23
03	**Design criteria**	**31**
	3.1. Geometric conditions	31
	3.2. Loads	33
	3.3. Dynamics	34
04	**Cable analysis**	**41**
	4.1. Single cable	41
	4.2. Bending of the cable	47
	4.3. Natural modes and frequencies	52
05	**Effects of prestressing**	**55**
06	**Creep and shrinkage of concrete**	**63**
	6.1. Time-dependent analysis	63
	6.2. Redistribution of stresses between members of different age	66
	6.3. Redistribution of stresses in structures with changing static systems	69
07	**Stress ribbon structures**	**71**
	7.1. Structural arrangement	71
	7.2. Prestressed band	72
	7.3. Piers and abutments	79
	7.4. Transferring stress ribbon force to the soil	81
	7.5. Erection of the deck	86
	7.6. Static and dynamic analysis	90
	7.7. Special arrangements	101
	7.8. Stress ribbon supported by arch	106
	7.9. Structures stiffened by external tendons	113
	7.10. Static and dynamic loading tests	119
08	**Suspension structures**	**127**
	8.1. Structural arrangement	127
	8.2. Erection of the structures	134
	8.3. Static and dynamic analysis	141
09	**Cable-stayed structures**	**155**
	9.1. Structural arrangement	155
	9.2. Erection of the structures	159
	9.3. Static and dynamic analysis	164
10	**Curved structures**	**171**
	10.1. Suspension of the deck on both edges	171
	10.2. Suspension of the deck on an outer edge	172
	10.3. Suspension of the deck on an inner edge	173
	10.4. Curved stress ribbon bridge	175
11	**Examples**	**179**
	11.1. Stress ribbon structures	179
	11.2. Suspension structures	213
	11.3. Cable-stayed structures	240
	Index	**257**

Foreword

I started my professional career in 1969, one year after Russian troops occupied Czechoslovakia and ruined our hopes of a free and creative life. Some of my colleagues decided to emigrate, while many others became resigned and passive. I was lucky to have collaborated with engineers who tried to prove that, even under difficult circumstances, we were still able to develop and build progressive bridge structures. We were supported by some professors who mostly had to leave the universities because of their opinions and who contributed to our group of similarly thinking people.

In the firm Dopravni stavby Olomouc we have developed our own prestressing system and our own segmental technology for constructing highway overpasses and viaducts. We have also built cable-stayed bridges and – more importantly – designed stress ribbon and suspension pedestrian bridges. These structures correspond to my design philosophy: they are light and transparent, their structural systems express the function of bridging of the site, they can be built without any influence on the terrain under the bridge and their dimensions have a human scale. Although they were developed from one of the oldest structural systems, they require a deep understanding of the structural systems, the function of prestressing and structural material.

We have not only built these structures, but we have also verified their function using detailed static and dynamic tests. We have also published the structural solutions and results of the tests in outstanding technical magazines and have achieved international awards. As a result of our publications, I was invited to contribute to the design of the first US stress ribbon bridge that was built across the Sacramento River in Redding, California in 1987.

Following the 'Velvet Revolution' in 1989, a lot of things have changed. I have started to work in the USA, my colleagues and I have launched our own design firm and I have been teaching bridges at the Technical University of Brno. For eight years I was a member of the presidium of the FIP (Fédération Internationale de la Precontrainte) that has been changed to the *fib* (Fédération Internationale du Béton – the International Federation for Structural Concrete). I also had the opportunity to participate in the design of large-scale projects in California and Japan.

I am extremely grateful of the fact that my design work has allowed me to become acquainted with many engineers who are not only outstanding experts, but also extremely good and cultural people. Due to my design work, I am now a member of a family of professionals who combine technical knowledge with aesthetic feeling and who create bridges from perfect structures. We build clean, economical and timeless structures while protecting our environment.

I am grateful to ICE Publishing for the opportunity to publish my experience in design and construction via *Stress Ribbon and Cable-supported Structures*. First of all I would like to express my thanks to my partner Ilja Husty for assisting with the design of our first structures, my younger colleagues for the effort to further develop them and to Professor Miros Pirner and Dr Marie Studnickova for assistance with their dynamic analysis and tests.

I would also like to thank my colleagues in the USA and UK, in particular Charles Redfield, Bill Hall, Gary Rayor, Joe Tognoli and Cezary Bednarski, for the opportunity to participate in their projects. Thanks are also due to Dr Akio Kasuga for the opportunity to visit Japanese structures and providing information on them.

In the book I present results of parametric studies and tests performed by my former and present PhD students: Dr Radim Necas, Dr Tomas Kulhavy, Dr Robert Broz, Tomas Romportl, Richard Novak, Ivan Belda, Michaela Hrncirova, Petr Kocourek, Michal Jurik and Jan Kolacek. It was a pleasure to work with them all. I also thank my colleague Dr Tomas Kompfner for checking the text and Jarek Baron for preparing images of our new structures. Most of all, I would like to thank my wife Nada Straska for drawing all the figures and for her support and understanding of my work.

I was delighted by the request of ICE Publishing to prepare a new edition of the book, in which I have included the latest developments. It is highly encouraging to know that there are engineers who find my book interesting and useful for their work. It is also heartening to know that there are still designers who favour bridges whose beauty is expressed in pure structural function and structural efficiency.

Jiri Strasky

Stress Ribbon and Cable-supported Pedestrian Bridges
ISBN 978-0-7277-4146-2

ICE Publishing: All rights reserved
doi: 10.1680/srcspb.41462.001

Chapter 1
Introduction

What would be the best bridge? Well, the one which could be reduced to a thread, a line, without anything left over; which fulfilled strictly its function of uniting two separated distances.
                                                        Pablo Picasso

Cable-supported bridges have been designed from the beginning of human history. Their fascinating development corresponds to the development of civil engineering (Brown, 1996; Gimsing, 1998; Leonhardt, 1984; Troyano, 2003; Wittfoht, 1972). The beauty of these structures comes from their clear static function that determines their architectural expression.

The architecture of bridges is not some kind of treatment added to or performed on the structural design of a bridge. The architecture should emerge from and is given by the bridge's basic function. The function of a bridge is to cross a particular space in order that some load or traffic may be conveyed over a natural obstacle or another man-made thoroughfare; the form of the bridge must express this basic function. The best structural solution should be some form inherent in the site which best fulfils the function of bridging the site. The task of the structural designer is to discover that form which can be realised in a way that is economical and efficient (Strasky, 1994).

Pedestrian bridges should be light and transparent. Of course, a bridge structure must be safe, should invite use, be comfortable for the user and should be designed and constructed to human scale. Vibrations of the deck under excitation of pedestrians walking or from wind must not produce feelings of discomfort in the users.

Architects and engineers generally agree that the whole structure and structural members forming the bridge should express – by their shape – the flow of internal forces through the structural system, which is integrated into the social surrounding, historical/time, technological and physical environments (Pearce and Jobson, 2002).

The structures described in this book are dedicated to helping us to fulfil those environmental, cultural and social objectives within the constraints of available economic resources. These structures use as a main load-bearing member a hanging cable, whose shape (funicular to resisting load) is given by its pure static function.

The decks of the cable-supported structures are formed of steel and concrete. In many designs, the stiffness of the structure is given by the stiffness of the space net of the cables and/or from the bending and torsion stiffness of the deck. In several recent designs, the stiffness of the structures comes from the tension or compression (normal) stiffness of the stress ribbon deck.

Stress ribbon bridge is the term used to describe structures formed by a directly walked prestressed concrete deck in the shape of a catenary. The conception was first introduced by Ulrich Finsterwalder, who repeatedly proposed such a structure for bridging large spans. Among the bridges suggested were those over Bosporus (Figure 1.1), Lake Geneva and Zoo in Köln (Finsterwalder, 1973).

The bearing structure consists of slightly sagging tensioned cables, bedded in a concrete slab that is very thin compared to the span. This slab serves as a deck, but apart from distributing the load locally and preserving the continuity it has no other function. It is a kind of suspension structure where the cables are tended so tightly that the traffic can be placed directly on the concrete slab embedding the cables. Compared with other structural types, the structure is extremely simple. On the other hand, the force in the cables is very large, making the anchoring of the cable very expensive.

The stress ribbon structures combine a structural form of primitive bridges formed by ropes from liana or bamboo (Figures 1.2 and 1.3) with a structural arrangement of prestressed suspended roofs (Leonhardt, 1964; Lin and Burns, 1981). For illustration, two architecturally successful roofs are presented in Figures 1.4 and 1.5.

The first roof was built for the Dulles International Airport terminal in Washington, DC, US in 1962 (Figure 1.4). The

Figure 1.1 Bosporus Bridge (built by U. Finsterwalder in 1958)

Figure 1.2 Liana Bridge

Figure 1.3 Mejorada Bridge across the Pari River, Peru (print of L. Angrand, 1838)

Figure 1.4 Dulles International Airport terminal in Washington, DC, USA

Figure 1.5 Portuguese National Pavilion for EXPO '98 in Lisbon, Portugal (courtesy of Segadaes Tavares & Partners)

roof of the span of 51.50 m is assembled of precast members suspended and post-tensioned by tendons anchored in cast-in-place curved-edge beams. The tension force from the roof is resisted by nicely shaped inclined columns passing through openings of the edge beams. The roof was designed by Aero Saarinen and Ammann & Whitney Engineers. When the terminal was extended in 1995, the same structure was constructed beside the original one.

The second roof was built for the Portuguese National Pavilion for the Lisbon World Exposition (EXPO '98) in Portugal (Figure 1.5). The roof forms a singly curved shell forming a canopy that covers an area of 65×50 m, and is formed of lightweight concrete which was cast in place. It is supported by bearing tendons anchored in an adjacent reinforced concrete structure and is prestressed by prestressing tendons anchored in the shell. There is a gap between the anchor slab and the shell. The roof was designed by Alvaro Siza and Segadaes Tavares & Partners (STA).

The technology used in the construction of precast singly curved shells was also adopted in the design of double-curved shells. Figure 1.6 shows a 'Saddledome' structure built in 1983 for the 1984 Calgary Winter Olympic Games (Bobrowski, 1984).

The roof structure, which was created from the projection of a hyperbolic paraboloid from a sphere of radius 67.5 m, is assembled from precast members supported and post-tensioned by a space net of prestressing tendons.

The failure of part of the Berlin Congress Hall suspended roof built in 1957 reminded engineers of the importance of the analysis of the local bending and execution of structural details of the stress ribbon at supports (Schlaich et al., 1980). The stability of the outer inclined arches of the Berlin roof was guaranteed by tension members formed by stress ribbon strips anchored at the central ring (Figure 1.7). The collapse of the roof in 1980 was caused by the corrosion of prestressing steel at anchor members. A new structure of a similar shape but of a different structural solution was built in 1986 again from prestressed concrete (Figure 1.8).

The characteristic feature of stress ribbon structures is a variable slope, which disqualifies the use of this structural type for highway bridges. It is difficult to imagine that the structure presented in Figure 1.1 would be acceptable for highway agencies. A stress ribbon structure might only represent the correct solution in special cases when the highway is straight and the elevation is a concave (sag) curve.

Figure 1.6 Saddledome, Calgary, Canada

On the other hand, the variable slope is acceptable or even advantageous for pedestrian bridges built in an environment of no straight lines.

The first stress ribbon bridge for public use was built according to the design of Professor Walther across the N3 motorway near Pfäfikon, Switzerland in 1965 (see Section 11.1; Walther, 1969).

From that first success, stress ribbon bridges were built in many countries all over the world.

Stress ribbon structures can be designed with one or more spans and are characterised by successive and complementary smooth curves. The curves blend into the rural environment and their forms, the most simple and basic of structural solutions

Figure 1.7 Berlin Congress Hall: structural arrangement of the original structure

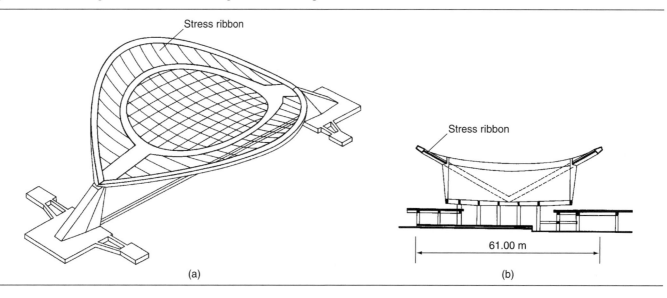

Figure 1.8 Berlin Congress Hall: new structure

(Figures 1.9 and 1.10), clearly articulate the flow of internal forces. Their fine dimensions also correspond to a human scale.

From Figure 1.11, it is evident that the stress ribbon structure represents the simplest structural form. Both the engineering and aesthetic beauty of this type of bridge lies in the fact that the suspended walkway itself is the structure. It carries itself without the need for any arms, props, masts, cables or dampers. Its stiffness and stability result from its geometry.

Such structures can be either cast in situ or formed from precast units. In the case of precast structures, the deck is assembled from precast segments that are suspended on bearing cables and shifted along them to their final position (Figure 1.12).

Figure 1.9 Grants Pass Bridge, Oregon, USA

Figure 1.10 Redding Bridge, California, USA

Figure 1.11 Redding Bridge: structural types, California, USA

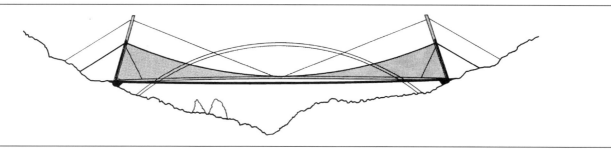

Prestressing is applied after casting the joints between the segments to ensure sufficient rigidity of the structures.

The main advantage of these structures is that they have minimal environmental impact because they use very little material and can be erected independently of the existing terrain. Since they do not need bearings or expansion joints, they only need minimal long-term maintenance.

Figure 1.12 Redding Bridge, California, USA: erection of a segment

They are able to resist not only uniformly distributed loads but also large concentrated loads created by the wheels of heavy trucks. Extremely large floods that occurred in the Czech Republic in summers 1997 and 2002 also confirmed that they are able to resist a large ultimate load. Although two bridges were totally flooded, their static function remains without any changes.

Even though stress ribbon structures have low natural frequencies, our experience confirmed that the speed of motion of the deck created by walking is within acceptable limits (Figure 1.13). Our detailed dynamic test also confirmed that vandals cannot damage these structures.

Over the course of time, people have realised that the deck can be suspend on the cables of larger sag; in this way it is possible to reduce tension in the cables (Figure 1.14). The first suspension structures had a flexible deck that was, in some cases, stiffened by a net of additional cables. J. Finley first identified the major components of suspension bridges: a system comprising main cables from which a deck was hung braced by trusses (Figure 2.12). The concept behind this system was

Figure 1.13 Brno-Komin Bridge, Czech Republic

Figure 1.14 Bridge in Nepal

developed in Britain. For main cables, suspension chains of circular eyebars were used (Figure 1.15). The French engineers substituted the chains with wire cables (Figure 1.16). J. Roebling developed cable spinning and started the development of construction of modern suspension structures from Brooklyn to Golden Gate Bridge (Figures 1.17 and 1.18).

The problem of vibration and even overturning of light pedestrian suspension structures due to wind load has been well documented (Podolny and Scalzi, 1976). However, the failure of the Tacoma Bridge (Washington, DC) has drawn the attention of engineers to the aerodynamic stability of suspension structures (Scott, 2001). New suspension structures therefore have decks formed of an open stiffening steel truss of sufficient torsional and bending stiffness (Endo *et al.*, 1994; Figure 1.19(a)) or use a streamlined steel box girder (Leonhardt and Zellner, 1980; Figure 1.19(b)).

To guarantee the aerodynamic stability of pedestrian bridges, solutions used in the design of highway bridges can be adopted. However, these solutions can appear to be too heavy, inappropriate and expensive for pedestrian bridges.

Another approach to providing stiffness is to construct a slender concrete deck and stiffen the structure by a system of inclined suspenders (Figure 1.19(c)). This approach has been successfully developed and erected in designs by Professor J. Schlaich. However, the maintenance of so many suspenders is not an easy task.

Slenderising the deck could be combined with techniques often used to stabilise utility bridges (Figure 1.19(d)). In such cases, the concrete deck is stiffened by the post-tensioning of external cables with an opposite curvature to that of the main suspension cables. Similar effects can be achieved by eliminating the longitudinal movement of the deck.

Figure 1.15 Menai Straits Bridge, Wales, UK

Figure 1.16 Fribourg Bridge, Switzerland

The combination of the stiffening of the deck by external cables situated within the deck and elimination of the horizontal movement was used in a design of the Vranov Lake Bridge built in 1993 in the Czech Republic (Figure 1.20). The deck of the span $l = 252\,\text{m}$ is assembled of precast concrete segments of depth of only $d = 0.40\,\text{m}$ (Figure 1.21). The bridge deck with the ratio $d/l = 1/630$, which is one of the most slender ever built, was erected without any effect on the environment (Figure 1.22).

From Figure 1.23, which shows possible bridge options, it is evident that suspension structures have a minimum impact on the environment. A tied arch is too dominating of the lake and the pylons towered above the trees in the stay-cable solutions. The modest suspension bridge produced the right proportions for the setting (Figure 1.20) and also proved economical to construct.

The bridge deck can also be supported by suspension (Figure 1.24) or stay cables (Figure 1.25).

While suspension structures have been widely designed both in the nineteenth and twentieth century, the development of modern cable-supported structures started after the Second World War. Professor F. Dishinger first emphasised the importance of high initial stress in the stay cables and designed the first modern cable-stayed bridge built in Strömsund, Sweden in 1955. Since then, many cable-stayed structures with both

Figure 1.17 Brooklyn Bridge, New York, USA

Figure 1.18 Golden Gate Bridge, San Francisco, USA

Figure 1.19 Suspension structures

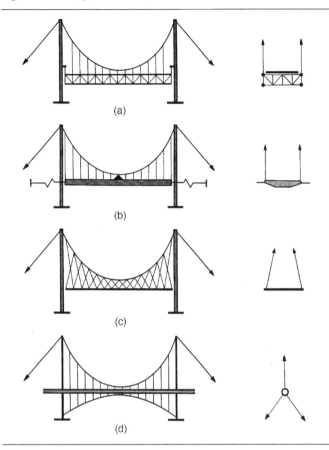

Figure 1.21 Vranov Lake Bridge, Czech Republic: deck

Figure 1.22 Vranov Lake Bridge, Czech Republic: erection of the deck

steel and concrete decks have been constructed (Gimsing, 1998; Leonhardt and Zellner, 1980; Liebenberg, 1992; Mathivat, 1983; Menn, 1990; Podolny and Muller, 1982; Wells, 2002). The development of the prestressed concrete technology has significantly contributed to the development of the cable-stayed structures.

Figure 1.20 Vranov Lake Bridge, Czech Republic

Figure 1.23 Vranov Lake Bridge, Czech Republic: structural types

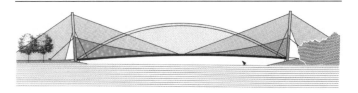

At present, a concrete bridge deck is widely used for the majority of bridges and the stay cables developed from the prestressing cables are used both for concrete and steel structures. Since the 1970s, the deck of many pedestrian bridges is formed by a slender concrete slab. Professor F. Leonhardt first used a slender concrete slab for a large-span cable-stayed pedestrian bridge (Völkel et al., 1977). Professor R. Walther proved that a deck of average depth of 450 mm can be safely used for bridges with spans up to 200 m (Figure 1.26; Walther et al., 1998).

Cable-supported pedestrian bridges can have timber, steel, concrete or composite decks and the stiffness of the structures can be given by the bending stiffness of the deck, space arrangement of cables or by tension stiffness of the prestressed concrete deck.

The concept of prestressing, a product of the twentieth century, gave designers the ability to control structural behaviour. At the same time, it enabled (or forced) them to think more deeply about the construction. While reinforced concrete combines concrete and steel bars by simply putting them together and letting them act together as required, prestressed concrete combines high-strength concrete with high-strength steel in an active manner. The prestressing allows us to balance the load, change boundary conditions and create supports within the structure. Prestressing is a radical step from passive reinforcement to creative thinking and development (Collins and Mitchell, 1987; Hampe, 1978; Leonhardt, 1964; Lin and Burns, 1981).

Figure 1.24 Design of the Almond River Bridge in Scotland (built by R. Stevenson, 1821)

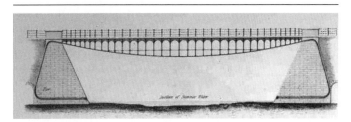

Figure 1.25 Varrugas Bridge, Peru

To focus on that fact and to narrow the scope of the book, cable-supported structures formed by prestressed stress ribbon decks are primarily described in the book.

Although the above structures have a very simple shape, their design is not straightforward. It requires a deep understanding of structural forms, function of structural details and the behaviour of prestressed concrete structures post-tensioned by internal and/or external tendons. The static and dynamic analysis also requires understanding of the function of cables and resolves various problems regarding both geometric and material non-linearity.

Figure 1.26 Model test of a cable-stayed structure with slender concrete deck (courtesy of R. Walther)

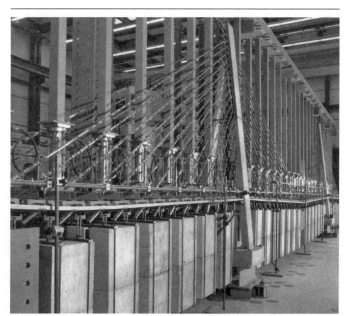

1.1. Eurocode

Since 1910, bridges designed in countries that are members of the European Union should comply with rules given by Eurocodes that are implemented in national standards. The philosophy of the new codes is explained in the *Designers' Guide to EN 1900 – Eurocode: Basis of Structural Design* (Gulvanessian *et al.*, 2002) and the problems of designing concrete bridges are explained in the *Designers' Guide to EN 1992-200 – Eurocode 2: Design of concrete structures. Part 2: Concrete Bridges* (Hendy and Smith, 2007).

The basis for verification of safety and serviceability of the structures is the partial factor method. Each structure has to be checked at the ultimate and serviceability limit state and must comply with the requirements for durability and detailing.

The initial aim of this book was to explain the problems of design and construction of cable-supported structures with stress ribbon decks, rather than to follow the standards and codes. For the second edition, however, the new Eurocodes will be implemented. The majority of bridges described in this book utilise the slender concrete deck. It is evident that the new structures have to comply with the standards explained in the above-cited literature. The structures should be designed on the basis of both the global and local analysis in order to cover possible geometric imperfections and redistribution of stresses due to cracks and shrinkage and creep of concrete.

While previous codes were concerned with average values of the load and material properties, new Eurocodes are based on their characteristic values. In the analysis of typical concrete structures it is possible to find an equilibrium of forces on un-deformed structures. This significantly simplifies analysis and allows us to use a superposition of static effects; it is therefore quite easy to use load factors to obtain specific load combinations.

The structures described in this book utilise the slender concrete deck, which exhibits large deformations. A non-linear analysis, in which an equilibrium of forces on deformed structures has to be found, is therefore mandatory. It is not possible to analyse these structures for different loading cases (such as dead load and live load) as their results are multiplied by different coefficients. It is necessary to prepare load combinations and analyse their structure. Since the deformations of the structure depend on the integral material characteristics, the calculations utilise the average property values.

The Eurocodes use the term *action* to mean a load and/or an imposed deformation. Actions are classified according to their variation in time as either a

- permanent action (G)
- variable action (Q), or
- accidental action (A).

A permanent action includes self-weight, additional dead load, prestressing force, effects of creep and shrinkage of concrete, and effects of settlement of supports. A variable action includes imposed load (live load), erection load, wind action and snow load, etc. To ensure this book is understandable to designers from other countries, well-known terms such as self-weight, dead load, live load and wind load, etc. are used.

REFERENCES

Bobrowski J (1984) *The Saddledome: the Olympic ice stadium in Calgary (Canada)*. L'Industria Italiana del Cemento 5/1984.

Brown DJ (1996) *Bridges. Three Thousand Years of Defying Nature*. Reed International Books Ltd., London.

Collins MP and Mitchell D (1987) *Prestressed Concrete Basics*. CPCI, Ottawa.

Endo T, Tada K and Ohashi H (1994) *Development of suspension bridges: Japanese experience with emphasis on the Akashi Kaikyo Bridge*. AIPC-FIP Int. Conf. on Cable-stayed and Suspension Bridges, Proceedings Vol. 1, Deauville, France.

Finsterwalder U (1973) *Festschrift: 50 Jahre für Dywidag*. Verlag G. Braun, Karlsruhe.

Gimsing NJ (1998) *Cable Supported Bridges: Concept & Design*. John Wiley & Sons, Chichester.

Gulvanessian H, Calgaro JA and Holicky M (2002) *Designers' Guide to EN 1900 – Eurocode: Basis of Structural Design*. Thomas Telford, London.

Hampe E (1978) *Spannbeton*. VEB Verlag für Bauwesen, Berlin.

Hendy CR and Smith DA (2007) *Designers' Guide to EN 1992-200 – Eurocode 2: Design of concrete structures. Part 2: Concrete Bridges*. Thomas Telford, London.

Leonhardt F (1964) *Prestressed Concrete: Design and Construction*. Wilhelm Ernst & Sons, Berlin.

Leonhardt F (1984) *Bridges: Aesthetics and Design*. Deutsche Verlags-Anstalt, Stuttgart.

Leonhardt F and Zellner W (1980) *Cable-stayed bridges*. IABSE Surveys, S-13/80.

Liebenberg AC (1992) *Concrete Bridges: Design and Construction*. John Wiley & Sons, New York.

Lin TY and Burns NH (1981) *Design of Prestressed Concrete Structures*. John Wiley & Sons, New York.

Mathivat J (1983) *The Cantilever Construction of Prestressed Concrete Bridges*. John Wiley & Sons, New York.

Menn C (1990) *Prestressed Concrete Bridges*. Birkhäuser Verlag, Basel.

Pearce M and Jobson R (2002) *Bridge Builders*. Wiley-Academy, John Wiley & Sons, Chichester, UK.

Podolny W Jr and Scalzi JB (1976) *Construction and Design of Cable Stayed Bridges*. John Wiley & Sons, New York.

Podolny W and Muller J (1982) *Construction and Design of Prestressed Concrete Bridges.* John Wiley & Sons, New York.

Scott R (2001) *In the Wake of Tacoma.* ASCE Press, Resno.

Schlaich J, Kordina K and Engell H (1980) Teileinsturz der Kongreßhalle Berlin: Schadensursachen Zusammenfassendes Gutachten. Beton-und Stahlbetonbau 1980, Heft 12, S. 281.

Strasky J (1994) Architecture of bridges as developed from the structural solution. *Proceedings of FIP '94: International Congress on Prestressed Concrete*, Washington, DC.

Völkel E, Zellner W and Dornecker A (1977) Die Schrägkabelbrücke für Fussgänger über den Neckar in Mannheim. *Beton- und Stahlbetonbau* 72, 29–35, 59–64.

Walther R (1969) Spannbandbrücken. *Schweizerische Bauzeitung* 87, S. 133–137.

Walther R, Houriet B, Walmar I and Moïa P (1998) *Cable Stayed Bridges.* Thomas Telford Publishing, London.

Wells M (2002) *30 Bridges.* Laurence King Publishing, London.

Wittfoht H (1972) *Triumph der Spannweiten.* Beton-Verlag GmbH, Düsseldorf.

Chapter 2
Structural systems and members

2.1. Structural systems

The beauty of the suspension and arch structures comes from their economic structural shape. Their economy is evident from Figure 2.1(a), which shows the trajectories of principal stresses in a uniformly loaded simply supported beam. From the figure, it is evident that the maximum stresses occur only at mid-span section and only in the top and bottom fibres. The beam has a significant amount of dead mass that does not contribute at all to resistance of the external loads.

From the figure, it is clear that if we want to reduce the weight of the beam we have to eliminate as much dead mass as possible and utilise the tension or compression capacities of the structural members. From the beam, we can derive a suspension cable or an arch in which the horizontal force is resisted by an internal strut or tie (Figure 2.1(b)). In the case of foundations, which are capable of resisting horizontal forces, we can substitute the strut or the tie by stiff footings (Figures 2.1(c), 2.2 and 2.3).

A uniformly loaded concrete arch can span several kilometres and a suspension cable can span even more. Their shape has to be funicular to the given load, however, and the structures need to have an economic rise or sag.

Figure 2.1 Cable and arch structure: (a) trajectories of principal stresses; (b) self-anchored cable and arch; and (c) cable and arch

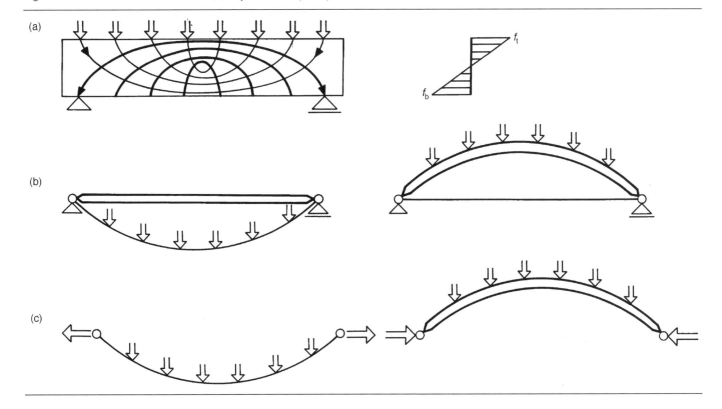

Figure 2.2 Natural Arch, Utah, USA

The layout of pedestrian bridges is influenced by two requirements. In cable-supported structures where the deck follows the shape of the cables, only a limited slope with corresponding sag can be accepted. Furthermore, these bridges need to have sufficient stiffness that guarantees comfortable walking and stability of the shape (Figure 2.3); it is therefore necessary to stiffen them.

The deformation of the suspension structures (Figure 2.4(a)) can be reduced by stiffening the cables using dead load (Figure 2.4(b)), external cables (Figure 2.4(c)) or by creating a prestressed concrete band with a certain amount of bending stiffness to guarantee the distribution of local loads and the stability of the overall shape (Figures 2.4(d) and 2.5).

Alternatively, cable-supported structures can be stiffened by beams that distribute the load and give stability to the system (Figure 2.6). In this way, a suspension system formed by beams and suspension cables of economic sags can be developed (Figures 2.6(a), 2.6(c), 2.7 and 2.9). The suspension cables could

Figure 2.3 Hukusai Bridge, Japan

Figure 2.4 Cable stiffening

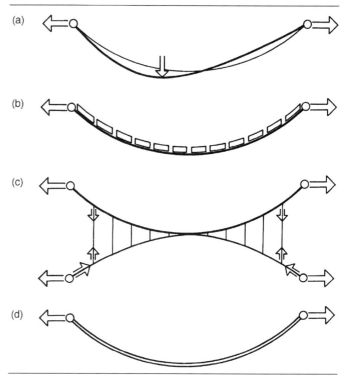

also be substituted by a system of stay cables (Figures 2.6(b), 2.6(d), 2.8 and 2.10).

When we think about a suspension structure, we usually assume that the cable is loaded at several points between the anchor points; when we imagine a cable-stayed structure, we assume that the cables are loaded only at their anchor points. The stay cables can have different arrangements, discussed in greater detail in Chapter 9.

Figure 2.5 Loading test of the Prague-Troja Bridge, Czech Republic

Figure 2.6 Beam stiffening of the cable

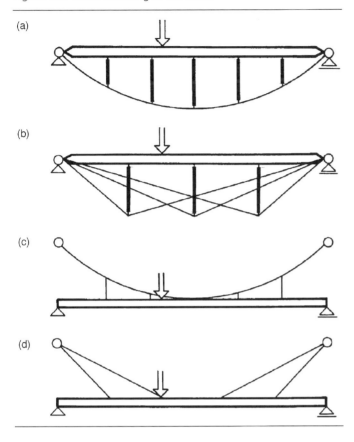

If the suspension cables are situated above the deck, the load is transferred from the deck into the cables by tensioned hangers. In structures with cables situated under the deck, the load is transferred from the deck into the cables by compression struts.

Figure 2.7 Werrekuss Bridge, Germany (courtesy of Schlaich Bergermann and Partners)

Figure 2.8 Virginia Railroad Bridge, Virginia, USA

The suspension and stay cables can be anchored outside the stiffening beam in anchor blocks, transferring the force from the cables to the soil (Figures 2.11(a) and 2.11(c)) or they can be anchored in the beam and create a so-called self-anchored

Figure 2.9 Golden Gate Bridge, California, USA

Figure 2.10 Elbe Bridge, Czech Republic

system (Figures 2.11(b) and 2.11(d)). In the latter case, the footings are stressed by vertical reactions only.

Typical arrangements of cable-supported structures with cables situated above or under the deck are presented in Figure 2.11. Figure 2.12 shows structural systems in which cables are situated both above and under deck. It is clear that it is possible to combine the suspension and stay cables situated below or above the deck. Systems with suspension or stay cables anchored in anchor blocks can also include cables anchored in the deck. Of all the possible variations, two solutions for long-span structures are presented in Figure 2.13.

Figure 2.13(a) shows a bridge where the central portion of the deck is supported on suspension cables; the portion of the deck close to the towers is supported on stay cables. This

Figure 2.11 Cable-supported structures

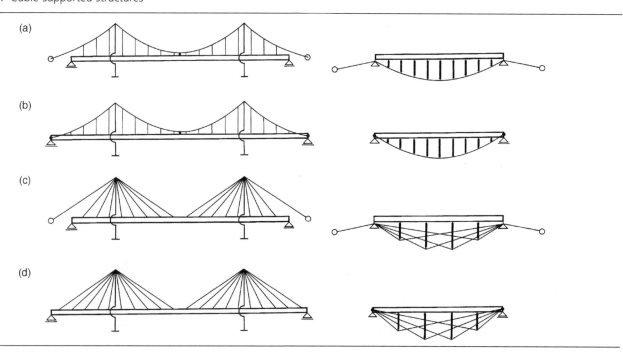

Figure 2.12 James Finley's chain bridge

arrangement, used in the past on several structures, has the modern equivalent of a proposed cable-stayed structure referred to as a 'bi-stayed bridge' (Muller, 1994; Figure 2.13(b)). In this structure, the longest back-stay cables are anchored in abutments that also serve as anchor blocks. The longest stays anchored in the main span are overlapped (coupled) by span tendons. Since the horizontal component of the stay cable is the same as the force in the span tendon the stay cable, together with the span tendon, behaves as a suspension cable in that it does not stress the structure with a compression force. This relief is critical for the design of support sections of long-span cable-stayed structures. From the above, it is clear that this structure functions similarly to the structure presented in Figure 2.13(a).

It is interesting to note that the famous Brooklyn Bridge (Figure 2.14), completed in 1883 across the East River in New York, combined suspension and stay cables. In the side spans, the suspension cables are situated both above and below the deck.

The decision on what type of structure can be used is also strongly influenced by the construction possibilities. In suspension structures, the cables have to be erected first. The erection of the deck then proceeds without consideration of the terrain under the bridge (see Section 8.2). For the self-anchored suspension bridge, the erection of the deck has to be completed first and the suspension cables can then be erected and tensioned. The fact that the erection of the deck depends on the terrain under the bridge means that use of this system is not possible in some cases.

Contrary to the self-anchored suspension structure, the self-anchored cable-stayed structure can be erected without consideration of the terrain under bridge. The deck can be erected progressively in cantilever and supported on symmetrical stay cables (see Section 9.2).

2.1.1 Stiffening of the structures

The importance of the stiffening of directly walked suspension structures, shown in Figure 2.4, is evident from Figure 2.15 in which four stress ribbon structures are compared.

Figure 2.15 shows a bridge of 99 m span with a maximum dead load slope of 8%, which yields maximum sag at mid-span of 1.98 m. The bridge is formed

1. by two cables on which timber boards are placed. The total area of the cables is $A_s = 0.0168\,\mathrm{m}^2$. The dead load $g = 5\,\mathrm{kN/m}$ and the horizontal force $H_g = 3094\,\mathrm{kN}$
2. by two cables that support concrete panels of 125 mm thickness. The total area of the cables is $A_s = 0.0252\,\mathrm{m}^2$

Figure 2.14 Brooklyn Bridge, New York, USA

Figure 2.13 Long-span cable-supported bridges

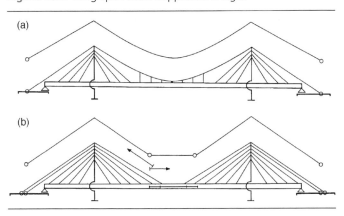

Figure 2.15 Stiffness of the stress ribbon structure

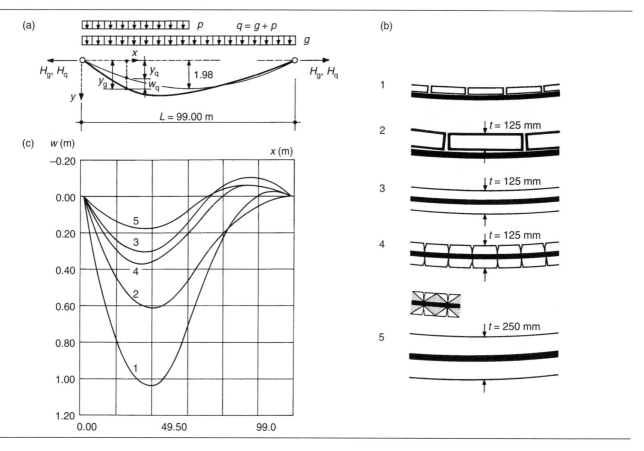

3. by a concrete band of 125 mm thickness that is supported by two cables. The total area of the cables is $A_s = 0.0252\,\text{m}^2$. The cables are embedded in the band that is fully prestressed and therefore uncracked
4. by a concrete band of 125 mm thickness that is supported by two cables. The total area of the cables is $A_s = 0.0252\,\text{m}^2$. The cables are embedded in the band that is partially prestressed and therefore cracks can form in the concrete. It is assumed that the crack spacing is 125 mm and that the concrete between cracks resists the tension. The area of concrete that resists the tension is taken as

$$A_{c,cr} = \frac{A_c}{2}$$

as shown in Figure 2.15(b)(4).

5. by a concrete band of 250 mm thickness that is supported by two cables. The total area of the cables is $A_s = 0.0392\,\text{m}^2$. The cables are embedded in the band that is fully prestressed and therefore uncracked.

In Figure 2.15(b), structures 2–4 are stressed by a dead load $g = 17\,\text{kN/m}$ with corresponding horizontal force $H_g = 10\,519\,\text{kN}$. Structure 5 is stressed by a dead load $g = 33.35\,\text{kN/m}$ with corresponding horizontal force $H_g = 20\,635\,\text{kN}$. The contribution of the prestressing tendons and reinforcing steel to the tension stiffness of the band is neglected in structures 3–5.

The above structures were analysed for the effects of the dead load g and live load $p = 20\,\text{kN/m}$ placed on one half of the structure. The analysis was performed for the structure modelled as a cable of effective tension stiffness $E_s A_e$ and zero bending stiffness. The effective stiffness $E_s A_e$ was determined from the modulus of elasticity of the cable $E_s = 195\,000\,\text{MPa}$ and an effective area A_e that depends on the area of the cable A_s and concrete band A_c (or $A_{c,cr}$, respectively) as well as the ratio of the modulus of elasticity of steel E_s and concrete E_c. Effective area is defined:

$$A_e = A_s + \frac{E_c}{E_s} A_c = A_s + \frac{36}{195} A_c.$$

Resulting deformations of the structures are presented in Figure 2.15(c) and Table 2.1. From the results it is evident that the fully prestressed band of thickness of 250 mm has the smallest deformations. Fully and partially prestressed structures also

Table 2.1 Maximum deformation of structures depicted in Figure 2.15

		Structure				
		1	2	3	4	5
$E_s A_e$	(MN)	3276	4914	27 414	16 164	51 233
	%	6	10	54	32	100
w	(m)	1.039	0.609	0.298	0.372	0.179
	%	580	340	290	360	100

Figure 2.16 Studied structures

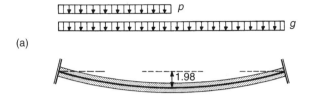
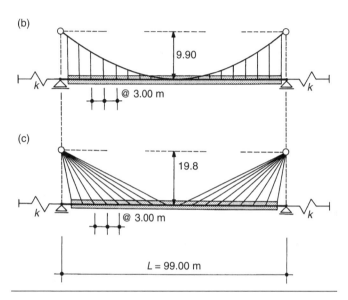

have reasonable deformations. Although the deformations of structures 1 and 2 could be reduced by substituting the cables with strips of structural steel that have lower allowable stresses and therefore require a larger cross-section, the deformations are still significant.

It is evident that the prestressed concrete deck (fully or partially prestressed) has a superior behaviour. The monolithic concrete deck not only gives the structure a sufficient tension stiffness but its membrane stiffness also guarantees the transverse stiffness of the bridge. These structures are referred to as *stress ribbon* structures.

Stiffening of the suspension structures by a tension cable of opposite curvature (Figure 2.4(c)) has a similar effect as stiffening of the structure by dead load (Figure 2.4(b)). If the cables create a space truss (e.g. as shown in Figure 11.39), the stiffness of the system can be further improved. In both these cases, however, the structures require complicated connections of the cables and good maintenance.

The above analysis was completed for the structure without considering the bending stiffness. To understand the influence of the bending stiffness of the prestressed band, extensive studies on the one-span structure were performed (Figure 2.16(a)). The structure of span $L=99.00$ m and sag $f=1.98$ m ($f=0.02L$) was developed from Structure No. 5 shown in Figure 2.15. The deck of constant area but different bending stiffness was analysed for several positions of the live load and for effects of temperature changes. Deformation, bending moments and normal force for two loadings are depicted in Figures 2.17 and 2.18.

A concrete band of width 5 m and depth 0.25 m was assumed with a modulus of elasticity $E_c = 36\,000$ MPa. The area of the concrete band A_c is calculated as:

$$A_c = bh = 5.00 \times 0.25 = 1.25$$

and the moment of inertia I_c as:

$$I_c = \tfrac{1}{12}bh^3 = \tfrac{1}{12}5.00 \times 0.25^3 = 0.00651.$$

The analysis was performed for the following values of the moment of inertia:

$$I_c(1) = 6.51 \times 10^{-3}\,\text{m}^4$$
$$I_c(2) = 1.00 \times 10^{-2}\,\text{m}^4$$
$$I_c(3) = 5.00 \times 10^{-2}\,\text{m}^4$$
$$I_c(4) = 1.00 \times 10^{-1}\,\text{m}^4$$
$$I_c(5) = 5.00 \times 10^{-1}\,\text{m}^4$$
$$I_c(6) = 1.00\,\text{m}^4$$
$$I_c(7) = 5.00\,\text{m}^4$$
$$I_c(8) = 1.00 \times 10^{1}\,\text{m}^4$$
$$I_c(9) = 5.00 \times 10^{1}\,\text{m}^4$$
$$I_c(10) = 1.00 \times 10^{2}\,\text{m}^4$$

The band was suspended on bearing tendons of total area $A_s = 0.0392\,\text{m}^2$ and modulus of elasticity $E_s = 195\,000$ MPa. The band had a constant area and therefore the dead load was also constant. Its value, including the weight of bearing tendons, was $g = 33.35$ kN/m.

Figure 2.17 Deflection, moments and normal forces in stress ribbon structure due to live load

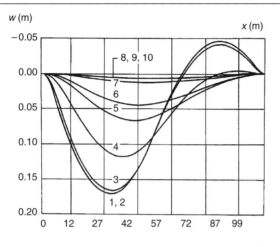
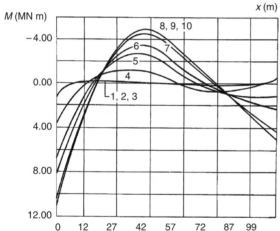
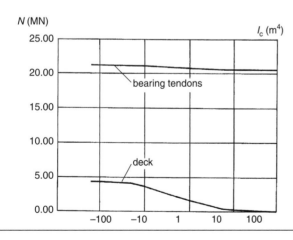

Figure 2.18 Deflection, moments and normal forces in stress ribbon structure due to temperature drop

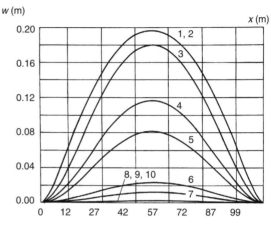
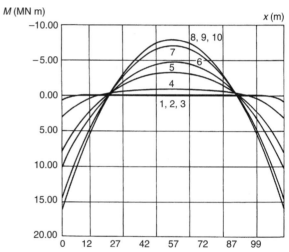
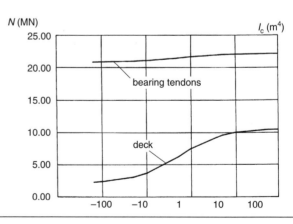

The initial stresses in the bearing tendons were derived from the horizontal force (kN):

$$H_g = \frac{gL^2}{8f} = \frac{33.35 \times 99.00^2}{8 \times 1.98} = 20\,635.$$

The structure was analysed as a two-dimensional (2D) geometrically non-linear structure with the program ANSYS. Both large deformations and the so-called tension stiffening were considered in the analysis (see Section 7.3). Figure 2.17 shows the deformation $w(x)$ and bending moments $M(x)$ along the length of the bridge for loading by dead load and live load

$p = 20.00$ kN/m situated on only one half of the structure. The values of the normal force in the bearing tendons and in the concrete deck at the mid-span section as a function of its bending stiffness are also presented.

The same results for loading by dead load and temperature drop of $\Delta_t = -20°C$ are presented in Figure 2.18. From the figures it can be seen how the deformations and bending moments depend on the stiffness of the deck. It is evident that, in the range of bending stiffness from 6.51×10^{-3} m^4 to 5.00×10^{-2} m^4, the deformation and bending moments are nearly the same. The structure resists the load by its tension stiffness and not by its bending stiffness. The deformations and normal forces are the same as those for the cable without bending stiffness.

It also evident that the bending stresses in slender structures are very low and originate only close to the end anchor blocks. Along the whole length, the structure is stressed by normal forces only.

A similar analysis was performed for the suspension (Figure 2.16(b)) and cable-stayed structures (Figure 2.16(c)). The deck was supported on movable bearings and connected to the abutments with horizontal springs.

The suspension structure is formed by a suspension cable of span $L = 99.00$ m and sag $f = 0.1L = 9.90$ m on which a concrete deck is suspended. The total area of the cables is $A_s = 0.00862$ m^2 and modulus of elasticity is $E_s = 195\,000$ MPa. It is assumed that the suspenders are pin-connected stiff bars. At the mid-span the suspension cable is connected to the deck.

The deck has the same material and sectional characteristics as the deck in the previous example. Since the deck has a constant area, the dead load $g = 31.25$ kN/m^2 is also constant. The initial stresses in the suspension cables are derived from the horizontal force (kN):

$$H_g = \frac{gL^2}{8f} = \frac{33.35 \times 99.00^2}{8 \times 1.98} = 20\,635.$$

The structure was analysed as a 2D geometrically non-linear structure with the program ANSYS for (a) zero and (b) infinite stiffness of the horizontal springs.

Figure 2.19 shows the deformation $w(x)$ and bending moments $M(x)$ along the length of the bridge for loading by dead load and live load $p = 20.00$ kN/m situated on only one half of the

Figure 2.19 Deflection and moments in suspension structure due to live load

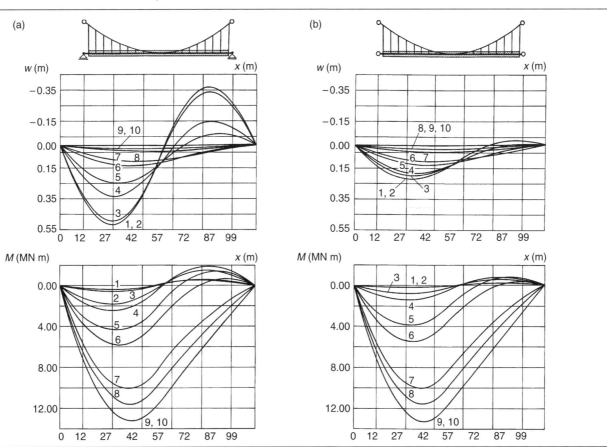

Figure 2.20 Deflection and moments in cable-stayed structure due to live load

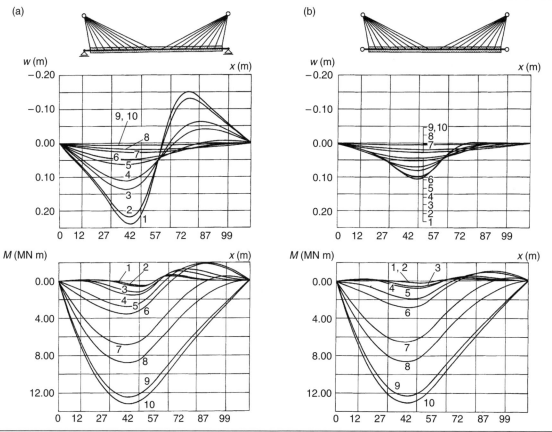

structure. From the figures, it is evident how deformation and bending moments depend on the stiffness of the deck and boundary conditions. It is clear that the deformation and bending moment in the deck can be significantly reduced by restraining the horizontal movement of the deck.

Similar results were obtained from the analysis of the cable-stayed structure (Figure 2.20). This structure was also analysed as a 2D geometrically non-linear structure with ANSYS. The initial forces in the stay cables were determined in such a way that they exactly balanced the dead load. The horizontal springs were connected to the deck after the erection.

The cable-stayed structure is usually demonstrated as superior; Figure 2.21 shows relative deformation of (a) a suspension bridge, (b) a cable-stayed bridge with twin towers and (c) a cable-stayed bridge with A-towers (Podolny and Muller, 1982). The relative deformations are caused by a live load applied as a chess board pattern. From the results it is evident that the higher stiffness of the cable-stayed structure, together with the A-towers eliminating the out-of-plane deflections of the legs, enhances the resistance of the system to torsional oscillation.

Figure 2.21 Relative deformations of suspension and cable-stayed structures

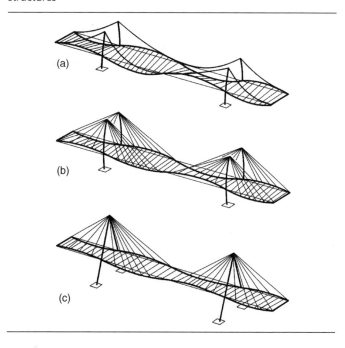

Figure 2.22 Deformations of the suspension structure

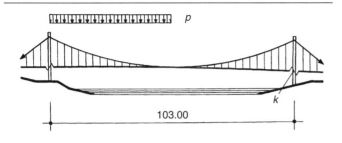

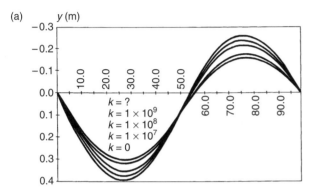

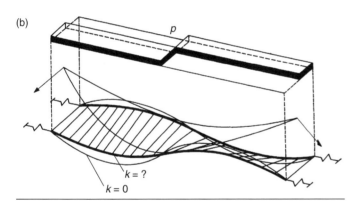

Figure 2.23 Tension stiffening

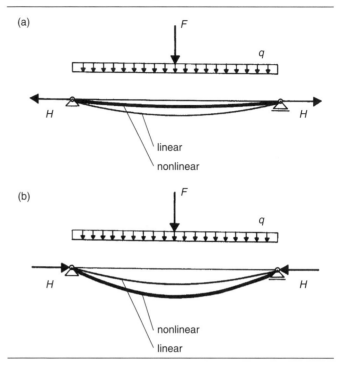

The reduction of the deflection of the deck for the same loading can also be achieved by eliminating the horizontal movement of the deck. Figure 2.22(a) shows the vertical deformations of the deck of the central span of the Willamette River Bridge (see Section 11.2) for a loading situated on one half of the main span, and for different values of the horizontal springs representing the flexible connection of the deck. A similar reduction of deflections and corresponding stresses occurs for loads situated in the main span in a chess board pattern that causes maximum distortions of the deck (Figure 2.22(b)).

Eliminating the horizontal movement of the deck generates tension stresses in the deck. In slender structures, tension stresses stabilise the structures and reduce the deflections and corresponding stresses (Figure 2.23). This phenomenon is utilised in the prevailing portion of the structures described in this book.

2.2. Structural members

The deck of the described pedestrian bridges is usually formed from concrete. In stress ribbon structures, the bearing and prestressing tendons are usually uniformly distributed in the deck (Figures 2.24(a) and 2.24(b)). The suspension and cable-stayed structures are usually suspended on the edges of the deck. The deck is therefore formed by a solid slab (Figure 2.24(c)) or two edge girders and deck slab that are stiffened by transverse diaphragms (Figure 2.24(d)). If the deck is supported by cables situated under the deck, they have a similar arrangement as the stress ribbon structures (Figures 2.24(e) and 2.24(f)). The structures suspended in the bridge axis require a torsionally stiff deck. Short-span structures can be designed as a solid slab (Figure 2.24(g)) and the long-span structures as a composite or concrete box girder (Figure 2.24(h)).

The curved decks can be suspended on only one side. To balance the transverse moments they require sufficient depth and radial prestress by tendons situated in the deck. Decks of short-span structures can be made of solid slab and the compressed member mutually connected by transverse diaphragms (Figure 2.24(i)); longer span structures require a composite or concrete box girder (Figure 2.24(j)).

Figure 2.24 Deck

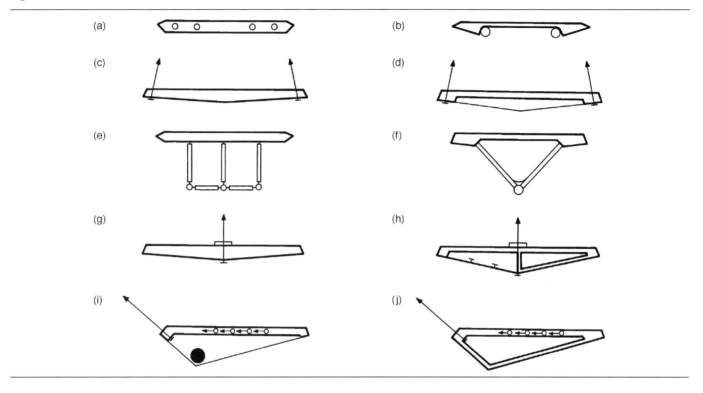

The towers can be constructed from concrete or steel; a possible arrangement is shown in Figure 2.25(a). They can have the shape of the letter H, V or A (Figure 2.26) or they can be formed by single columns situated in the bridge axis (Figure 2.27) or be located on one side of the deck (Figure 2.28). The suspension and cable-stayed structures that are supported on the edges can have cables arranged in vertical or inclined

Figure 2.25 Towers: (a) cross section; and (b) elevation

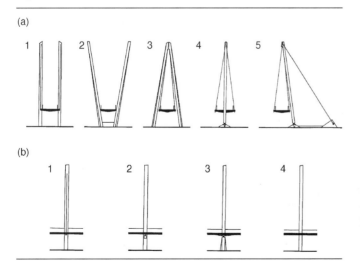

Figure 2.26 Willamette River Bridge, Oregon, USA

Figure 2.27 Max-Eyth-See Bridge, Stuttgart, Germany (courtesy of Schlaich Bergermann and Partners)

Figure 2.29 Shiosai Bridge, Japan (courtesy of Sumitomo Mitsui Construction Co. Ltd)

planes. The second option gives the structures larger transverse stiffness and increases their aerodynamic stability.

If the deck is suspended in the bridge axis, the towers can have a similar arrangement. The suspension of the deck on one side by an inclined column requires anchoring of the column by transverse cables, and should therefore only be used if necessary.

The deck can be totally suspended on the stay cables or can be supported at towers. The deck can be also frame connected with tower legs (Figure 2.26(b)).

The compression struts of cable-supported structures can also be made from either concrete or steel (Figures 2.29 and 2.30).

Figure 2.28 Hungerford Bridge, London, UK

Figure 2.30 Tobu Bridge, Japan (courtesy of Oriental Construction Co. Ltd)

Figure 2.31 Tension members: (a) cables developed by steel industry; and (b) cables developed by prestressed concrete industry

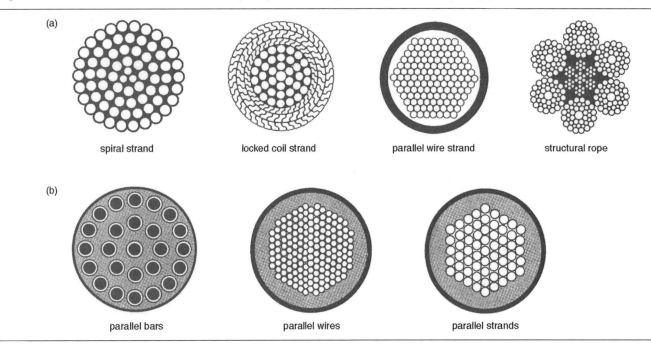

The tension members (stay or suspension cables) can be made from structural steel or cables developed by the steel or prestressed concrete industries. They can also be composite or prestressed concrete members.

Tension members of structural steel can be rolled shapes (Figure 2.30), plates, bars, tubes or pipes. The cables developed by the steel industry are formed by spiral strands, locked coil strands, parallel wires or ropes (Figure 2.31(a)). They are factory fitted with a combination of socket types to enable load transmission between the structure and the cable (Figure 2.32). To eliminate the construction stretch it is important to prestress the cables. Prestressing is conducted using a series of cyclic loadings, typically between 10% and 50% of the strand breaking force, until an apparent stable modulus of elasticity is achieved. Typical arrangements of the cables and structural details can be found in brochures prepared by the producers.

Cables developed from the prestressing tendons are formed by parallel prestressing bars, parallel wires or prestressing strands (Figure 2.31(b); Post-Tensioning Institute, 1993). A typical arrangement of a prestressing tendon and a stay cable formed by prestressing strands is shown in Figure 2.33. Cables were traditionally made with prestressing strands that were grouted in steel or polyethylene (HDPE) tubes. More recently, they have also been made with epoxy coated strands, polyethylene (PE)-sheathed and greased strands (monostrands) or PE-coated galvanised strands with wax fill. All types of strands can be grouted in steel or PE tubes.

Figure 2.32 Anchoring of the cables: (a) plain cylindrical socket; (b) threaded cylindrical socket with bearing nut; and (c) open socket

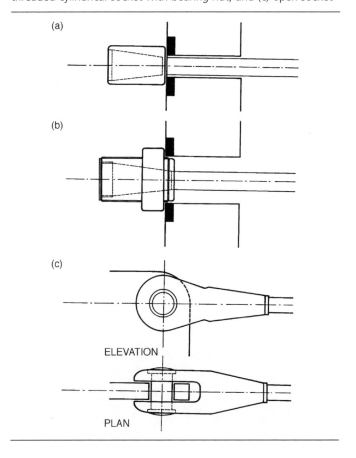

Figure 2.33 Anchoring of the cables: (a) prestressing tendon; and (b) stay cable

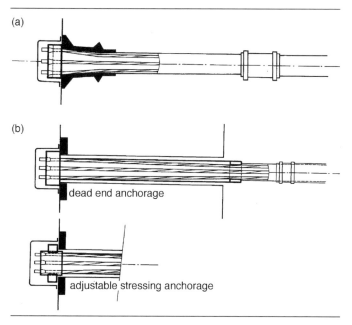

Figure 2.34 Hungerford Bridge, London, UK

The arrangement of cables varies according to the prestressing system used; this information is published in brochures by the producers.

Since the suspension cables are stressed by relatively low fatigue stresses, they are usually designed as non-replaceable structural members. In contrast, the suspenders and stay cables are stressed by large fatigue stresses. It is therefore necessary to design their connections to towers, struts, suspension cables and the deck in such a way that the forces in the cables can be adjusted and allow the replacement of the suspenders of cables (Figures 2.34–2.36).

The suspension and stay cables can be anchored at saddles situated on the towers (Figure 2.37) or struts (uprights), or can be bent at such saddles (Figures 2.27 and 2.38). The anchors can be attached to tension plates (Figure 2.39) or they can be made to overlap there (Figure 2.26).

The structural details of the cables at saddles or at their anchors are influenced by local bending moments that originate there (Chapter 9).

The tension members can also be formed by composite or prestressed concrete members (Figure 2.40). The prestressed members can form ties or stress ribbons (Figure 2.29). The composite members can be formed not only by a prestressed composite section of structural steel and concrete, but also by strands grouted in steel tubes. The author used this arrangement in his designs of suspension and stay cables. Through the construction process, compression stresses in both the cement mortar and the tubes were created. In this way the tension stiffness of the cables, and consequently the whole structure, was significantly increased. The process was as follows.

The grouting of the suspension cables was done after the erection of the deck was completed (Figure 2.41). The deck was

Figure 2.35 Nordbahnhof Bridge, Stuttgart, Germany (courtesy of Schlaich Bergermann and Partners)

Figure 2.36 McKenzie River Bridge, Oregon, USA

Figure 2.38 Tobu Bridge, Japan (courtesy of Oriental Construction Co. Ltd)

Figure 2.39 Hungerford Bridge, London, UK

Figure 2.37 Anchoring of the cables at towers and struts: (a) saddle; (b) tension plate; and (c) overlapping

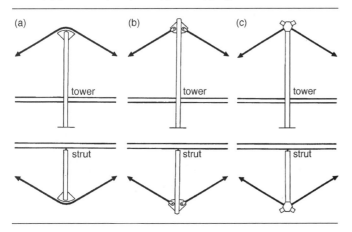

Figure 2.40 Composite and prestressed concrete tension members

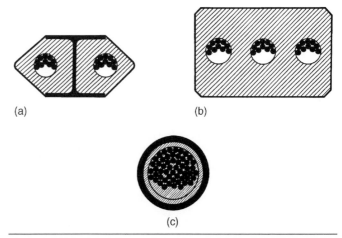

Figure 2.41 Grouting of suspension cables: (a) elevation; (b) cross-section; and (c) stresses in the strands, tubes and cement mortar

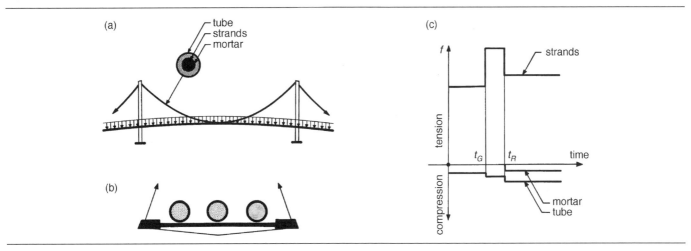

temporarily loaded before the cables were grouted; the load was created with plastic tubes filled with water. When the mortar reached a sufficient strength the loading was released: the water was drained from the plastic tubes. In this way, compression stresses in both the steel tubes and cement mortar were obtained.

For a cable-stayed structure the process was similar (Strasky, 1993). During the erection of the structure the strands were tensioned to their design stress. Before grouting, the whole cable was tensioned to higher tension (Figure 2.42). When grouting was completed, a closure weld between the steel

Figure 2.42 Grouting of stay cables: (a) tensioning of the cable; (b) grouting; and (c) release of stresses

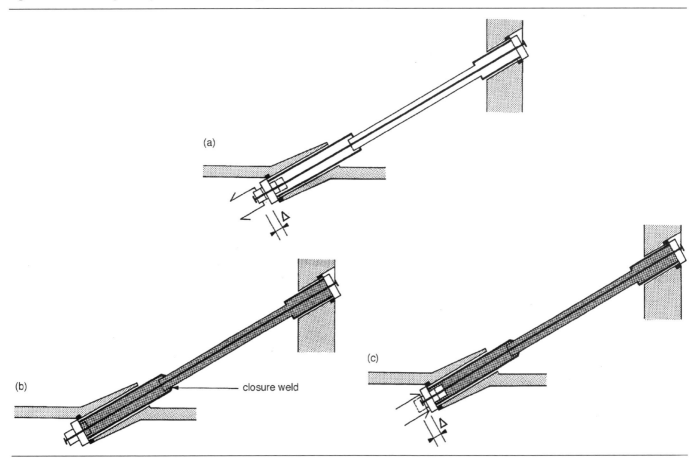

tubes was done. As soon as the grout reached sufficient strength, the length of the completed stay cable was adjusted by loosening the rectification nut. In this way, compression stresses in both the steel tubes and the cement mortar were created. The value of the compression stress in the steel tubes was as large as 220 MPa; in cement mortar this would be 14 MPa.

Cables made from parallel wires of carbon-fibre reinforced polymer (CFRP) or bars of carbon-fibre composite were recently built in Switzerland and France. These cables are seven times lighter than steel cables with the same breaking load. Since the cables are resistant to corrosion and chemical attack, no corrosion-inhibiting compound or grout is necessary. However, to protect the wires against wind erosion and ultra-violet radiation, polyethylene sheaths are used.

REFERENCES

Muller J (1994) Reflections on cable-stayed bridges. Revue generale des routes et des aerodromes. Paris.

Podolny W and Muller J (1982) *Construction and Design of Prestressed Concrete Bridges*. John Wiley & Sons, New York.

Post-Tensioning Institute (1993) Recommendations for stay-cable design, testing and installation. Post-Tensioning Institute, Committee on Cable-stayed Bridges, August 1993.

Strasky J (1993) Design and construction of cable-stayed bridges in the Czech Republic. *PCI Journal* **38(6)**: 24–43.

Chapter 3
Design criteria

The design criteria including the dynamics of pedestrian bridges are widely discussed in the book *Guidelines for the Design of Pedestrian Bridges* which was prepared by *fib* Task Group 1.2 Bridges, Working Party Pedestrian Bridges (2005). The criteria for the design of pedestrian bridges are given by Eurocode and National Standards (Gulvanessian *et al.*, 2002; Department of Transport, 1988; AASHTO, 1997). Only key issues are discussed in this chapter.

3.1. Geometric conditions
3.1.1 Deck width

From Chapter 11, in which 50 examples of pedestrian bridges are presented, it is evident that width of pedestrian bridges varies from 0.85 to 6.60 m. The width depends on the local conditions and expected density of pedestrians. The width is different if the bridge is situated on a trail, park or in a city. The width of Japanese pedestrian bridges is usually smaller than the width of bridges built in Europe and North America.

If the bridge is used by pedestrians only, the minimum width W_1 of common bridges should be at least 3.00 m. If the bridge is used by both pedestrians and bicycles, the width W_2 should be at least 3.50 m (Figure 3.1).

3.1.2 Grades

The maximum permissible longitudinal grade also depends on the location. If the bridge is situated on a trail in the mountains, the maximum slope can reach as much as 20%. If the bridge is situated in a city, the slope has to accommodate the disabled.

It is usually required that the slope of ramps and separated pathways should not exceed 20:1 (5%). A maximum slope of 12:1 (8.33%) is acceptable for a rise of no more than 0.75 m if a level landing of at least 1.5 m is provided at each end. Some specifications allow the maximum slope of 1:8 (12.50%) to be used for short ramps of maximum length of 3.00 m (Figure 3.2).

The bridges described in this book are mostly of a variable longitudinal slope. While suspension and cable-stayed bridges are usually in convex (crest) elevation, stress ribbon bridges are in concave (sag) elevation which is characteristic of this structural type. Their shape is close to the shape of the second-degree parabola.

The author is convinced that the value of the permissible slope should not be given by the value of the maximum slope at one point, but it should be derived from the conditions of the same energy that a disabled person has to overcome.

It is important to realise that the average slope of the second-degree parabola $p_{avr} = 0.50 p_{max}$. For the parabola of the length L and sag $f = 0.02L$ the maximum slope $p_{max} = 10\%$

Figure 3.1 Deck width and railings height

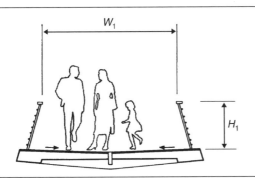
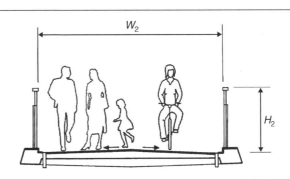

Figure 3.2 Longitudinal grade

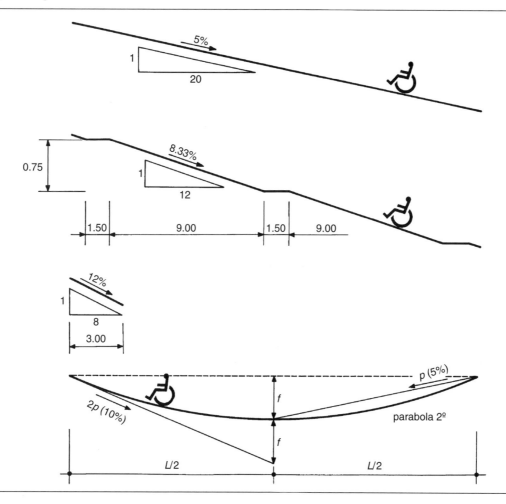

and average slope $p_{avr} = 5\%$. A disabled person who moves on this parabolic deck has to exert the same energy as if climbing a constant slope of 1:20. Of course, the length of the slope larger than 1:20 has to have a reasonable length that should be discussed with local authorities.

3.1.3 Surfacing, railing and lighting

The deck of the prevailing portion of structures described in this book is formed by a slender concrete ribbon that can provide a smooth and durable finish (Figure 3.3). Its waterproofing can be provided by epoxy coating spread by crushed silicon stone to prevent slippage (Figure 3.4). Another type of overlay formed by polymer concrete or asphalt can also be used.

The cross-slope of 1% is usually sufficient for drainage (Figure 3.1). The deck can be drained to the edge curb or to the central channel (Figure 3.5). If the bridge is situated in the countryside, it is possible to drain the bridge by many small pipes or openings covered by stainless grid (Figure 3.4).

Figure 3.3 Grants Pass Bridge, Oregon, USA

Figure 3.4 Maidstone Bridge, Kent, UK

Horizontal loads for which the railings of footbridges are designed are less than those for road bridges; the shape and the materials for railings can therefore be chosen more freely. Usually the required height of the railings H_1 is 1.00–1.10 m for pedestrians. If the bridge is designed for both pedestrians and cyclists, the required height H_2 is 1.30–1.40 m (Figure 3.1).

There are many possibilities for railing designs. If horizontal bars are used, railings inclined to the deck which prevent

Figure 3.5 Freiburg Bridge, Germany

Figure 3.6 Maidstone Bridge, Kent, UK

climbing should be used. Steel wire meshes that are very transparent (Figure 3.6) have been successfully used in recent years.

Lampposts can be avoided if the lighting can be included into the handrail or the railing posts. Modem LEDs emit sufficient light to be used for this purpose. They fit into the narrow spaces of rails and posts, do not heat up and consume very little energy.

The deck walkway can also be lit by airport runway lights, fitted with frosted lenses and powered by reduced voltage current. The lamps shed dispersed light onto the parapet mesh, causing it to glow and create an ambient light aura at night.

3.2. Loads

Pedestrian bridges are designed for a uniform live load whose intensity is similar to, or even higher than, those for highway bridges. The basic uniform load of 4–5 kN/m² is usually reduced according to the length and width applied. This means that all structural members should be designed for a load of higher intensity applied to a smaller area and for a load of reduced intensity applied to a larger area.

However, if the bridge is situated in a vicinity of, e.g., a football stadium, reduction of the live load should not be used. The designer should also consider the possibility of unbalanced loading caused by pedestrians standing only on one side.

Pedestrian bridges should also be designed for load of emergency, police or maintenance vehicles. The weight of such a vehicle distributed onto its corresponding plan area will generally be less than $5\,\text{kN/m}^2$. Nevertheless, the corresponding wheel loads should be applied.

3.2.1 Eurocode 1: Action on structures
Part 2 of Eurocode 1 states that traffic loads on bridges must have a uniform load (kN/m^2):

$$q_\text{fk} = 2.0 + \frac{120}{L+30}$$

$$q_\text{fk} \geqslant 2.5; \quad q_\text{fk} \leqslant 5.0$$

where L is the loaded length (m).

The characteristic value of the concentrated load Q_fwk should be assumed equal to $10\,\text{kN}$ acting on a square surface of sides $0.10\,\text{m}$.

If a service vehicle is specified, Q_fwk should not be considered. The characteristics of service vehicle (axle weight and spacing, contact area of wheels), the dynamic amplifications and all other appropriate loading rules should be specified for the individual project.

3.3. Dynamics
The pedestrian bridges described in this book have low natural frequencies and damping; it is therefore necessary to check their dynamic behaviour from the point of view of the physiological effect of vibrations and response of the structure to wind. In seismic regions, a seismic analysis of the structure also has to be provided.

3.3.1 Physiological effect of vibrations
Some pedestrian bridges have exhibited unacceptable performance due to vibration caused by people walking or running on them. Wind can also cause an unpleasant movement. Dynamic actions induced by people result from the rhythmical body motions (Bachmann et al., 1997; Bachmann, 2002; Kreuzinger, 2002; Roberts, 2003).

3.3.1.1 Vertical vibration
Typical pacing frequencies when a person is walking or running and frequencies when jumping on the spot are given in Table 3.1 (Bachmann, 2002). Approximate mean values are $f_s = 2\,\text{Hz}$ for walking and $f_r = 2.5\,\text{Hz}$ for running and jumping.

Table 3.1 Pacing and jumping frequencies: Hz

	Total range	Slow	Normal	Fast
Walking	1.4–2.4	1.4–1.7	1.7–2.2	2.2–2.4
Running	1.9–3.3	1.9–2.2	2.2–2.7	2.7–3.3
Jumping	1.3–3.4	1.3–1.9	1.9–3.0	3.0–3.4

To avoid the resonance, some standards specify that pedestrian bridges with fundamental frequencies below $3\,\text{Hz}$ should be avoided (AASHTO, 1997). However, all stress ribbon and suspension pedestrian bridges designed by the author have these fundamental frequencies below $2\,\text{Hz}$. Although they were built from 1979 onwards, no complaints about their dynamic behaviour have been made.

It is obvious that, rather than checking the natural modes and frequencies, the speed of motion or acceleration of the bridge deck caused by forced vibration (which represents the effects of moving people) should be checked.

According to the Department of Transport (1988) the maximum vertical acceleration should be calculated assuming that the dynamic loading applied by a pedestrian can be represented by a pulsating point load F moving across the main span of the superstructure at a constant speed v as follows.

$$F = 180 \times \sin 2\pi f_0 T$$

$$v = 0.9 f_0$$

where F is point load (N), T is the time (s) and v is speed (m/s).

The maximum vertical acceleration (m/s^2) should be limited to

$$0.5\sqrt{f_0}.$$

Similar critical values of the acceleration are presented in Figure 3.7 (Walther et al., 1998). The author usually checks that the speed of motion is no greater than $24\,\text{mm/s}$.

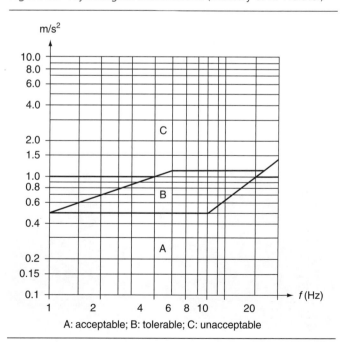

Figure 3.7 Psychological classifications (courtesy of R. Walther)

A: acceptable; B: tolerable; C: unacceptable

3.3.1.2 Horizontal vibration

With every step, there is also a horizontal power that interacts with the bridge. While the vertical power has a downwards effect by each footstep, the horizontal power sends our force to the right and left alternately (Figure 3.8) which is why we are dealing with a case of resonance if:

Vertical vibration: $f_V = f_s$
Horizontal vibration: $f_H = f_{s/2}$

Step frequencies f_s of about 2.0 Hz will affect bridges with vibrations of 1.0 Hz with substantial horizontal deformations. Circumstances where $f_V = 2f_H$ should be avoided.

According to Bachmann (2002), amplitude spectra of the horizontal forces of a person walking at a pacing frequency of 2 Hz show considerable scatter. In general, the amplitudes increase with increasing vibrations of the deck. In the transverse direction, maximum values of $\Delta G/G$ (where G is dead load) of up to 0.07 in the case of a stationary deck and up to 0.14 in the case of a vibrating deck were measured. Although the horizontal forces from walking and running are relatively small compared to the vertical forces, they are sufficient to produce strong vibrations in the case of horizontally soft and hence low-frequency structures.

3.3.1.3 Lock-in effect

The so-called lock-in effect may also be of substantial importance. A walking or running person adapts to and synchronises his/her motions in frequency and phase (ϕ_i) with a vibrating deck if the displacement amplitude exceeds a certain threshold value.

The threshold value depends on the direction of vibration and the person's age and fitness. For vertical vibrations of ≈ 2 Hz, this value falls within the range 10–20 mm. For horizontal vibrations with a frequency of ≈ 1 Hz, some persons begin to adapt their motion when the amplitude exceeds 2–3 mm.

Figure 3.8 Vertical and horizontal load

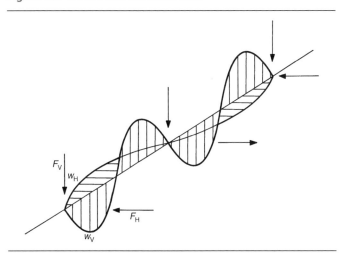

If the individual threshold value of a person is exceeded due to the synchronisation, an 'impulse into every wave trough of the bridge vibration' occurs which is a much more adverse dynamic action than described above. As a consequence, the vibration amplitude increases and more persons are locked into synchronisation. In certain cases, synchronisation of more than 80% of the persons involved was observed. A relatively simple analysis of these effects was published by Roberts in 2003.

According to the Eurocode, basic structural design (Gulvanessian et al., 2002) of pedestrian bridges requires checking for feelings of discomfort which arise for basic natural frequencies smaller than 5 Hz. A maximum acceptable acceleration is specified (0.7 m/s^2); however, the method of analysis is not given.

In checking dynamic behaviour of pedestrian bridges, the authors utilises the method described by the Department of Transport (1988). This procedure was also accepted in a draft of Eurocode 2 (Design of Concrete Structures, Part 2: Concrete Bridges from 1998). Unfortunately, for reasons unknown to the author, this procedure has not been included in the final version of Eurocode 2 from 2007.

The author has used the above approach to check the dynamic response of 15 pedestrian bridges and the results of the analysis are summarised in the following. All structures were analysed by the program ANSYS as geometrically non-linear structures for a static load. The analyses described their progressive erection and actual boundary conditions. Tension stiffening of tension members has also been considered.

During the dynamic analyses, the natural modes and frequencies were first determined. Depending on the static system, arrangement and stiffness of supports and mass, the first bending modes had the shape as depicted in Figure 3.9. In the analysis of pedestrian bridges of a greater number of

Figure 3.9 Typical natural modes

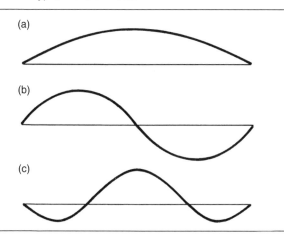

Table 3.2 Pedestrian bridges: results of analysis

Bridge, structural type and date of completion/design	Cross-section, reference depth and dead load	Natural frequency (Hz) and mode shape	Checking
1. Prague-Troja, Czech Republic Stress ribbon 1984	$h_s = 0.25$ m; $g = 27.0$ kN/m	$f_{(0)} = f_{(1)} = 0.490$ (A) $f_{(2)} = 0.609$ (A) $f_{(3)} = 0.966$ (A) $f_{(4)} = 1.010$ (A) & (B) $f_{(H)} = f_{(9)} = 2.204$	$u_{max} = 5.330$ mm $v_{max} = 0.020$ m/s $a_{max} = 0.078$ m/s^2 $a_{lim} = 0.350$ m/s^2 $f_{(0)}/2 = 0.245$ Hz
2. Lake Hodges, San Diego, California, USA Stress ribbon 2009	$h_s = 0.28$ m; $g = 38.6$ kN/m	$f_{(0)} = f_{(1)} = 0.574$ (A) $f_{(2)} = 0.587$ (A) $f_{(3)} = 0.796$ (A) $f_{(5)} = 1.188$ (B) $f_{(6)} = 1.195$ (B) $f_{(7)} = 1.197$ (B) $f_{(H)} = f_{(4)} = 1.087$	$u_{max} = 3.630$ mm $v_{max} = 0.013$ m/s $a_{max} = 0.047$ m/s^2 $a_{lim} = 0.379$ m/s^2 $f_{(0)}/2 = 0.287$ Hz
3. Kent Messenger, Maidstone, Kent, UK Stress ribbon 2001	$h_s = 0.23$ m; $g = 19.8$ kN/m	$f_{(0)} = f_{(1)} = 1.171$ (A) $f_{(2)} = 1.964$ (A) & (B) $f_{(H)} = f_{(4)} = 2.619$	$u_{max} = 1.730$ mm $v_{max} = 0.021$ m/s $a_{max} = 0.263$ m/s^2 $a_{lim} = 0.541$ m/s^2 $f_{(0)}/2 = 0.586$ Hz
4. Olse River, Bohumin, Czech Republic Stress ribbon Design	$h_s = 0.14$ m; $g = 15.3$ kN/m	$f_{(0)} = f_{(1)} = 0.888$ (A) $f_{(2)} = 1.116$ (A) & (B) $f_{(3)} = 1.332$ (B) $f_{(H)} = f_{(4)} = 1.350$	$u_{max} = 3.710$ mm $v_{max} = 0.031$ m/s $a_{max} = 0.260$ m/s^2 $a_{lim} = 0.471$ m/s^2 $f_{(0)}/2 = 0.444$ Hz
5. R35 Expressway, Olomouc, Czech Republic Stress ribbon supported by arch 2008	$h_s = 0.14$ m; $g = 17.8$ kN/m	$f_{(0)} = f_{(1)} = 1.530$ (A) $f_{(2)} = 1.746$ (B) $f_{(3)} = 2.149$ (B) $f_{(H)} = f_{(1)} = 0.961$	$u_{max} = 1.570$ mm $v_{max} = 0.015$ m/s $a_{max} = 0.145$ m/s^2 $a_{lim} = 0.490$ m/s^2 $f_{(0)}/2 = 0.765$ Hz
6. Svratka River, Brno, Czech Republic Stress ribbon supported by arch 2008	$h_s = 0.18$ m; $g = 22.6$ kN/m	$f_{(0)} = f_{(1)} = 1.912$ (B) $f_{(2)} = 2.163$ (A) $f_{(3)} = 3.819$ (C) $f_{(H)} = f_{(4)} = 4.627$	$u_{max} = 0.880$ mm $v_{max} = 0.012$ m/s $a_{max} = 0.162$ m/s^2 $a_{lim} = 0.691$ m/s^2 $f_{(0)}/2 = 0.956$ Hz
7. Vltava River, Ceske Budejovice, Czech Republic Stress ribbon suspended on arch 2007	$h_s = 0.10$ m; $h_i = 0.16$ m; $g = 16.0$ kN/m	$f_{(0)} = f_{(2)} = 1.711$ (B) $f_{(3)} = 1.851$ (A) $f_{(4)} = 2.863$ (C) $f_{(H)} = f_{(1)} = 1.264$	$u_{max} = 1.890$ mm $v_{max} = 0.026$ m/s $a_{max} = 0.218$ m/s^2 $a_{lim} = 0.380$ m/s^2 $f_{(0)}/2 = 0.856$ Hz

Table 3.2 Continued

Bridge, structural type and date of completion/design	Cross-section, reference depth and dead load	Natural frequency (Hz) and mode shape	Checking
8. McLouglin Blvd, Portland, Oregon, USA Stress ribbon suspended on arch 2006	$h_s = 0.25$ m; $g = 31.7$ kN/m	$f_{(0)} = f_{(1)} = 1.021$ (B) $f_{(5)} = 1.791$ (C) $f_{(H)} = f_{(2)} = 1.282$	$u_{max} = 2.540$ mm $v_{max} = 0.016$ m/s $a_{max} = 0.104$ m/s^2 $a_{lim} = 0.505$ m/s^2 $f_{(0)}/2 = 0.511$ Hz
9. Vranov Lake, Czech Republic Suspension with a stress ribbon deck 1993	$h_s = 0.16/0.14$ m; $g = 27.4/35.7$ kN/m	$f_{(0)} = f_{(1)} = 0.298$ (B) $f_{(2)} = 0.360$ (C) $f_{(H)} = f_{(4)} = 0.431$	$u_{max} = 10.96$ mm $v_{max} = 0.025$ m/s $a_{max} = 0.056$ m/s^2 $a_{lim} = 0.273$ m/s^2 $f_{(0)}/2 = 0.149$ Hz
10. Willamette River Eugene, Oregon, USA Suspension with a stress ribbon deck 2002	$h_s = 0.22$ m; $g = 36.6$ kN/m	$f_{(0)} = f_{(1)} = 0.541$ (B) $f_{(2)} = 0.613$ (C) $f_{(3)} = 0.888$ (C) $f_{(H)} = f_{(5)} = 1.136$	$u_{max} = 7.040$ mm $v_{max} = 0.027$ m/s $a_{max} = 0.105$ m/s^2 $a_{lim} = 0.367$ m/s^2 $f_{(0)}/2 = 0.271$ Hz
11. Harbor Drive, San Diego, California, USA Suspension with a girder deck 2010	$h_s = 0.23$ m; $g = 35.4$ kN/m	$f_{(0)} = f_{(2)} = 0.955$ (B) $f_{(5)} = 2.039$ (C) $f_{(H)} = f_{(1)} = 0.731$	$u_{max} = 1.650$ mm $v_{max} = 0.010$ m/s $a_{max} = 0.059$ m/s^2 $a_{lim} = 0.489$ m/s^2 $f_{(0)}/2 = 0.478$ Hz
12. Johnson Creek, Portland, Oregon, USA Stress ribbon supported by a suspension cable Design	$h_s = 0.26$ m; $g = 36.3$ kN/m	$f_{(0)} = f_{(1)} = 0.838$ (A) $f_{(3)} = 1.774$ (B) $f_{(5)} = 2.965$ (C) $f_{(H)} = f_{(2)} = 1.674$	$u_{max} = 3.534$ mm $v_{max} = 0.019$ m/s $a_{max} = 0.098$ m/s^2 $a_{lim} = 0.458$ m/s^2 $f_{(0)}/2 = 0.419$ Hz
13. I-5 Gateway, Eugene, Oregon, USA Cable stayed with a stress ribbon deck 2009	$h_s = 0.26$ m; $g = 36.3$ kN/m	$f_{(0)} = f_{(4)} = 1.654$ (A) $f_{(10)} = 3.295$ (A) $f_{(17)} = 4.633$ (A) & (B) $f_{(H)} = f_{(1)} = 1.131$	$u_{max} = 0.810$ mm $v_{max} = 0.008$ m/s $a_{max} = 0.083$ m/s^2 $a_{lim} = 0.639$ m/s^2 $f_{(0)}/2 = 0.827$ Hz
14. Delta Pond, Eugene, Oregon, USA Cable stayed with a stress ribbon deck 2010	$h_s = 0.26$ m; $g = 36.3$ kN/m	$f_{(0)} = f_{(1)} = 1.267$ (A) $f_{(7)} = 2.207$ (A) & (B) $f_{(9)} = 2.584$ (A) & (B) $f_{(H)} = f_{(4)} = 1.403$	$u_{max} = 2.020$ mm $v_{max} = 0.016$ m/s $a_{max} = 0.128$ m/s^2 $a_{lim} = 0.477$ m/s^2 $f_{(0)}/2 = 0.634$ Hz

Table 3.2 Continued

Bridge, structural type and date of completion/design	Cross-section, reference depth and dead load	Natural frequency (Hz) and mode shape	Checking
15. D47 Freeway, Bohumin, Czech Republic Cable stayed with a girder deck 2010	$h_s = 0.31$ m; $g = 63.2$ kN/m	$f_{(0)} = f_{(2)} = 1.420$ (A) $f_{(3)} = 2.199$ (A) $f_{(8)} = 4.115$ (B) $f_{(H)} = f_{(4)} = 2.746$	$u_{max} = 0.530$ mm $v_{max} = 0.005$ m/s $a_{max} = 0.260$ m/s^2 $a_{lim} = 0.596$ m/s^2 $f_{(0)}/2 = 0.710$ Hz

spans, the natural modes A, B or C have sometimes occurred simultaneously.

After determining natural modes and frequencies, the forced vibration was calculated. The parameters maximum amplitude (u_{max}, mm), maximum speed of motion (v_{max}, m/s), maximum acceleration (a_{max}, m/s^2) and limited acceleration (a_{lim}, m/s^2) were determined.

Structures 1–6 and 8–15 listed in Table 3.2 are also described in Chapter 11. The pedestrian bridge across the Vltava River (structure 7) is formed by a tied arch of span 53.20 m. A composite deck that is formed by two steel pipes and slender concrete slab is suspended on one side of an inclined arch with a rise of 8.00 m.

Results of the analyses are summarised in Table 3.2 which lists cross-section of the deck, depth of a reference deck $h_S = A/w$ (area A divided by width w), dead load of the deck g (kN/m),
values and basic shapes of the first bending frequencies and maximum and limited accelerations. Since the structures are also described in Chapter 11, the cross-sections are schematic. For comparison, however, they are drawn to the same scale.

A comparison of one-half of the vertical and horizontal frequency is included, i.e. $f_V/2 = f_{(0)}/2 \neq f_H$. For the pedestrian bridge across the Vltava River in Ceske Budejovice (structure 7) that has a composite deck, a reference depth of concrete slab and an effective depth h_i that includes an area of steel multiplied modular ratio is defined as $n = E_s/E_c$ (where E_s and E_c are the modulus of elasticity of steel and concrete).

Results of the analyses are depicted in Figure 3.10, where maximum accelerations and criteria for acceleration checking in terms of the first bending frequency f_0 are presented.

All of the studied structures have a very slender deck with reference depth h_S from 0.10 to 0.31 m. Thirteen of the fifteen

Figure 3.10 Acceleration of analysed footbridges

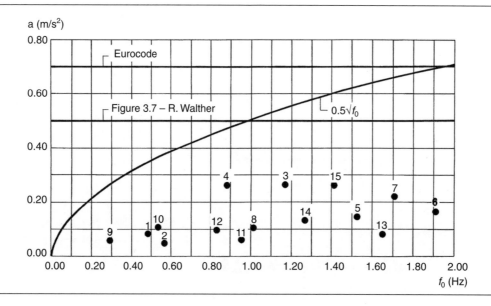

described pedestrian bridges have been built, and (so far) no complaints about their dynamic behaviour have been recorded. The author therefore believes that the presented approach can be used to estimate the dynamic response of new structures.

3.3.1.4 Vibration due to wind
Pedestrian bridges should also be checked for possible vibration caused by wind. Since walking on the bridge at wind speeds larger than 20 m/s (72 km/hour) is very difficult, the speed of motion or acceleration should be checked for reasonable wind speed in which the bridge is usually used by ordinary pedestrians.

3.3.2 Aerodynamic behaviour
Depending on the geographical location of the bridge, the deck may be exposed to crosswinds (Institution of Civil Engineers, 1981; Walther *et al.*, 1998). A flow of air tends to induce torsional and bending oscillations in the structure which, under the effect of small variations in the angle of the wind, modify the lifting effects. This phenomenon, known as flutter, was illustrated in 1940 by the total collapse of the Tacoma Narrows Bridge (USA) (Scott, 2001). From studies carried out since, it has been seen that the torsional and bending frequencies must be sufficiently far apart. Mathivat (1983) has shown that a ratio of torsional to bending frequency of 2.5 is adequate.

Flutter can be induced by the phenomenon of vortex shedding (Figure 3.11). The shape of the deck governs the air flow and, if the section is badly streamlined, vortices form as the air passes and can introduce the danger of vortex shedding.

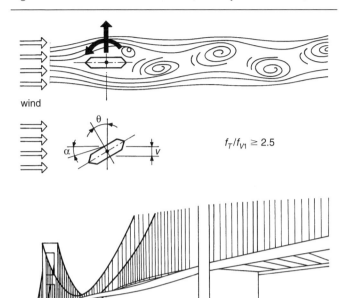

Figure 3.11 Phenomenon of flutter (courtesy of R. Walther)

Flutter can be also caused by auto-excited oscillations that are caused by the movement itself. Figure 3.12 provides a simplified description of this mechanism in the case of a phase difference of $\pi/2$ between bending and torsion. Beyond a certain wind

Figure 3.12 Phenomenon of auto-excited oscillations (courtesy of R. Walther)

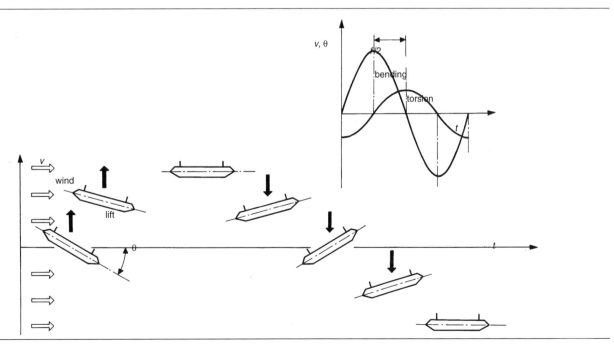

speed, known as the critical speed V_{crit}, the deck receives more energy than can be dissipated by damping. The result is to produce combined bending and torsional movements due to the aerodynamic forces, with rapidly increasing amplitudes with no other limit other than the destruction of the bridge.

The stay cables should also be checked for the possibility of their oscillation. If the stay cables or hanger are formed by several parallel members, they can be subjected by galloping oscillations induced by the turbulent wake. The problem can be easily solved by mutual connection of parallel members. The connection should be made outside the nodes of the first natural modes of vibration.

3.3.3 Seismic design

Since the stress ribbon and cable-supported pedestrian bridges have low natural frequencies, they are not sensitive to seismic load. Although the stress ribbon deck is fixed into the end abutments, the horizontal force due to live load is usually larger than the force created by seismic load. The seismic load therefore does not usually control the design of these structures for longitudinal effects.

However, the dynamic analysis based on the application of a response spectrum of ground acceleration of a space structure is mandatory for multi-span stress ribbon structures and for cable-supported structures. When designing the structural members and their connections, it is necessary to follow the recommendations published by Priestly *et al.* (1996).

The design of pedestrian bridges has to be based on a careful balance between the stiffness of supporting members. On the one hand, the structure should be sufficiently flexible to reduce the seismic effects; on the other hand, the structure should be sufficiently stiff to guarantee comfortable walking.

REFERENCES

AASHTO (1997) Guide specification for design of pedestrian bridges. American Association of State Highway and Transportation Officials.

Bachmann H (2002) 'Lively' footbridges – a real challenge. *Proceedings of Footbridge 2002: Design and Dynamic Behaviour of Footbridges*, OTUA, Paris.

Bachmann H, Ammann W and Deischl F (1997) *Vibration Problems in Structures: Practical Guidelines*, 2nd edn. Birkhäuser Verlag, Basel, Berlin, Boston.

Department of Transport (1988) *Design Criteria for Footbridges*. Department of Transport, UK.

fib (2005) *Guidelines for the Design of Footbridges*. Guide to good practice prepared by Task Group 1.2. Fédération Internationale du Béton, Lausanne.

Gulvanessian H, Calgaro JA and Holicky M (2002) *Designers' Guide to EN 1900 – Eurocode: Basis of Structural Design*. Thomas Telford, London.

Institute of Civil Engineers (1981) Bridge aerodynamics: proposed British design rules. *Proceedings of the Bridge Aerodynamics Conference*, 25–26 March. Institution of Civil Engineers, London.

Kreuzinger H (2002) Dynamic design strategies for pedestrian and wind action. *Proceedings of Footbridge 2002, Design and Dynamic Behaviour of Footbridges*, Paris. OTUA.

Mathivat J (1983) *The Cantilever Construction of Prestressed Concrete Bridges*. John Wiley & Sons, New York.

Priestly JN, Seible F and Calvi GM (1996) *Seismic Design and Retrofit of Bridges*. John Wiley & Sons, New York.

Roberts TM (2003) Synchronised pedestrian excitation of footbridges. *Bridge Engineering* **156**: 155–160.

Scott R (2001) *In the Wake of Tacoma*. ASCE Press, Resno.

Walther R, Houriet B, Walmar I and Moïa P (1998) *Cable Stayed Bridges*. Thomas Telford Publishing, London.

Chapter 4
Cable analysis

Analysis of the stress ribbon and cable-supported structure is based on the understanding of the static and dynamic behaviour of the single cable.

4.1. Single cable

We assume that a cable of area A and modulus of elasticity E acts as a perfectly flexible member that is able to resist the normal force only. Under this assumption, the cable curve will coincide with the funicular curve of the load applied to the cable and to the chosen value of the horizontal force H (Figure 4.1).

Consider a cable supported at two fixed hinges a and b loaded by a vertical load $q(x)$. We have:

$$l = X(b) - X(a) = x(b),$$
$$h = Y(b) - Y(a) = y(b),$$
$$\tan \beta = \frac{h}{l}.$$
(4.1)

For the given load $q(x)$ and chosen horizontal force H the cable curve is determined by coordinate $y(x)$, sag $f(x)$, the slope of the tangent $y'(x) = \tan \varphi(x)$ and radius of the curvature $R(x)$. These values are derived from the general equilibrium conditions on the element ds. See Figure 4.1 for definitions of notation.

The cable is stressed by a normal force $N(x)$ that has vertical and horizontal components $V(x)$ and $H(x)$, defined:

$$N(x)^2 = H(x)^2 + V(x)^2$$
$$H(x) = N(x) \cos \phi(x)$$
$$V(x) = N(x) \sin \phi(x).$$
(4.2)

For a vertical load $H = \text{const}$, we have

$$V(x) = H \tan \phi(x) = H \frac{dy}{dx} = Hy'(x)$$
$$\frac{dV(x)}{dx} = \frac{H\,dy}{dx\,dx} = Hy''(x)$$
$$dV(x) = Hy''(x)\,dx.$$
(4.3)

The equilibrium of forces in the vertical direction can be defined

$$V(x) - q(x)\,dx - (V(x) + dV(x)) = 0$$
$$V(x) - q(x)\,dx - V(x) - dV(x) = 0$$ (4.4)
$$q(x)\,dx = -dV(x) = -Hy''(x)\,dx$$

where

$$y''(x) = -\frac{q(x)}{H}$$
$$y''(x) = -\frac{1}{R(x)}$$ (4.5)
$$q(x) = \frac{H}{R(x)}$$

and

$$q(x) = -Hy''(x)$$
$$y'(x) = \frac{Q(x)}{H} + C_1$$ (4.6)
$$y(x) = \frac{M(x)}{H} + C_1 x + C_2$$

where $Q(x)$ and $M(x)$ are shear force and bending moment on a simple beam of span l. The constants C_1 and C_2 are determined from the boundary conditions:

$$x = 0; \quad y = y(a); \quad y(a) = 0 + C_1 0 + C_2; \quad C_2 = y(a)$$
$$x = l; \quad y = y(b); \quad y(b) = 0 + C_1 l + C_2 = C_1 l + y(a)$$

hence

$$C_1 = \frac{y(b) - y(a)}{l} = \frac{h}{l}.$$

Figure 4.1 Basic characteristics of the single cable

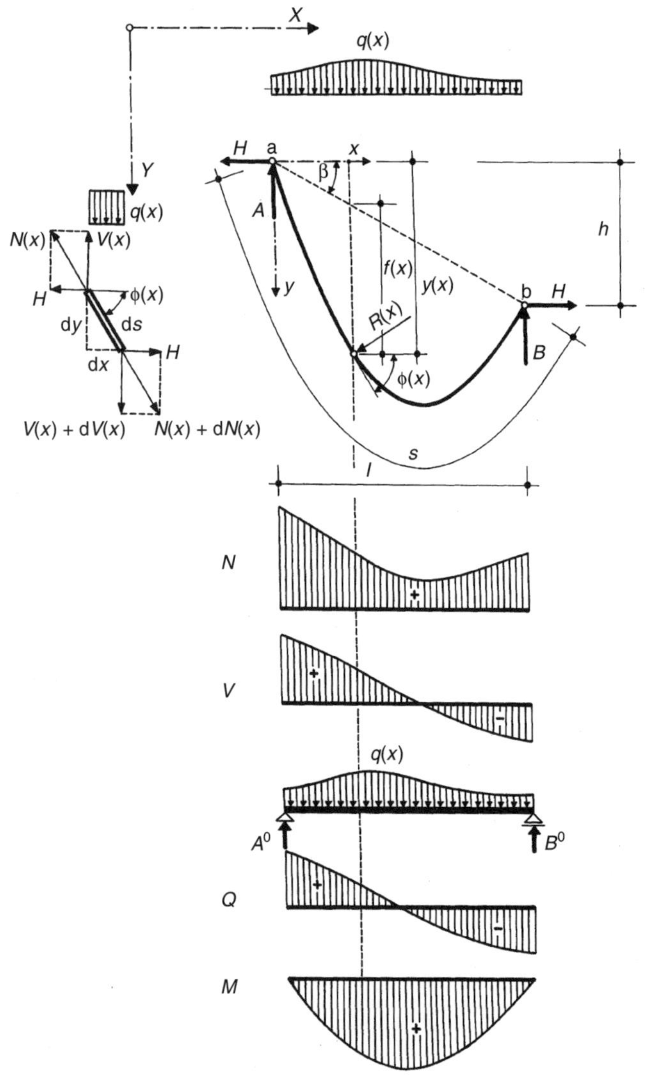

We then have

$$p^0(x) = \frac{Q(x)}{H}$$

$$p(x) = y'(x) = p^0(x) + \frac{h}{l} = p^0(x) + \tan\beta$$

$$f(x) = \frac{M(x)}{H}$$

$$y(x) = \frac{M(x)}{H} + \frac{h}{l}x = f(x) + x\tan\beta$$

(4.7)

for $q(x)$ constant (Figure 4.2). From Figure 4.2, we have

$$p^0(x) = \frac{1}{H}\left(\frac{1}{2}ql - qx\right) = \frac{q}{2H}(l - 2x)$$

$$p(x) = p^0(x) + \frac{h}{l} = p^0(x) + \tan\beta$$

$$f(x) = \frac{M(x)}{H} = \frac{1}{H}\left(\frac{1}{2}qlx - \frac{1}{2}qx^2\right) = \frac{q}{2H}x(l - x)$$

$$y(x) = f(x) + \frac{h}{l}x = f(x) + x\tan\beta$$

(4.8)

$$x = 0$$

$$\max p^0 = \frac{ql}{2H}$$

(4.9)

$$x = \frac{l}{2}$$

$$\max f = \frac{ql^2}{8H}$$

(4.10)

The length of the cable (Figure 4.3) is defined

$$s = \int_0^s ds = \int_0^l \sqrt{dx^2 + dy^2}$$

(4.11)

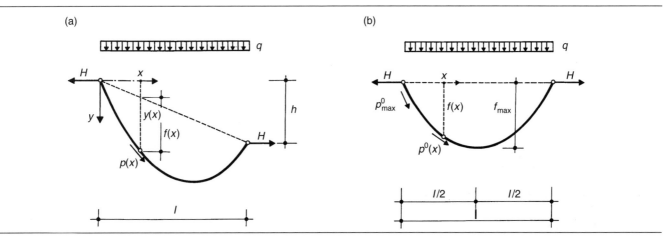

Figure 4.2 Uniformly loaded cable

Figure 4.3 Non-tension length of the cable: (a) initial stage and (b) final stage

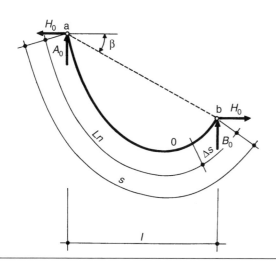
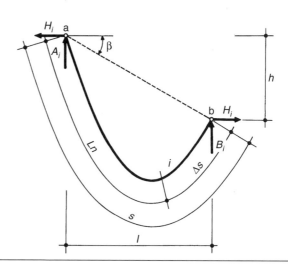

since

$$\tan \phi(x) = y'(x) = \frac{dy}{dx} \quad \text{and} \quad dy = y'(x)\, dx.$$

We then have

$$ds^2 = dx^2 + dy^2 = dx^2(1 + y'^2(x))$$

and

$$s = \int_0^s ds = \int_0^l \sqrt{dx^2 + (1 + y'^2(x))}$$
$$= \int_0^l \sqrt{1 + y'^2(x)}\, dx. \quad (4.12)$$

Since

$$y'(x) = \frac{Q(x)}{H} + \frac{h}{l} = \frac{Q(x)}{H} + \tan \beta$$

and

$$\cos^2 \beta = \frac{1}{1 + \tan^2 \beta}$$

$$\tan \beta = \frac{h}{l}$$

$$\frac{1}{\cos^2 \beta} = 1 + \left(\frac{h}{l}\right)^2,$$

then

$$s = \int_0^l \sqrt{1 + y'^2(x)}\, dx = \int_0^l \sqrt{1 + \left(\frac{Q(x)}{H} + \frac{h}{l}\right)^2}\, dx$$

$$= \int_0^l \sqrt{\frac{1}{\cos^2 \beta}\left(1 + \frac{Q^2(x)\cos^2 \beta}{H^2} + \frac{2Q(x)h\cos^2 \beta}{Hl}\right)}$$

$$= \int_0^l \frac{1}{\cos \beta}\sqrt{1 + \frac{Q^2(x)\cos^2 \beta}{H^2} + \frac{2Q(x)h\cos^2 \beta}{Hl}}\, dx. \quad (4.13)$$

We can express s as

$$s = \frac{1}{\cos \beta} \int_0^l (1 + B)^{1/2}\, dx \quad (4.14)$$

where

$$B = \frac{Q^2(x)\cos^2 \beta}{H^2} + \frac{2Q(x)h\cos^2 \beta}{hl}. \quad (4.15)$$

Since

$$|B| < 1,$$

it is possible to use the binomial formula:

$$(1 + B)^p = 1 + \binom{B}{1}x + \binom{B}{2}x^2 + \binom{B}{3}x^3 + \ldots \quad (4.16)$$

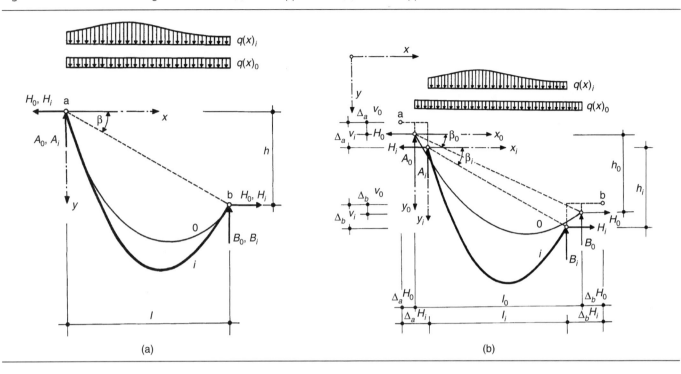

Figure 4.4 Initial and final stage of the cable: (a) fixed supports and (b) flexible supports

If we use only the first two terms, we have

$$(1+B)^p = 1 + \binom{p}{1}B$$

$$(1+B)^{1/2} = 1 + \frac{1}{2}B$$

$$= 1 + \frac{1}{2}\left(\frac{Q^2(x)\cos^2\beta}{2H^2} + \frac{2Q(x)h\cos^2\beta}{Hl}\right)$$

and hence

$$s = \frac{1}{\cos\beta}\left(\int_0^l dx + \int_0^l \frac{Q^2(x)\cos^2\beta}{2H^2} dx + \int_0^l \frac{2Q(x)h\cos^2\beta}{Hl} dx\right). \quad (4.17)$$

Since

$$\int_0^l Q(x)\,dx = 0$$

i.e. there is an equilibrium of forces in the vertical direction, then

$$\int_0^l \frac{2Q(x)h\cos^2\beta}{Hl}\,dx = 0$$

and

$$s = \frac{1}{\cos\beta}\left(l + \frac{\cos^2\beta}{2H^2}\int_0^l Q^2(x)\,dx\right)$$

$$= \frac{l}{\cos\beta} + \frac{\cos\beta}{2H^2}\int_0^l Q^2(x)\,dx \quad (4.18)$$

The term

$$D = \int_0^l Q^2(x)\,dx = \int_0^l Q(x)Q(x)\,dx \quad (4.19)$$

is usually determined by a Verescagin rule. The area $Q(x)$ is multiplied by the value of Q_t, which occurs at the centre of the gravity of the area (Figure 4.4).

For example, for uniform load $q(x) = \text{const}$ (Figure 4.5(a)), we have

$$D = \int_0^l Q^2(x)\,dx = 2 \times \left[\left(\frac{1}{2}\left(\frac{1}{2}ql\right)\frac{l}{2}\right) \times \left(\frac{2}{3} \times \frac{1}{2}ql\right)\right] = \frac{q^2l^3}{12}.$$

For the cable of length l (Figure 4.2(b)), sag f, horizontal force H and uniform load q, the length of the cable is defined:

$$s = \frac{l}{1} + \frac{1}{2H^2}\frac{q^2l^3}{12} = l + \frac{q^2l^3}{24H^2}$$

Figure 4.5 Determination of D for: (a) a uniform load and (b) an arbitrary load

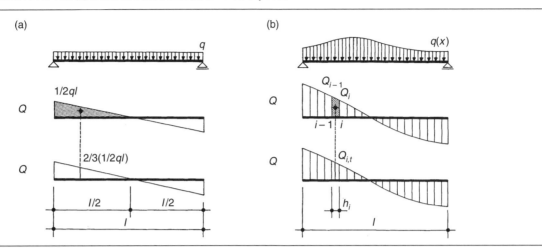

for

$$H = \frac{ql^2}{8f} \quad \text{and} \quad s = l + \frac{8f^2}{3l}.$$

For a general load (Figure 4.5(b)) it is possible to divide the length of the girder into the elements of the length h and substitute the course of $Q(x)$ by a polygon. Then:

$$D = \int_0^l Q^2(x)\,dx = \sum_{i=1}^n D_i = \sum_{i=1}^n \int_0^h Q^2\,dx$$

It is possible to calculate the values of D_i for each element:

$$D_i = \frac{Q_{i-1} + Q_i}{2} \times h \times Q_{i,t}$$

where $Q_{i,t}$ is the value of $Q(x)$ at the centre of gravity of the element.

4.1.1 Elastic elongation of the cable
Since

$$\frac{N(s)}{H} = \frac{ds}{dx}, \quad \text{i.e. } N(s) = H\frac{ds}{dx}$$

we have

$$\Delta s = \int_0^s \frac{N(s)}{EA}\,ds = \int_0^s \frac{H}{EA}\frac{ds^2}{dx} \qquad (4.20)$$

and

$$ds = \int_0^l (1 + y'^2(x))\,dx. \qquad (4.21)$$

For the vertical load,

$$\Delta s = \frac{H}{EA}\int_0^s \frac{ds^2}{dx} = \frac{H}{EA}\int_0^l \frac{\left(\sqrt{1+y'^2(x)}\,dx\right)^2}{dx}$$

$$= \frac{H}{EA}\int_0^l \left[1 + \left(\frac{Q(x)}{H} + \frac{h}{l}\right)^2\right] dx \qquad (4.22)$$

$$= \frac{H}{EA\cos^2\beta}\left(l + \frac{\cos^2\beta}{H^2}\int_0^l Q(x)\,dx\right)$$

$$= \frac{H}{EA}\left(\frac{l}{\cos^2\beta} + \frac{1}{H^2}\int_0^l Q(x)\,dx\right).$$

Since

$$s = \frac{l}{\cos\beta} + \frac{\cos\beta}{2H^2}\int_0^l Q^2(x)\,dx$$

we have

$$\Delta s = \frac{H}{EA\cos^2\beta}\left(l + \frac{\cos^2\beta}{H^2}\int_0^l Q^2(x)\,dx\right) \qquad (4.23)$$

and hence

$$\Delta s = \frac{2H}{EA\cos\beta}\left(\frac{l}{\cos\beta} + \frac{\cos^2\beta}{2H^2\cos\beta}\int_0^l Q^2(x)\,dx\right)$$

$$= \frac{2H}{EA\cos\beta}\left(\frac{l}{\cos\beta} + \frac{\cos^2\beta}{2H^2\cos\beta}D\right) \qquad (4.24)$$

$$= \frac{2H}{EA\cos\beta}\left(s - \frac{l}{2\cos\beta}\right).$$

The elastic elongation of the cable (Figure 4.2(b)) is

$$\Delta s = \frac{2H}{EA}\left(s - \frac{l}{2}\right) = \frac{2H}{EA}\left(l + \frac{8f^2}{3l} - \frac{l}{2}\right) = \frac{H}{EA}\left(l + \frac{16f^2}{3l}\right).$$

4.1.2 Determining the horizontal force H_i

For the load $q(x)_0$, horizontal force H_0 and temperature t_0 the length of the non-tension cable (Figure 4.4(a)) is defined:

$$Ln = s_0 - \Delta s_0$$
$$= \frac{l}{\cos\beta} + \frac{\cos\beta}{2H_0^2}D_0 - \frac{H_0 l}{EA\cos^2\beta} - \frac{1}{H_0 EA}D_0 \quad (4.25)$$

$$D_0 = \int_0^l Q_{x,0}^2\, dx.$$

For the load $q(x)_i$, unknown horizontal force H_i and temperature t_i, the length of the non-tension cable (Figure 4.4(b)) is defined:

$$Ln_i = s_i - \Delta s_i$$
$$Ln_i = Ln(1 + \alpha_t \Delta t_i) \quad (4.26)$$
$$D_i = \int_0^l Q_{x,i}^2\, dx$$

where the temperature change $\Delta t_i = t_i - t_0$ and α_t is a coefficient of thermal expansion. Since

$$s_i = \frac{l}{\cos\beta} + \frac{\cos\beta}{2H_i^2}D_i$$

and

$$\Delta s_i = \frac{H_i l}{EA\cos^2\beta} + \frac{1}{H_i EA}D_i,$$

we have

$$Ln_i = s_i - \Delta s_i$$
$$= \frac{l}{\cos\beta} + \frac{\cos\beta}{2H_i^2}D_i - \frac{H_i l}{EA\cos^2\beta} - \frac{D_i}{EAH_i} \quad (4.27)$$
$$\left(Ln_i - \frac{l}{\cos\beta}\right) - \frac{\cos\beta}{2H_i^2}D_i + \frac{H_i l}{EA\cos^2\beta} + \frac{D_i}{EAH_i} = 0.$$

If we denote

$$a = \frac{l}{EA\cos^2\beta},$$
$$b = Ln_i - \frac{l}{\cos\beta},$$

$$c = \frac{D_i}{EA}$$

and

$$d = \frac{\cos\beta}{2}D,$$

we can describe H_i in terms of a cubic equation, i.e.

$$b + \frac{d}{H_i^2} + aH_i + \frac{c}{H_i} = 0$$
$$aH_i^3 + bH_i^2 + cH_i + d = 0. \quad (4.28)$$

from which the unknown horizontal force H_i can be easily determined.

4.1.3 Influence of deformation of supports and elongation of the cable at the anchor blocks

In actual structures it is necessary to include possible deformations of supports and elongations of the cable at the anchor blocks.

Deformations of supports for load 0 and i are defined:

$$\begin{aligned}
\Delta_{ai}^V &= A_i \delta_a^V & \Delta_{a0}^V &= A_0 \delta_a^V \\
\Delta_{ai}^H &= H_i \delta_a^H & \Delta_{a0}^H &= H_0 \delta_a^H \\
\Delta_{bi}^V &= B_i \delta_b^V & \Delta_{b0}^V &= B_0 \delta_b^V \\
\Delta_{bi}^H &= H_i \delta_b^H & \Delta_{b0}^H &= H_0 \delta_b^H
\end{aligned} \quad (4.29)$$

and depend on values of the reactions and positive unit deformations δ_a^V, δ_a^H, δ_b^V and δ_b^H. For load 0,

$$l_0 = X_b - X_a - \Delta_{a0}^H - \Delta_{b0}^H$$
$$h_0 = Y_b - Y_a - \Delta_{a0}^V - \Delta_{bi0}^V \quad (4.30)$$
$$\tan\beta_0 = \frac{h_0}{l_0}.$$

For load i,

$$l_i = X_b - X_a - \Delta_{ai}^H - \Delta_{bi}^H$$
$$h_i = Y_b - Y_a - \Delta_{ai}^V - \Delta_{bi}^V \quad (4.31)$$
$$\tan\beta_i = \frac{h_i}{l_i}.$$

The elastic deformations of the cable in the anchor blocks a and b are defined for load 0:

$$\Delta_{an,0} = kH_0 \quad (4.32)$$

Figure 4.6 Elastic deformations of the cable at the anchor blocks

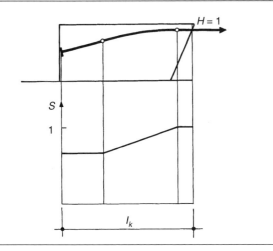

and load i:

$$\Delta_{an,i} = kH_i \quad (4.33)$$

where $k = k_a + k_b$ (Figure 4.6) expresses the elongation of the cable at the anchor blocks a and b due to unit horizontal force $H = 1$. k_a and k_b are defined:

$$k_a = \int_0^{l_{ka}} \frac{S_{ka}}{EA}\,\mathrm{d}s \qquad k_b = \int_0^{l_{kb}} \frac{S_{kb}}{EA}\,\mathrm{d}s \quad (4.34)$$

For the load $q(x)_0$, horizontal force H_0 and temperature t_0, the length of the non-tension cable (Figure 4.4(a)) is defined:

$$\begin{aligned}
Ln &= s_0 - \Delta s_0 - \Delta_{an,0} \\
&= \frac{l_0}{\cos \beta_0} + \frac{\cos \beta_0}{2H_0^2} D_0 - \frac{H_0 l_0}{EA \cos^2 \beta_0} \\
&\quad - \frac{1}{H_0 EA} D_0 - kH_0
\end{aligned} \quad (4.35)$$

$$D_0 = \int_0^{l_0} Q_{x,0}^2\,\mathrm{d}x$$

For the load $q(x)_i$, unknown horizontal force H_i and temperature t_i, the length of the non-tension cable (Figure 4.4(b)) is defined:

$$\begin{aligned}
Ln_i &= Ln(1 + \alpha_t \Delta t_i) \\
Ln_i &= s_i - \Delta s_i - \Delta_{an,i} \\
&= \frac{l_i}{\cos \beta_i} + \frac{\cos \beta_i}{2H_i^2} D_i - \frac{H_i l_i}{EA \cos^2 \beta_i} - \frac{1}{H_i EA} D_i - kH_i
\end{aligned}$$

$$D_i = \int_0^{l_i} Q_{x,i}^2\,\mathrm{d}x$$

where

$$\left(Ln_i - \frac{l_i}{\cos \beta_i}\right) - \frac{\cos \beta_i}{2H_i^2} D_i + \frac{H_i l_i}{EA \cos^2 \beta_i} + kH_i + \frac{D_i}{EAH_i} = 0 \quad (4.36)$$

If we set

$$a = \frac{l_i}{EA \cos^2 \beta_i} + k, \qquad b = Ln_i - \frac{l_i}{\cos \beta_i}, \qquad c = \frac{D_i}{EA}$$

and

$$d = \frac{\cos \beta_i}{2} D_i,$$

we obtain a cubic equation to determine H_i:

$$aH_i^3 + bH_i^2 + cH_i + d = 0. \quad (4.37)$$

Since the parameters a, b, c and d depend on the span l_i and vertical difference h_i (which depends on horizontal force H_i), it is not possible to determine the unknown H_i directly by solving Equation (4.37); it is therefore necessary to determine H_i by iteration. First, the unknown H_i is determined for zero deformation of supports and zero elongation of the cable at the anchor blocks. For this force, the vertical reactions A_i and B_i, span length l_i, vertical difference h_i, parameters a, b, c and d and new horizontal force H_i are utilised. The computation is repeated until the difference between the subsequent solutions is smaller than the required accuracy.

4.2. Bending of the cable

The bending of the cable is derived from the analysis of the single cable which is stressed by a known horizontal force H (Eibel et al., 1973).

Figure 4.7 shows a single cable of the area A, moment of inertia I and modulus of elasticity E that is fixed to the supports a and

Figure 4.7 Geometry and internal forces at the cable

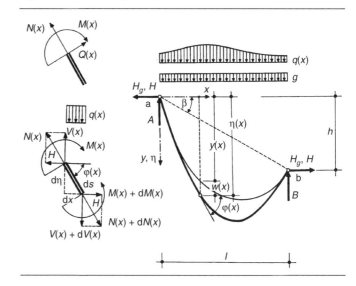

b. The cable is loaded by load $g(x)$ and $q(x)$. Corresponding horizontal forces are H_g and H.

It is assumed that erection of the cable is done is such a way that the load g does not cause any bending of the cable. For constant g, the shape of the cable given by $y(x)$ is the second-degree parabola:

$$y(x) = f(x) + \frac{h}{l}x = \frac{g}{2H_g}x(l-x) + \frac{h}{l}x$$

$$= \frac{g}{2H_g}xl - \frac{g}{2H_g}x^2 + \frac{h}{l}x \tag{4.38}$$

$$y'(x) = \frac{g}{2H_g}l - 2\frac{g}{2H_g}x + \frac{h}{l} = \frac{g}{H_g}\left(\frac{l}{2} - x\right) + \frac{h}{l}$$

$$y''(x) = -\frac{g}{H_g} = \frac{1}{R_g}.$$

The shape of the cable for load $q(x)$ is given by the coordinate

$$\eta(x) = y(x) + w(x)$$
$$d\eta(x) = dy(x) + dw(x) \tag{4.39}$$

where $w(x)$ is the deformation of the cable due to load $q(x) - g(x)$.

The cable is stressed by normal force $N(x)$, shear force $Q(x)$ and by bending moment $M(x)$:

$$N(x) = H\cos\phi(x) + V\sin\phi(x)$$
$$Q(x) = -H\sin\phi(x) + V\cos\phi(x). \tag{4.40}$$

These values are derived from the equilibrium conditions on the element ds.

For vertical load with constant H,

$$\Sigma V = 0$$
$$V(x) - q(x)\,dx - (V + dV) = 0$$
$$-q(x)\,dx = dV \tag{4.41}$$
$$-q(x) = \frac{dV}{dx}$$

and

$$\Sigma M = 0$$
$$V(x)\,dx - H\,d\eta(x) + M(x) - \tfrac{1}{2}q(x)\,dx^2 \tag{4.42}$$
$$- (M(x) + dM(x)) = 0.$$

Since $dx^2 \ll dx$, dx^2 can be neglected. We then have

$$V(x)\,dx - H\,d\eta(x) - dM(x) = 0$$
$$V(x) - H\frac{d\eta(x)}{dx} - \frac{dM(x)}{dx} = 0$$
$$\frac{dV(x)}{dx} - H\frac{d^2\eta(x)}{dx^2} - \frac{d^2M(x)}{dx^2} = 0$$
$$-q(x) = \frac{dV(x)}{dx} \tag{4.43}$$
$$M(x) = -EI\frac{d^2w(x)}{dx^2}$$
$$-q(x) - H\frac{d^2\eta(x)}{dx^2} - \frac{d}{dx^2}\left(-EI\frac{d^2w}{dx^2}\right) = 0$$
$$EI\frac{d^4w(x)}{dx^4} - H\frac{d^2\eta(x)}{dx^2} = q(x).$$

We then have

$$\eta(x) = y(x) + w(x)$$
$$EI\frac{d^4w(x)}{dx^4} - H\frac{d^2\eta(x)}{dx^2} = q(x)$$
$$EI\frac{d^4w(x)}{dx^4} - H\frac{d^2y(x)}{dx^2} - H\frac{d^2w(x)}{dx^2} = q(x) \tag{4.44}$$
$$EI\frac{d^4w(x)}{dx^4} - H\frac{d^2w(x)}{dx^2} = q(x) + H\frac{d^2y}{dx^2}$$
$$= q(x) + H\left(-\frac{g}{H_g}\right) = q(x) + H\left(\frac{1}{R_g}\right)$$
$$= q(x) + \frac{H}{R_g} = q(x) - \frac{H}{H_g}g$$

Equation (4.44) can be easily extended to express the elastic support of the portion of the cable by Winkler's springs (Figure 4.8):

$$\Sigma V = 0$$
$$V(x) - q(x)\,dx + kw(x)\,dx - (V + dV) = 0$$
$$-q(x)\,dx + kw(x) = dV \tag{4.45}$$
$$-q(x) + kw(x) = \frac{dV}{dx}.$$

The characteristic of the spring $k(x)$ is a 'stress' that corresponds to its unit deformation.

$$\Sigma M = 0$$
$$V(x)\,dx - H\,d\eta(x) + M(x) - \tfrac{1}{2}q(x)\,dx^2 \tag{4.46}$$
$$+ kw(x)\tfrac{1}{2}dx^2 - (M(x) + dM(x)) = 0$$

Figure 4.8 Geometry and internal forces at flexibly supported cable

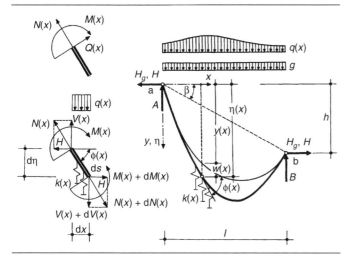

Figure 4.9 Bending moments: (a) at support and (b) under point load

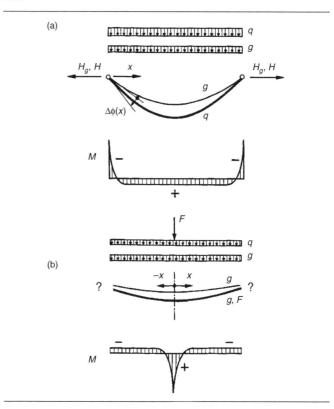

Since $dx^2 \ll dx$, dx^2 can be neglected and we then have:

$$EI\frac{d^4w(x)}{dx^4} - H\frac{d^2w(x)}{dx^2} + kw(x) = q(x) + \frac{H}{R_g}$$

$$= q(x) - \frac{H}{H_g}g \quad (4.47)$$

The solution of Equation (4.45) can be written as:

$$w(x) = w_h(x) + w_p(x) \quad (4.48)$$

where the particular solution $w_p(x)$ corresponds to deformation of the cable without bending stiffness; the homogenous solution can be written:

$$w_h(x) = Ae^{\lambda x} + Be^{-\lambda x} + C + Dx$$

$$\lambda = \sqrt{\frac{H}{EI}}. \quad (4.49)$$

A direct solution is possible only for special cases (Figure 4.9), for example the course of the bending moment $M(x)$ in the vicinity of the support of the cable loaded by uniform loads g and q and corresponding horizontal forces are H_g and H is solved for an infinitively long cable.

The particular solution is:

$$w_p(x) = \eta(x) - y(x) = \frac{q}{2H}(l-x)x - \frac{g}{2H_g}(l-x)x$$

$$= \frac{qlx}{2H} - \frac{qx^2}{2H} - \frac{glx}{2H_g} + \frac{gx^2}{2H_g}$$

$$w'_p(x) = \frac{ql}{2H} - \frac{qx}{H} - \frac{gl}{2H_g} + \frac{gx}{H_g}$$

$$w''_p(x) = -\frac{q}{H} + \frac{g}{H_g} = -\Delta q$$

$x = 0$

$$w'_p(0) = \frac{ql}{2H} - \frac{gl}{2H_g} = \phi_q(0) - \phi_g(0) = \Delta\phi \quad (4.50)$$

The homogenous solution is:

For $x = \infty$, $M = 0$, $w = 0$, $e^{\lambda x} \to \infty$, and therefore $A = 0$ and $D = 0$. Thus

$$w_h(x) = Be^{-\lambda x} + C$$
$$w'_h(x) = -\lambda Be^{-\lambda x}$$
$$w''_h(x) = \lambda^2 Be^{-\lambda x}$$
$$w'(x) = w'_h(x) + w'_p(x)$$

For $x = 0$

$$w'(x) = 0 = -\lambda Be^0 + \Delta\phi$$

$$B = \frac{\Delta\phi}{\lambda} \quad (4.51)$$

$$w''(x) = w_h''(x) + w_p''(x) = \lambda^2 \frac{\Delta\phi}{\lambda} e^{-\lambda x} - \Delta q$$

$$M(x) = EJw''(x)$$

$$\lambda = \sqrt{\frac{H}{EI}} \quad (4.52)$$

$$M(x) = \Delta\phi\sqrt{H \cdot EI}\, e^{-\lambda x} + \Delta q EI$$

The bending moment at an infinitively long cable that is loaded by point load F and by uniform load q (Figure 4.9(b)) can be derived similarly. For uniform load g that does not cause the bending, the cable is stressed by horizontal force H_g; for uniform load q and point load F the cable is stressed by horizontal force H. The course of bending moment is defined:

$$M(x) = \frac{F}{2\lambda} e^{-\lambda x} + EI\Delta q$$

$$\lambda = \sqrt{\frac{H}{EI}} \quad (4.53)$$

$$\Delta q = \frac{q}{H} - \frac{g}{H_g}$$

Rather then solving the equations for different loading conditions, the author developed a program in which the deformation and corresponding shear forces and bending moments were solved using the *finite difference method* (Tomishenko and Goodier, 1970). This approach enables us to describe a local stiffening of the cable and supporting of portion of the cable by Winkler springs.

In the analysis, the cable was divided into short elements of length h (in the actual structure, a cable of length 100 m was divided into 10 000 elements of length of 0.010 m). The elements can have different stiffness values given by EI and can be supported by springs of different stiffness.

The derivations were substituted by well-known formulae (Figure 4.10)

Figure 4.10 Finite difference method

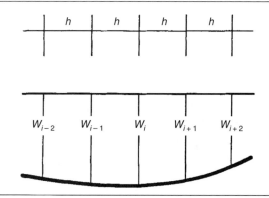

$$\frac{d^2w(x)}{dx^2} = \frac{1}{h}(w_{i-1} - 2w_i + w_{i+1})$$

$$\frac{d^4w(x)}{dx^4} = \frac{1}{h^4}(w_{i-2} - 4w_{i-1} + 6w_i - 4w_{i+1} + w_{i+2}) \quad (4.54)$$

In this way, a solution of Equation (4.47) was substituted by a solution of a system of linear equations. From the discrete values of the deflections, the bending moments and shear forces were determined.

From the discrete values of deformations, the bending moment and shear forces were determined from:

$$M_i = \frac{EI_i}{h^2}(w_{i-1} - 2w_i + w_{i+1})$$

$$T_i = \frac{M_{i+1} - M_{i-1}}{2h}. \quad (4.55)$$

The described analysis of the bending of the cable was verified during the loading tests of the cables that were developed for the Elbe Bridge (Figure 2.10). The cable was formed from 18 strands of 15.5 mm diameter that were grouted in steel tubes. The tubes and mortar are composite with the strands and contribute to the resistance of the strands to the live load (Figure 2.42). The tested cable was loaded with a point load situated at its mid-span (Figure 4.11). During the test, the strain in the steel tubes at the mid-span and support sections was carefully measured. The bending moment was calculated from these strains. Figure 4.12, which depicts the arrangement and results of the test, highlights that a good agreement of results has been achieved.

During the analysis, the cable is initially analysed as a perfectly flexible member. The initial state is chosen as the state in which erection guarantees that the cable is without bending. To analyse loading, the unknown horizontal force is determined and then used to determine the deformation, shear forces and bending moments.

Figure 4.11 Loading test of the stay cable

Figure 4.12 Deformation and bending moments of the tested cable

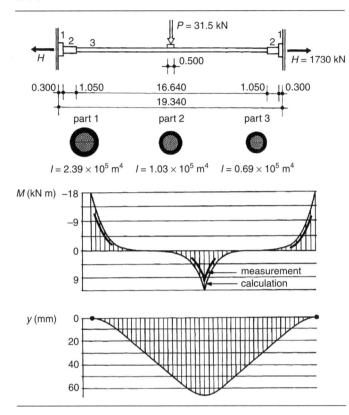

In actual cables loaded by a uniform load, the bending moment is nearly zero along the length of the cable. Significant values of bending moments originate only close to the supports and under the point load (Figure 4.9); their course is exponential, which must be taken into consideration when we analyse the structures by modern non-linear programs by finite element methods. To cover the concentration of stresses, a very fine mesh of elements has to be used close to supports and point loads.

To understand the difference between the behaviour of the beam and of the cable, a stress ribbon and beam structure of span $L=33$, 66 and 99 m loaded by uniform load and by vertical deflection of supports are examined here. The stress ribbon was modelled as a cable and analysed by the above process.

The structure has area $A=1.25\,\text{m}^2$, moment of inertia $I=0.0065104\,\text{m}^4$ and modulus of elasticity $E=36\,000$ MPa. The stress ribbon structures have sag at mid-span $f_{L/2}=0.02L$, corresponding horizontal force $H_g=gL^2/8f_{L/2}=6.25(gL)=195.3125L$. The dead load $g=31.25$ kN/m does not cause bending of the structure.

Figure 4.13(a) and Table 4.1 show the deflection and bending moment in the beam and cable for load $p=20$ kN/m. It is evident that the deflections and bending moments in the stress ribbon are very low.

Figure 4.13 Deformation and bending moments at beam and stress ribbon: (a) uniform load and (b) vertical deflection of support

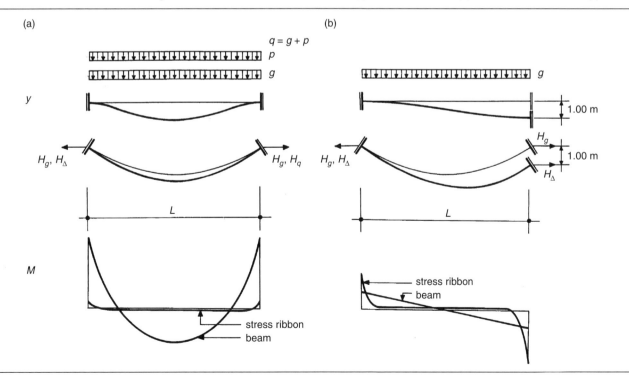

Table 4.1 Comparison of the static effects

		L: m		
		33.00	66.00	99.00
M_p	Beam (MNm)	−1.815	−7.262	−16.335
$M_{L/2}$	Beam (MNm)	0.908	3.631	8.168
$y_{L/2}$	Beam (m)	0.5271	8.4332	42.6933
H_q	Stress ribbon (MN)	10.180	19.768	28.939
M_p	Stress ribbon (MNm)	−0.106	−0.339	−0.594
$M_{L/2}$	Stress ribbon (MNm)	0.034	0.039	0.036
$y_{L/2}$	Stress ribbon (m)	0.115	0.0726	0.1679

Figure 4.13(b) and Table 4.2 show the deflection and bending moment in the beam and stress ribbon stressed by a vertical deflection of support $\Delta = 1\,\text{m}$. With the increasing span length, the bending moments in the beams are reduced proportionally to the square of their length.

On the contrary, the bending moments in the stress ribbon have significant values that in the longer spans are even higher that in the beam. It is therefore necessary to carefully analyse the bending of cables (or prestressed bands) in structures where significant vertical deformations can occur (e.g. in the cables of a cable-supported structure).

Note: the above-described analysis is hypothetical since a beam of the above dimensions would fail.

4.3. Natural modes and frequencies

The natural modes and frequencies of a single cable (Figure 4.14) can be determined according the following formules (Strasky and Pirner, 1986):

$$f_{(1)} = \frac{1}{2}\sqrt{\frac{1}{\mu}\left(\frac{H}{l^2} + \frac{EAf^2\pi^2}{2l^4} + \frac{EI\pi^2}{l^4}\right)} \quad (4.56)$$

Table 4.2

		L: m		
		33.00	66.00	99.00
$M_{p,a}$	Beam (MNm)	−1.291	−0.323	−0.143
$M_{p,b}$	Beam (MNm)	1.291	0.323	0.143
H_q	Stress ribbon (MN)	8.266	13.542	19.730
$M_{p,a}$	Stress ribbon (MNm)	−1.429	−0.787	−0.637
$M_{p,b}$	Stress ribbon (MNm)	2.485	1.166	0.839

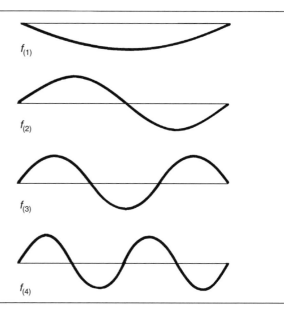

Figure 4.14 Natural modes

$$f_{(n)} = \frac{1}{2}\sqrt{\frac{1}{\mu}\left(\frac{Hn^2}{l^2} + \frac{EI\pi^2 n^2}{l^4}\right)} \quad (4.57)$$

where H is horizontal force, μ is mass of the cable per unit length, f is sag of the cable, E is the modulus of elasticity, A is area and I is moment of inertia.

The term

$$\frac{EAf^2\pi^2}{2l^4}$$

in Equation (4.56) describes the normal stiffness of the cable that has to elongate when vibrating in the first mode. This is the reason why, in some cases, the first mode is higher than the second mode.

The term

$$\frac{EI\pi^2 n^2}{l^4}$$

in Equation (4.57) describes the bending stiffness of the cable which, in engineering calculations, is insignificant.

Natural modes of vibrations are defined as:

$$w(x,t) = \sum_{i=1}^{n} A_i \cos(2\pi f_i)t + B_i \sin(2\pi f_i)t \sin\frac{i\pi\left(x - \frac{l}{2}\right)}{l}$$

where A_i, B_i are determined by the progression of the right-hand side of equations:

$$w(x,0) = g(x)$$
$$w'(x,0) = h(x).$$

REFERENCES

Eibel J, Pelle K and Nehse H (1973) *Zur Berechnung von Spannbandbrücken – Flache Hängebänder*. Verner, Verlang, Düsseldorf.

Strasky J and Pirner M (1986) *DS-L Stress ribbon footbridges*. Dopravni stavby, Olomouc, Czechoslovakia.

Timoshenko SP and Goodier JN (1970) *Theory of Elasticity*. McGraw-Hill, New York.

Stress Ribbon and Cable-supported Pedestrian Bridges
ISBN 978-0-7277-4146-2

ICE Publishing: All rights reserved
doi: 10.1680/srcspb.41462.055

Chapter 5
Effects of prestressing

For a better understanding of the function of prestressing in stress ribbon bridges, some basic facts will be recalled. According to FIP (1998) prestress is applied through a construction control process (prestressing) by stressing tendons (prestressing reinforcement) relative to the concrete member. The prestress is exerted by tendons made of high-strength steel (either bars, wires or strands). The tendons can be pre-tensioned or post-tensioned.

Post-tensioned tendons can be situated inside (internal tendons; Figure 5.1(a)) or outside of the cross-section (external tendons). The internal tendons can be bonded to the structure by grouting or left provisionally or permanently unbonded.

In the structure with bonded tendons the cement mortar grouted between the prestressing steel and the ducts guarantees that, after post-tensioning, any additional strain is the same for the concrete and steel in all sections. In the ultimate limit state the force in the steel that resists the ultimate load therefore corresponds to the width of the cracks.

The internal tendons can also be formed by monostrands, greased and sheathed strands. Since there is no bond between the steel and concrete, the strains in the concrete and steel are different. In the 'ultimate limit state' the force in the steel that resists the ultimate load therefore depends on the total elongation of the tendon between the anchors. The force in the tendons is therefore smaller than for the case of the bonded tendons.

The external tendons can be situated inside (Figure 5.1(b)) or outside (Figure 5.1(c)) the depth of the section. The tendons are anchored in so-called anchor blocks and deviated at deviators. The deformations of the structure and the tendons are the same only at anchor blocks and deviators. The deviator can be rigidly connected to the tendons or it can allow the movement of the prestressing steel in the ducts. At the ultimate limit state the force in the external tendons corresponds to the elongation of the tendons between the points in which the tendon slippage is prevented. The actual geometry of the tendons has to be considered in the analysis; a sectional analysis alone is not sufficient.

During post-tensioning and during the service of the structure, the effects of prestress can be expressed by an equivalent load at anchors by the force $N=-P$ acting in the direction of the tendons and along the length of the duct or at deviators by radial forces r. These radial forces are the resultant of the normal and tangent forces k and t.

The normal and tangent forces can be determined from the theory of frictional loss of a cable around a curve. Consider an infinitesimal length ds of prestressing tendon whose centroid follows an arc of radius $\rho(\alpha)$ (Figure 5.2). The change in angle for a length ds is

$$d\alpha = \frac{ds}{\rho(\alpha)} \qquad (5.1)$$

where

$$ds = \rho(\alpha)\, d\alpha. \qquad (5.2)$$

Along the length ds the force $P(\alpha)$ is changed by $dP(\alpha)$. The forces $P(\alpha)$ and $P(\alpha) + dP(\alpha)$ act on this infinitesimal element.

The action of these forces can be substituted by normal forces $k(\alpha)$ and by tangent forces $t(\alpha)$. Their resultants are

$$dK(\alpha) = ds\, k(\alpha) \qquad (5.3)$$

and

$$dT(\alpha) = ds\, t(\alpha). \qquad (5.4)$$

These forces can be determined from the conditions for equilibrium. The equilibrium in the direction of the tangent is

$$P(\alpha)\cos\frac{d\alpha}{2} - (P(\alpha) + dP(\alpha))\cos\frac{d\alpha}{2} = dT(\alpha). \qquad (5.5)$$

Since

$$\cos\frac{d\alpha}{2} \cong 1,$$

Figure 5.1 Types of prestressing: (a) internal; (b) external within perimeter; and (c) external outside the perimeter

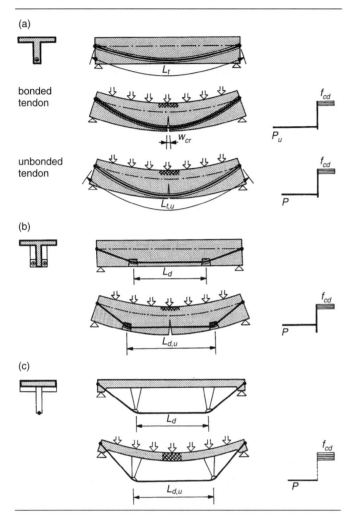

Figure 5.2 Equivalent forces

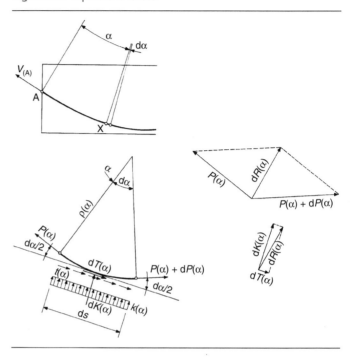

we have

$$P(\alpha)\,d\alpha = dK(\alpha). \tag{5.8}$$

In other words,

$$dK(\alpha) = P(\alpha)\frac{ds}{\rho(\alpha)}$$

$$k(\alpha)\,ds = P(\alpha)\frac{ds}{\rho(\alpha)} \tag{5.9}$$

$$k(\alpha) = \frac{P(\alpha)}{\rho(\alpha)}.$$

Friction $t(\alpha)$ is proportional to the normal compression $k(\alpha)$ and the coefficient of friction μ, thus

$$t(\alpha) = \mu k(\alpha). \tag{5.10}$$

The friction force always acts against the direction of movement of the tendon.

The value of the tendon force $P(\alpha)$ at the point X is given by the well-known formula

$$P(\alpha) = P_A\,e^{-\mu(\alpha + kx)} \tag{5.11}$$

where P_A is the prestressing force at the anchor, μ is the coefficient of friction, α is the sum of angular displacement along x, k

we have

$$dT(\alpha) = -dP(\alpha). \tag{5.6}$$

Equilibrium in normal direction is

$$P(\alpha)\sin\frac{d\alpha}{2} + (P(\alpha) + dP(\alpha))\sin\frac{d\alpha}{2} = dK(\alpha). \tag{5.7}$$

Since

$$\sin\frac{d\alpha}{2} \cong \frac{d\alpha}{2}$$

and

$$dP(\alpha)\frac{d\alpha}{2} \cong 0,$$

Figure 5.3 Modelling prestressing

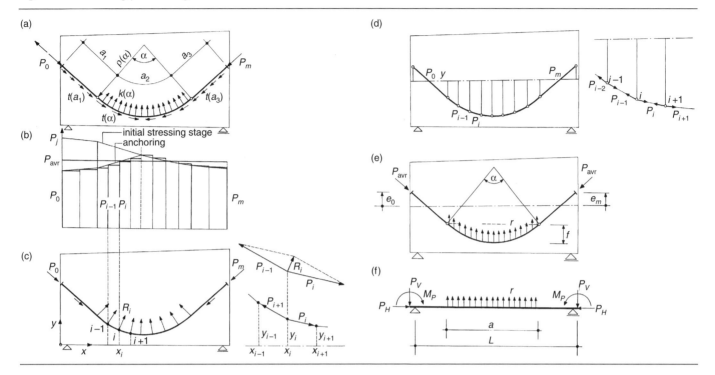

is the unintentional angular displacement or wobble (per unit length) and x is the length of the tendon from the anchor.

With $k(\alpha)$ and $t(\alpha)$ determined using equations (5.9) and (5.10), the resultant force $r(\alpha)$ can be determined

$$r(\alpha) = \sqrt{k^2(\alpha) + t^2(\alpha)}. \tag{5.12}$$

This force that acts along the length of the cable represents the equivalent load of the tendon. It can be transformed into the global coordinate axes X and Y.

Since most of the structures are currently analysed by the finite element method, it is not necessary to determine the function of forces $r(\alpha)$. It is more appropriate to determine the forces at several nodes along the length of the tendons. Figure 5.3(a) shows a typical layout of a tendon formed by two straight parts and one parabolic curve. The tendon is post-tensioned from the left anchorage.

Figure 5.3(b) depicts the prestressing force which is influenced by the friction losses and the wedge draw-in at the anchorage. The tendon is divided in several elements along the length of the structure in which we can substitute a continuous force diagram with a piecewise constant diagram. The length of the elements should be chosen in such a way that sufficient accuracy is ensured.

From the geometry of the tendon and the forces in the elements, we can determine the equivalent radial forces R:

$$\tan \alpha_{i-1} = \frac{y_{i-1} - y_i}{x_i - x_{i-1}}$$

$$\tan \alpha_i = \frac{y_i - y_{i+1}}{x_{i+1} - x_i}$$

$$R_{xi} = P_{i-1} \cos \alpha_{i-1} - P_i \cos \alpha_i \tag{5.13}$$

$$R_{yi} = P_{i-1} \sin \alpha_{i-1} - P_i \sin \alpha_i \tag{5.14}$$

$$R_i = \sqrt{R_{xi}^2 + R_{yi}^2}. \tag{5.15}$$

Although the above procedure is relatively simple, it requires some effort. It can be easier to model the tendon as a chain of straight, mutually connected elements. The elements are linked to the nodes of the analysed structure by rigid links. The tendon elements are pin-connected to the rigid links and have zero stiffness ($EA = 0$) during the post-tensioning phase. They are stressed by axial forces determined according to Figure 5.3(b). Since the members have zero stiffness, the resultant of forces in each node has the same value and direction as the forces determined according to Equation (5.15).

After the post-tensioning of the structure, it is possible to give the tendon elements their actual stiffness (E_s, A_s) and make the final link of the structure. In this way it is possible to

Figure 5.4 Curved beam – equivalent load: (a) section A–A; (b) section B–B; (c) elevation; and (d) plan

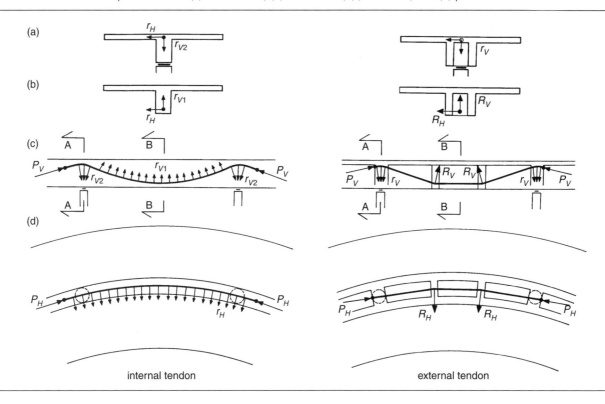

internal tendon   external tendon

correctly describe the function of the tendon in the structure: during the post-tensioning the tendon is not a part of the structure (it is not included in the stiffness matrix) but after the post-tensioning the tendon contributes to the stiffness of the structure (it is included in the stiffness matrix). Any loads will cause corresponding stresses in the tendons proportional to the stiffness, position and connection to the structure.

Modelling the tendons by a chain of straight members also allows us to easily describe the function of the draped tendon in structures curved in plan (Figure 5.4). Figure 5.5 shows the calculation model of the structure post-tensioned by internal and external tendons both during the post-tensioning operation (Figure 5.5(a)) and during service (Figure 5.5(b)).

For preliminary calculations we can assume that the structure is post-tensioned by a tendon for which the horizontal component of the prestressing force P_H is constant and is equal to the horizontal component of the average value of the prestressing force P_{avr} (Figure 5.3(b)):

$$P_H = P_{avr} \cos \alpha = \text{const.} \quad (5.16)$$

If the curve of the duct is a second-degree parabola (Figure 5.3(e)) the radial forces r are assumed to be equal to the normal forces k and are given by

$$f = \frac{a}{4} \tan \frac{\alpha}{2}, \quad r = \frac{8f}{a^2} P_H. \quad (5.17)$$

To understand the behaviour of the prestress, it is useful to review several basic examples of post-tensioned structures.

Figure 5.6 shows simply supported beams and beams with restrained ends. One end of the beam is rigidly fixed so it cannot undergo any rotation or displacement; the other end is prevented from rotating but is free to move longitudinally. These beams of a span length L are post-tensioned by straight and parabolic tendons. It is assumed that the horizontal component of the prestressing force is constant along the length of the cable.

Figures 5.6(1)–5.6(4) show the equivalent load, normal forces and bending moments in the beams that are post-tensioned by the straight tendons. In all these cases the effects of post-tensioning can be expressed by the equivalent load acting on the concrete at the anchorages:

$$N = -P_H.$$

Since the tendon has an eccentricity $e_a = e_b$ in cases 2 and 4, the normal force N introduces external bending moments at the

Figure 5.5 Modelling the curved beam: (a) during post-tensioning and (b) during service

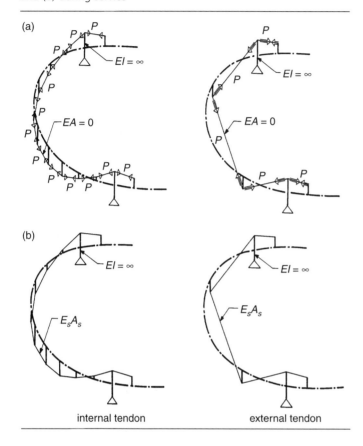

internal tendon external tendon

beam ends:

$$M_a = M_b = Ne_b.$$

In all cases, the girders are stressed by a constant normal force N. However, a bending moment

$$M = M_a = M_b$$

is seen only in the case of the simply supported beam. In the beam with restrained ends the bending moment is zero, since the equivalent bending moment acting at the beam ends cannot cause any rotation of the beam. In this case, the so-called primary and secondary bending moments have the same value but opposite sign:

$$M_a^0 = -M_a' = M_b^0 = -M_b' = Ne_b.$$

Figures 5.6(5)–5.6(8) show the equivalent load, normal forces and bending moments in the beams that are post-tensioned with parabolic tendons. In all cases the tendon has the same sag f. Similarly to the previous example, the effects of the post-tensioning can be described by an equivalent loading acting at the anchorages:

$$N = -P_H = -P_{avr} \cos \alpha.$$

Since the tendon has eccentricity $e_a = e_b$ in cases 5.6(6) and 5.6(8) the normal force N loads the beam ends with bending moments:

$$M_a = M_b = Ne_b.$$

Because the tendons are laid out as a parabolic curve, the beams are also loaded by radial forces:

$$r = \frac{8f}{a^2} P_H = \frac{8f}{L^2}.$$

In all cases, the girders are stressed by a constant normal force N.

Since the radial forces are constant, the bending moments have the shape of a second-degree parabola.

In the case of Figure 5.6(5), the bending moment at mid-span is

$$M_{L/2} = -\tfrac{1}{8} rL^2 = -P_H f.$$

In the case of Figure 5.6(6) the bending moment at mid-span is

$$M_{L/2} = -\tfrac{1}{8} rL^2 + M_b = -P_H f + P_H e_b = -P_H(f - e_b).$$

It is important to note that the values and shapes of bending moments are the same for the layout of prestressing steel shown in Figures 5.6(7) and 5.6(8). As also noted above, the equivalent bending moment acting at the beam ends cannot cause any rotation of the beam. The bending moments at the ends are therefore:

$$M_a = M_b = \tfrac{1}{12} rL^2 = \tfrac{2}{3} P_H f$$

and the bending moment at mid-span is

$$M_{L/2} = -\tfrac{1}{24} rL^2 = -\tfrac{1}{3} P_H f.$$

Figure 5.7 shows curved beams post-tensioned with curved tendons. The tendon is situated in the beam axis or parallel to the axis. The beams of span L are circular with radius ρ. The beams in Figures 5.7(1) and 5.7(2) are simply supported; the beams in Figures 5.7(3) and 5.7(4) have rigidly fixed ends that cannot undergo any rotation or displacements. It is assumed that the prestressing force P is constant along the length of the cable and the friction (tangent) forces are zero.

The effects of post-tensioning can be described with the equivalent loads acting at the anchorages and the radial

Figure 5.6 Equivalent load, normal forces and bending moments in straight beam

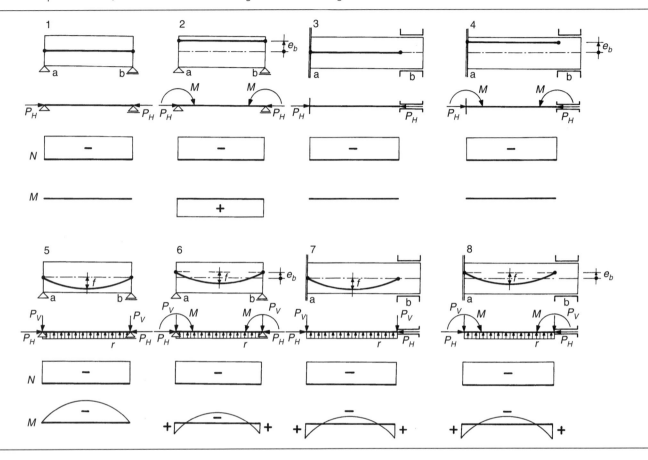

forces. Since the tangent forces are zero, the radial forces are constant and

$$k = \frac{P}{\rho}.$$

These uniform radial forces create a constant normal force in the beams:

$$N = -P = -\rho r.$$

Figure 5.7 Equivalent load, normal forces and bending moments in curved beam

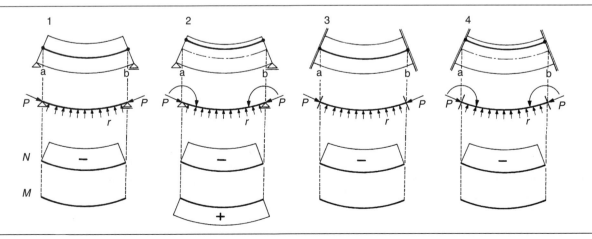

Figure 5.8 Radial forces acting on stiff and slender arch

This is also true for the curved beam in which both ends are fixed. Bending moments originate only in the simply supported beam (Figure 5.7(2)) that is post-tensioned with an eccentric tendon. The anchor forces create equivalent moments at the ends of the beam:

$$M_a = M_b = P_H e_b.$$

Since the reactions due to prestress are zero, the bending moments are constant along the length of the beam.

From Figures 5.7(3) and 5.7(4) it is evident that an arch structure, which is post-tensioned by a tendon that is parallel to the axis, is stressed only by normal forces. This is only true for structures in which deformations due to normal forces are not significant, however.

In slender structures analysed as geometrically non-linear structures in which the final state of stresses is determined on the deformed shape the bending stresses are significant. From Figure 5.8, which depicts radial forces and deformations of a typical and slender arch structure, it is evident that the radial forces change their value and direction. Any analysis therefore has to include this phenomenon.

This can be easily achieved with a model in which the prestressing tendons are modelled as a chain of members loaded by axial forces.

REFERENCE

FIP (1998) FIP Recommendations 1996: Practical design of structural concrete. *Proceedings of FIP Congress, Amsterdam.* Thomas Telford Publishing, London.

Chapter 6
Creep and shrinkage of concrete

During construction, stress ribbon and cable-supported structures utilise different static systems. Boundary conditions change, new structural members are added or cast, post-tensioning is applied and temporary support elements are erected and consequently removed. Structural elements of various ages are combined and the concrete is gradually loaded. During both construction and throughout the service life of concrete structures, account must be taken of the creep and shrinkage of concrete.

The deformation of concrete due to shrinkage and creep may vary considerably with the type of cement and aggregate, the climate (temperature and humidity), the member size and the time of loading.

Modern creep functions (see Figure 6.1) combine the theory of the delayed elasticity (the final value of creep does not depend on the age of concrete) and rate-of-creep theory (Smerda and Kristek, 1988).

For final design, a time-dependent analysis using CEB-FIP (1993) Model Code 90 (MC90) or specific tests should be used. Eurocode 2 also employs the MC90 approach. The function MC90 uses the following function for shrinkage (cs) and creep (cc) strains:

$$\varepsilon_{cs}(t, t_s) = \varepsilon_{cs,0}\beta_s(t - t_s) \tag{6.1}$$

$$\varepsilon_{cc}(t, t_0) = \frac{f_c(t_0)}{E_c}\phi(t, t_0) \tag{6.2}$$

where β_s is the coefficient to describe the development of shrinkage with time, $\phi(t, t_0)$ is the creep coefficient, f_c is normal stresses, E_c is modulus of elasticity at the age of 28 days, t_0 is time of loading (days), t_s is the age of concrete (days) at the beginning of shrinkage or swelling and t is the age of concrete (days).

For the preliminary considerations, the final values of shrinkage and concrete are listed in Tables 6.1 and 6.2 (FIP, 1998). The tables presents mean values and apply to concrete of grades 20–50 MPa subjected to a stress not exceeding $0.4f_c(t_0)$ at age t_0 of loading.

The development of the shrinkage strain and creep coefficient with age may be estimated from Figure 6.2.

6.1. Time-dependent analysis

For the time-dependent analysis of stress ribbon and cable-supported structures, specialised software should be used. The software needs to cover all aspects of the construction needs in order to take into account the fact that many structures are assembled from members of different ages. For design and for the parametric studies performed for this book, the program TDA was used. This program, which was recently included in the program system SCIA (ESA PrimaWin, 2000), was developed by Dr Navratil of the Technical University of Brno in collaboration with the author's design office Strasky, Husty and Partners, Ltd. (SHP) in Brno, Czech Republic (Strasky et al., 2001).

Figure 6.1 Creep functions: (a) delayed elasticity; (b) rate-of-creep; and (c) modified rate-of-creep

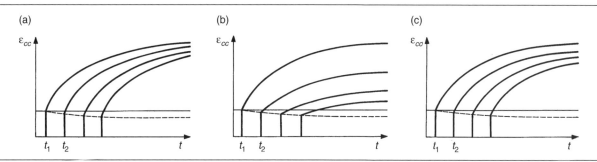

Table 6.1 Final value of shrinkage strain ε_{cs} ($\times 10^{-3}$), where A_c is the cross-sectional area of concrete and u is the exposed perimeter of A_c (RH: relative humidity)

Atmospheric conditions	Effective member size $2A_c/u$: mm		
	50	150	600
Dry, indoor (RH = 50%)	−0.53	−0.51	−0.36
Humid, outdoors (RH = 80%)	−0.30	−0.29	−0.20

The linear aging viscoelastic theory is applied for the analysis. This means that the creep prediction model is based upon the assumption of linearity between stresses and strains to ensure the applicability of linear superposition. The numerical solution

Table 6.2 Final value of the creep coefficient ϕ

Age at loading t_0: days	Dry, indoor (RH = 50%)			Humid, outdoors (RH = 80%)		
	Effective member size $2A_c/u$: mm					
	50	150	600	50	150	600
1	5.6	4.6	3.7	3.7	3.3	2.8
7	3.9	3.2	2.6	2.6	2.3	2.0
28	3.0	2.5	2.0	2.0	1.8	1.5
90	2.4	2.0	1.6	1.6	1.4	1.2
365	1.9	1.5	1.2	1.2	1.1	1.0

Figure 6.2 Shrinkage strain $\varepsilon_{cs}(t)$ and creep coefficient $\phi(t)$ at time t (days) divided by the ultimate shrinkage or creep coefficients listed in Tables 6.1 and 6.2

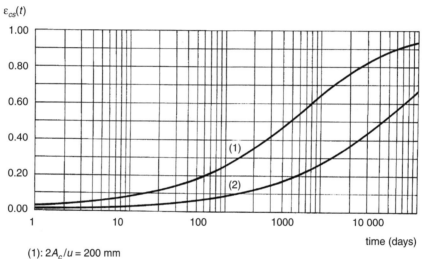

(1): $2A_c/u = 200$ mm
(2): $2A_c/u = 600$ mm

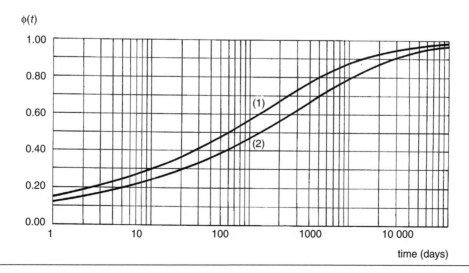

Figure 6.3 Time-dependent modelling of cable-supported structures: (a) partial elevation; (b) cross-section; (c) detail 'A'; and (d) finite-element degrees of freedom

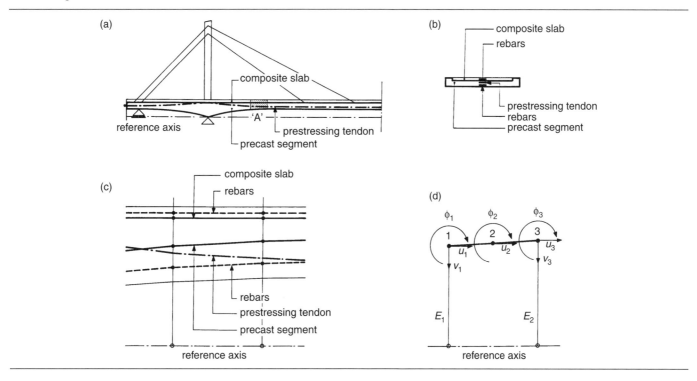

is based on the replacement of Stieltjes' hereditary integral by a finite sum. The general creep problem is thus converted into a series of elasticity problems. The shrinkage and creep of structural members is predicted through the mean properties of a given cross-section, taking into account the average relative humidity and member size. The development of the modulus of elasticity with time due to aging is also considered.

The method is based on a step-by-step computer procedure in which the time domain is subdivided by discrete times (time nodes) into time intervals (Navrátil, 1992). The finite element analysis is performed at each time node. The element stiffness matrix and load vector terms include the effect of axial, bending and shear deformations. The centroidal axis of the element can be placed eccentrically to the reference axis connecting the nodes. Six external and two internal degrees of freedom are used. The static condensation of internal node parameters is used, thus the full compatibility between eccentric elements is fulfilled. The elements represent, for example, precast segments, composite slab, stay cable, prestressing tendon or reinforcement (Figure 6.3). They can be installed or removed according to the construction scheme and take into account the influence of concrete of different ages in both the longitudinal and transverse directions of the structure. The various operations used in the construction such as the addition or removal of segments and prestressing cables, changes of boundary conditions, loads and prescribed displacements can be modelled.

The stress-produced strain consists of an elastic instantaneous strain $\varepsilon_e(t)$ and a creep strain $\varepsilon_c(t)$. The change of the modulus of elasticity with time due to ageing is also considered. The creep prediction model is based on the assumption of linearity between stresses and strains to allow the application of a linear superposition procedure. The numerical solution is based on the replacement of the Stieltjes hereditary integral by a finite sum. The general creep problem is thus converted to a series of elasticity problems:

$$\varepsilon_m(t) = \sum_{j=0}^{n} \left\{ [1 + \phi(t, t_j)] \frac{\Delta f(t_j)}{E(t_j)} \right\} \qquad (6.3)$$

where $\varepsilon_m(t)$ is the total strain at time t.

The method used for the time-dependent analysis is based on a step-by-step computer procedure in which the time domain is subdivided into time intervals by discrete time nodes t_i ($i = 1, 2, \ldots, n$). The solution at the time node i is as follows.

1. The increments of normal strains, curvatures and shear strains caused by creep during the interval $\langle t_{i-1}, t_i \rangle$ are calculated using the second summand of Equation (6.3). The corresponding shrinkage strains are also calculated.

2. A load vector $\mathrm{d}\boldsymbol{F}_p$, equivalent to the effects of the generalised strains calculated in step 1, is assembled.
3. The element stiffness matrices $\boldsymbol{K}$ are calculated for the time t_i and the global stiffness matrix $\boldsymbol{K}_g$ is assembled.
4. The system of equations $\boldsymbol{K}_g \mathrm{d}\Delta_g = \mathrm{d}\boldsymbol{F}_p$ is analysed. The vector of increments of nodal displacements $\mathrm{d}\Delta_g$ is added to the vector of total nodal displacements Δ_g.
5. The elements are analysed in the central coordinate system. The increments of internal forces and increments of elastic strains are calculated from the increments of displacements of the element nodes.
6. The increments of the structural configuration at the time node t_i are applied.
7. The increments of generalised strains of the elements that are prestressed (or loaded with temperature change) at time node t_i are calculated. The losses of prestressing due to the deformation of the structure are automatically included in the analysis through the inclusion of these increments of internal forces.
8. The global load vector $\mathrm{d}\boldsymbol{F}_z$ is assembled as equivalent to the internal forces calculated in step 7. The increments of other types of the long-term load applied at the time node t_i are added to the load vector $\mathrm{d}\boldsymbol{F}_z$.
9. The system of equations $\boldsymbol{K}_g \mathrm{d}\Delta_g = \mathrm{d}\boldsymbol{F}_z$ is analysed. The vector of increments of nodal displacements $\mathrm{d}\Delta_g$ is added to the vector of total nodal displacements Δ_g.
10. The increments of internal forces and of elastic strains are calculated from the increments of displacements of the element nodes.
11. The increments of internal forces calculated in steps 5 and 10 are added to the total internal forces. The increments of elastic strains calculated in steps 5 and 10 are added together and saved to the history of elastic strains as the increments for time node t_i.
12. Go to step 1 for time node $i+1$.

Considering the extensive amount of input and output data, a user-friendly graphic interface for pre- and post-processing has been developed. The pre-processor guides the user through the process of preparing and inputting data. Post-processors allow the user to select and view only the required data. In a graphical environment, it is possible to filter output data, switch individual parameters on and off (internal forces, stresses, deformations), zoom in to the structure, chose the time node to view and carry out other operations.

A modified numerical procedure was implemented for the time-dependent geometrically non-linear structural analysis. The computer procedure for creep and shrinkage analysis described above was adapted to collaborate with the finite-element software package ANSYS used at the Technical University of Brno, Czech Republic. The effects of creep and shrinkage calculated in steps 1 and 2, and the external load applied in steps 6 and 7 are converted to equivalent loads for the ANSYS structural model. After the full geometrical analysis is performed, the increments of displacements and internal forces are processed in the same way as described in steps 5, 10 and 11 before the analysis moves onto a new time node.

The importance of the time-dependent analysis will be demonstrated with the aid of the following examples.

6.2. Redistribution of stresses between members of different age

It is well known that a significant redistribution of stresses occurs in structural members composed of elements of different ages and in structures formed by steel and concrete. For the 1 m member depicted in Figure 6.4, the significance of this phenomenon is depicted in Figures 6.5–6.7 which display the redistribution of stresses between the two parts of different concrete ages.

The cross-section of the member is rectangular and of dimensions $5 \times 0.25\,\mathrm{m}$. The analysis was carried out for five cases that differ by the proportion of the area of the precast segment (A_{PS}) and composite slab (A_{CS}) part of the member (Table 6.3).

The precast segment (first part) was assumed to have a characteristic strength $f_c = 50\,\mathrm{MPa}$ ($E_c(28) = 38.5\,\mathrm{GPa}$, where $E_c(28)$

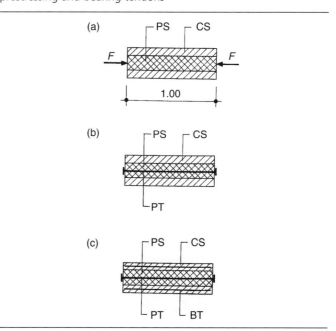

Figure 6.4 Progressively cast concrete member: (a) precast segment and composite slab; (b) precast segment, composite slab and prestressing tendon; and (c) precast segment, composite slab, prestressing and bearing tendons

Figure 6.5 Redistribution of forces of member depicted in Figure 6.4(a)

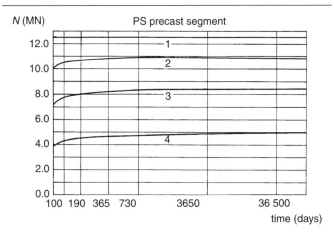

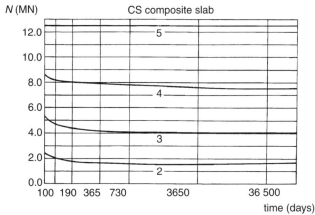

refers to the value of E_c at the age of 28 days) and the composite slab (second part) was assumed to have a characteristic strength of $f_c = 40$ MPa ($E_c(28) = 36.50$ GPa). The redistribution of stresses was determined by the program TDA using the MC90 rheological functions.

Figure 6.5 shows the redistribution of stresses between the two parts of the progressively cast concrete member that was loaded by a constant force. The process of casting and loading was as follows.

The first part of the member was cast, and after three days of curing it was placed on a support that allowed it to shrink. After 87 days (time 90 days) the second part of the member was cast and cured for three days. After an additional seven days (time 100 days) the element was loaded with a constant force $F = 12.5$ MN.

From Figure 6.5 it is evident that, at the time of loading, the force was distributed into the two parts of the member proportional to their stiffness given by the product of the area A_c and the modulus of elasticity $E_c(t)$. In the course of time the stresses (or portion of the force) were redistributed from the younger to the older part of the member.

Figure 6.6 shows the redistribution of stresses between the two parts of the progressively cast concrete member that was post-tensioned by a prestressing tendon (Figure 6.4(b)). The tendon has an area $A_{PT} = 0.01512$ m^2, with a modulus of elasticity $E_s = 195$ GPa. The process of casting and loading was the same as in the previous example. The prestressing force applied at day 100 was $P = 12.5$ MN. At the time of the post-tensioning the tendon had zero stiffness ($A_{PT}E_s = 0$); after post-tensioning the tendon is incorporated into the structure.

Similar to the previous example, the prestressing force was distributed to the components of the member proportional to their stiffness as given by the product of the area A_c at the modulus of elasticity $E_c(t)$ at the time of loading. With time, the prestressing force was redistributed. Due to the shortening of the member caused by the creep and shrinkage of the concrete, the tension force in the tendon was reduced. The compression stresses were also redistributed from the younger to the older concrete.

Figure 6.7 shows the redistribution of stresses between two parts of a progressively cast concrete member that was post-tensioned by a prestressing tendon (PT in Figure 6.4(c)). The tendon has an area $A_{PT} = 0.01512$ m^2 and a modulus of elasticity $E_s = 195$ GPa. Additionally, an unstressed tendon of area $A_{BT} = 0.02408$ m^2 and modulus of elasticity $E_s = 195$ GPa was placed in the member before casting the second portion. This tendon could represent the bearing tendons (BT) used in stress ribbon structures.

The prestressing force applied at day 100 was $P = 12.5$ MN. At the time of post-tensioning the prestressing tendon has zero stiffness ($A_{PT}E_s = 0$); after post-tensioning the tendon has become a part of the structure.

The prestressing force was distributed not only to the two parts of the concrete member but also to the unstressed tendon. The stresses were distributed proportionally to the stiffness given by the product of the area A_c, the modulus of elasticity $E_c(t)$ and the stiffness of the unstressed tendon given by the product of the area A_{PT} and modulus of elasticity E_s. With time the prestressing force is redistributed. Due to the shortening of the member due to the creep and shrinkage of concrete, the tension force in the tendon is reduced. The compression stresses are also redistributed from the concrete to the steel as well as from the younger concrete to the older concrete.

Figure 6.6 Redistribution of forces of member depicted in Figure 6.4(b)

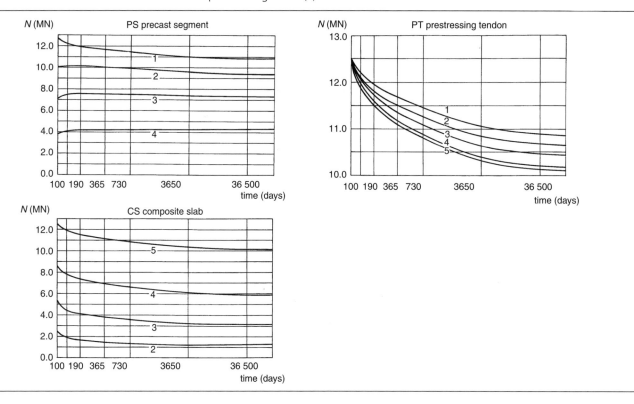

Figure 6.7 Redistribution of forces of member depicted in Figure 6.4(c)

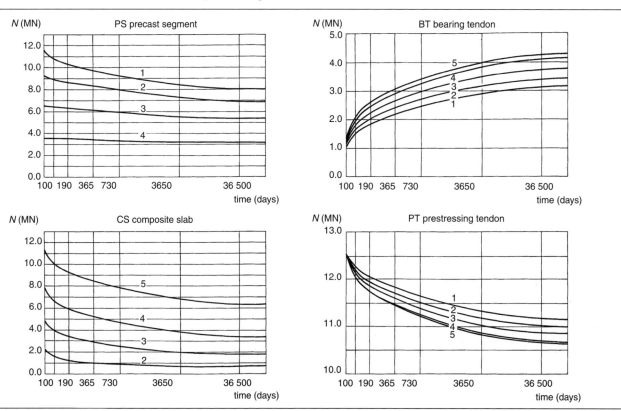

Table 6.3 Area of progressively cast concrete member: m²

Case	Part 1 (PS)	Part 2 (CS)	Total
1	1.2500	0.0000	1.2500
2	0.9375	0.3125	1.2500
3	0.6250	0.6250	1.2500
4	0.3125	0.9375	1.2500
5	0.0000	1.2500	1.2500

6.3. Redistribution of stresses in structures with changing static systems

It is also known that a significant redistribution of stresses can occur in structures that change their static system during construction. The significance of this phenomenon is illustrated with the example of a simple cable-supported structure.

Figure 6.8 shows a simple beam of 12 m span which, after 14 days of curing, is suspended at mid-span on a very stiff stay cable ($E_s A_s = \infty$). Before suspending, a force N was introduced in the cable. The values of N were 0, R and $2R$, where R is a reaction at the intermediate support of a two-span continuous beam under uniform loading.

Figure 6.8 Redistribution of bending moments in a two-span beam

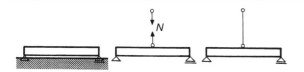
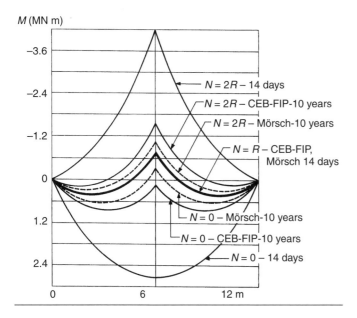

A time-dependent analysis was performed for the old Mörsch and CEB-FIP (MC90) creep functions (see Figure 6.1). With time, a significant redistribution of bending moments has occurred for $N = 0$ and $N = 2R$. The bending moment diagram is changing towards the diagram of a two-span beam. Concrete is a natural material and the structure therefore tries to behave naturally as a continuous beam. In this case, a larger redistribution of stresses is obtained for the old Mörsch creep function.

It is important to realise that for the force $N = R$ there is no redistribution for both creep functions. The structure keeps its shape and the stresses are constant in time. Their values do not depend on the adopted creep function.

Since it is difficult to design a structure in which the stresses are changing with time, it is very important to design an initial stage such that the redistribution of stresses is minimal.

This means that the geometry and forces in the internal prestressing tendons or external cables (situated inside or outside the perimeter of the deck) have to be determined in such a way that their effects in conjunction with a dead load create zero deflection at the deviators (Figure 6.9). This means that the dead load should be balanced by prestressing. Such a structure loaded only by normal force keeps its shape in time. This approach, which was developed by Leonhardt (1964) and Lin and Burns (1981), directs us to use partial, limited or full prestressing. The importance of load balancing was also demonstrated by Favre and Markey (1994).

Figure 6.9 Dead-load balancing at a two-span beam: (a) by internal tendons; (b) by external tendons situated within the depth of the cross-section; and (c) by external tendons situated outside the depth of the cross-section

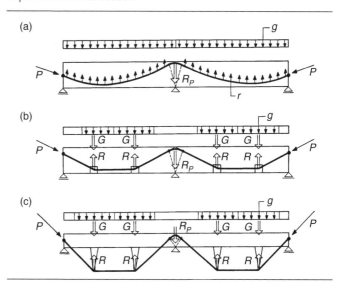

Figure 6.10 Dead-load balancing: (a) arch structure; (b) cable-stayed structure; (c) suspension structure; and (d) equivalent continuous beam

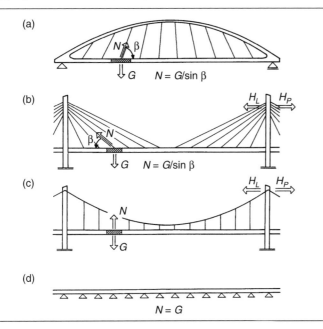

In case the deck is suspended on arches or pylons, the initial forces in the stay- or suspension cables have to be determined from the condition of zero deck deflection at the anchor points (Figure 6.10).

REFERENCES

CEB-FIP (1993) *CEB-FIP Model Code 1990.* Comité Euro-International du Béton. Thomas Telford, London.

ESA PrimaWin (2000) *Reference Manual, SCIA Software.* Scientific Application Group, Belgium.

Favre R and Markey I (1994) *Generalization of the load balancing method. Prestressed Concrete in Switzerland 1990–1994.* Proceedings of 12th FIP Congress, Washington, DC.

FIP Recommendations (1998) *Practical Design of Structural Concrete.* FIP Congress, Amsterdam.

Leonhardt F (1964) *Prestressed Concrete: Design and Construction.* Wilhelm Ernst & Sons, Berlin.

Lin TY and Burns NH (1981) *Design of Prestressed Concrete Structures.* John Wiley & Sons, New York.

Navrátil J (1992) Time-dependent analysis of concrete frame structures. *Building Research Journal* (Stavebnický časopis) **40(7)** (in Czech).

Smerda Z and Kristek V (1988) *Creep and Shrinkage of Concrete Elements and Structures.* Elsevier, Amsterdam.

Strasky J, Navratil J and Susky S (2001) Applications of time-dependent analysis in the design of hybrid bridge structures. *PCI Journal* **46(4)**: 56–74.

Chapter 7
Stress ribbon structures

As has been explained in Chapter 2, stress ribbon bridges with prestressed concrete decks have superior behaviour because of their increased stiffness. They are described in this chapter in greater detail.

7.1. Structural arrangement

Stress ribbon bridges can have one or more spans. A lightly draped shape characterises this structural type (Figure 7.1). The exact shape of the stress ribbon structure is the funicular line of the dead load. Since these structures usually have a constant cross-section and the sag is very small, they have the shape of a second-degree parabola. The typical geometry of a two-span structure is shown in Figure 7.2.

7.1.1 Geometry

The geometry of structures of multiple spans is derived from the condition that, for dead load, the horizontal force is constant along the length of the bridge. This guarantees that the intermediate piers are not stressed by horizontal forces and consequently by bending. This requirement is essential for all cable-supported structures and the construction should guarantee it.

In design we first determine the mid-span sag f_{max} for the longest span L_{max}. From the conditions of equilibrium in the horizontal direction, the mid-span sag $f_{i,max}$ for the span of the length L_i is

$$H = \frac{gL_i^2}{8f_i} = \frac{gL_{max}^2}{8f_{max}} \tag{7.1}$$

$$f_i = \frac{L_i^2}{L_{max}^2} f_{max}. \tag{7.2}$$

In the span i we have

$$f_i(x) = 4f_i \frac{x(L_i - x)}{L_i^2} \tag{7.3}$$

and

$$p_i(x) = 4f_i \frac{(L_i - 2x)}{L_i^2} \tag{7.4}$$

for a two-span structure as presented in Figure 7.2.

For span 1:

$$y_1(x) = h_1 - \frac{h_1}{L_1}x + f_1(x) = h_1 - \frac{h_1}{L_1}x + 4f_1 \frac{x(L_1 - x)}{L_1^2}$$

$$p_i(x) = -\frac{h_1}{L_1} + p_1^0 = -\frac{h_1}{L_1} + 4f_1 \frac{(L_1 - 2x)}{L_1^2}$$

Figure 7.1 Grants Pass Bridge, Oregon, USA: variable slope

Figure 7.2 Geometry of the deck

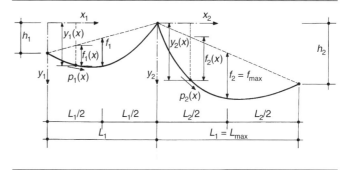

Figure 7.3 Koushita Bridge, Japan: structural members (courtesy of Oriental Construction Co., Ltd)

and for span 2:

$$y_2(x) = \frac{h_2}{L_2}x + f_2(x) = \frac{h_2}{L_2}x + 4f_2\frac{x(L_2 - x)}{L_2^2}$$

$$p_i(x) = \frac{h_2}{L_2} + p_2^0 = \frac{h_2}{L_2} + 4f_2\frac{(L_2 - 2x)}{L_2^2}.$$

7.1.2 Structural members

The deck of a stress ribbon structure can be formed by a monolithic band or can be assembled of precast segments. The band is fixed to the abutments and is supported by intermediate piers (Figure 7.3). Due to limits on the maximum slope, the ribbon is stressed by a large horizontal force that has to be transferred into the soil.

The deck cast in formwork is supported by a falsework only in special cases. Usually, stress ribbon structures are erected independently of the existing terrain. The formwork of precast segments is suspended on bearing tendons and shifted along them into their design position (Figure 7.4(a)). Prestressing applied after the casting of the whole band, or the joints between the segments, guarantees the structural integrity of the complete deck.

The structural arrangement of stress ribbon bridges is determined by their static function and by their process of construction. During erection the structure acts as a perfectly flexible cable (Figure 7.4(a)) during service as a prestressed band (stress ribbon) that is stressed not only by normal forces but also by bending moments (Figure 7.4(b)). However, the shape and the stresses in the structure at the end of the erection determine the magnitude of the stresses that will occur in the structure during service.

7.2. Prestressed band

Although a prestressed band can resist very heavy loads it can be very slender (Figure 7.5). In order to understand the behaviour of stress ribbon structures, the bending moments in the deck due to prestress, live load and temperature changes are shown in Figure 7.6. The structure has a span of 99.00 m and its sag is 1.98 m. The deck is the same as the deck of the structure depicted in Figure 2.16. The concrete band of modulus of

Figure 7.4 Static function: (a) erection stage and (b) service stage

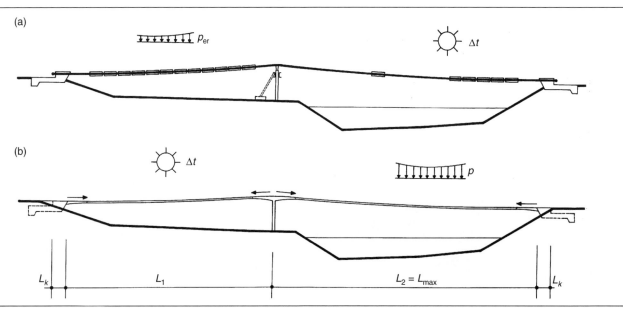

Figure 7.5 Brno-Komin Bridge, Czech Republic

elasticity of $E_c = 36\,000$ MPa had section properties taken from a basic rectangular section of width of 5.00 m and depth 0.25 m:

$A_c = bh = 5.00 \times 0.25 = 1.25$

$I_c = \frac{1}{12} bh^3 = \frac{1}{12} 5.00 \times 0.25^3 = 0.00651$.

The deck was suspended on bearing tendons of area $A_{s,bt} = 0.0196\,\text{m}^2$ and post-tensioning was applied with tendons of area $A_{s,pt} = 0.0196\,\text{m}^2$.

The horizontal force H_g due to the dead load $g = 33.345$ kN/m is 17.30 MN. The structure was loaded by (1) dead load, (2) dead load under temperature change $\Delta t = 20°$C, (3) dead load under temperature change $\Delta t = -20°$C, (4) live load $p = 20$ kN/m on the whole span, (5) live load $p = 20$ kN/m on the half span and (6) point load $F = 100$ kN at mid-span. The structure was analysed as a geometrically non-linear structure using the program ANSYS without and with post-tensioning of the deck with a prestressing force $P = 25.52$ MN. The process described in Section 7.6 was used in the analysis.

Figure 7.6 shows the position of the loadings, deformations and bending moments. Figure 7.6(a) shows the results of the analysis for the non-prestressed deck; Figure 7.6(b) for the prestressed deck.

From the presented results it is evident that significant bending moments originate only at the supports. The bending moments due to a point load that represents a typical maintenance car are relatively very low. Since the deck is always post-tensioned, the negative bending moments at the supports are very low.

Figure 7.6 Deformation and bending moments 1: (a) without prestressing and (b) with prestressing

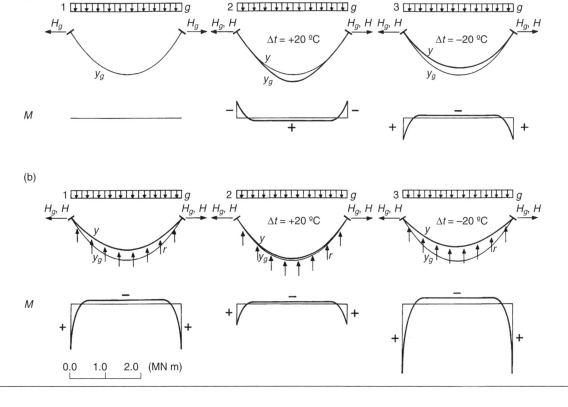

Figure 7.7 Deformation and bending moments 1: (a) without prestressing and (b) with prestressing

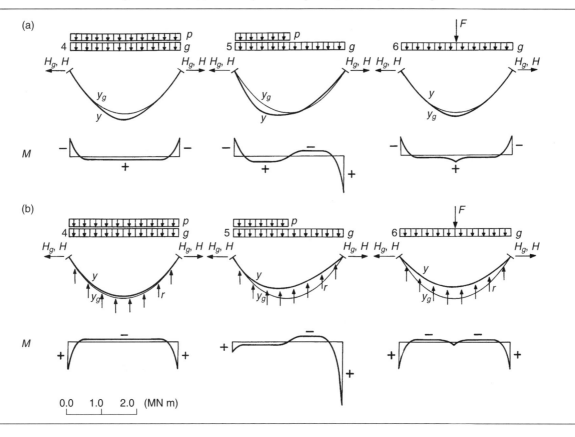

However, positive bending moments are very large and have a significant influence on the detailing of the stress ribbon structure at the supports.

Since the deck is stressed only by normal forces along the whole length of the structure, the deck can be formed with a very slender solid section that can be further reduced by *waffles* that create a coffered soffit (Figure 7.8). The minimum area of the deck is determined from the requirement that, under the various loading condition (including prestress), there are limited or zero tension stresses in the deck and that maximum compressive stresses are not exceeded. Since the bending moments due to the point load are low, the depth of the deck is essentially determined by the cover requirements of the prestressing steel. Usually the minimum depth guarantees a sufficient stiffness of the deck (see also Figure 2.15).

The deck of stress ribbon structures can be cast in a formwork that is suspended on the bearing cables (Figure 7.9(a)) or is assembled from precast segments. Usually the deck is suspended from bearing tendons and is prestressed by prestressing tendons. However, the function of bearing and prestressing tendons can also be combined.

Figure 7.8 Nymburk Bridge, Czech Republic: coffered soffit of the deck

Stress ribbon structures

Figure 7.9 Prestressed band: typical sections

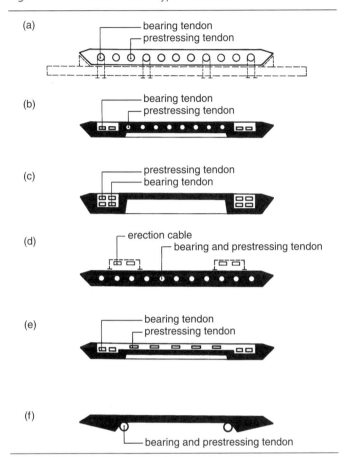

Figure 7.10 DS-L Bridge, Czech Republic: prestressed band

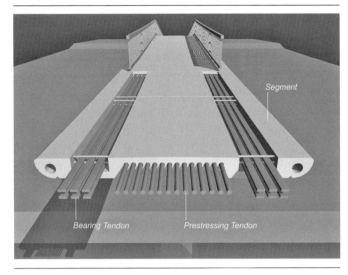

Figure 7.11 DS-L Bridge, Czech Republic: segment

Figure 7.12 Redding Bridge, California, USA: segment

Initially, the segments can be hung from bearing tendons situated in troughs. After erection, the deck is post-tensioned by the second group of cables situated either in the ducts within the segments (Figures 7.9(b), 7.10 and 7.11) or in the troughs (Figures 7.9(c) and 7.12). The bearing tendons are then protected by cast-in-place concrete poured simultaneously with the joints between the segments. Since longitudinal shrinkage cracks are likely to occur between the cast-in-place and precast concrete, it is recommended that the surface is protected with a waterproof overlay.

The deck can also be assembled from precast segments that are hung on temporary erection cables that are removed after post-tensioning of the deck with the internal tendons (Figure 7.9(d)). Another system utilises external tendons to support and pre-stress the deck (Figure 7.9(f)). These external tendons can either be uniformly distributed along the width of the segment or situated close to edges of the segments (Figure 7.17).

Another arrangement consists of precast segments with a composite slab (Figures 7.9(e) and 7.13–7.16). The segments are suspended on bearing tendons and serve as a falsework

Figure 7.13 Grants Pass Bridge, Oregon, USA: prestressed band

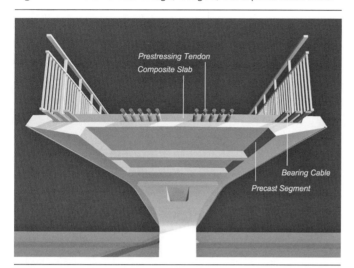

Figure 7.15 Maidstone Bridge, UK: prestressed band

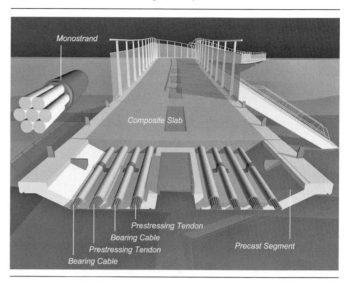

and formwork for the composite slab that is cast simultaneously with the segment joints. Both the precast segments and the composite slab are post-tensioned by tendons that are situated along the bearing tendons within the cast-in-place slab. A continuous deck slab without any joints provides an excellent protection of the prestressing steel and requires minimum maintenance.

Usually the bearing tendons consist of strands that are protected by the post-tensioned concrete of the troughs or composite slab (Figure 7.18(a)). The prestressing tendons are formed by strands grouted in a traditional duct (Figure 7.18(b)). If a higher degree of protection is required, the bearing tendons can also be made from strands grouted in ducts or both bearing and prestressing tendons can be made from monostrands additionally grouted in PE ducts (Figures 7.15 and 7.18(c)).

Figure 7.14 Grants Pass Bridge, Oregon, USA: segment

Figure 7.16 Maidstone Bridge, UK: segment

Figure 7.17 Olomouc Bridge, Czech Republic: prestressed band

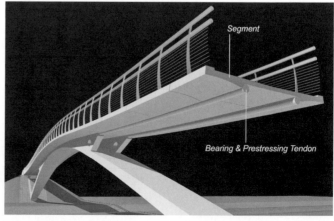

Figure 7.18 Bearing and prestressing tendons

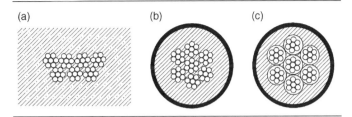

A typical section of stress ribbon is not able to resist the bending moments that occur at the supports (Figures 7.6 and 7.7). The support bending moments can be reduced by the following techniques.

- Creating a flexible support member close to the supports (Figure 7.19(b)). The function of this member is similar to the function of the neoprene rings usually designed for stay cables.
- Supporting the stress ribbon with a saddle from which the band can lift during post-tensioning and temperature drop, and to which the band can return for a temperature increase (Figure 7.19(c)).
- Strengthening the stress ribbon with a short support haunch (Figure 7.19(d)).

To understand the problem and quantify the above measures, extensive parametric studies were carried out. The structure depicted in Figure 7.19 was analysed for the different arrangements of the stress ribbon close to the supports. The analysis was completed for all loadings presented in Figures 7.6 and 7.7.

Figure 7.20 shows the bending moments in the stress ribbon near the support for different stiffnesses of the supporting member and for loading by prestressing force P and temperature changes $\Delta t = \pm 20°C$. The supporting member was modelled as a beam member of length $l = 1.00$ m and area $A = 1.00$ m^2, for which the modulus of elasticity E was varied. Stiffness k is a force that causes a deformation $\Delta l = 1$:

$$\Delta l = \frac{kl}{EI} = 1$$

$$k = \frac{EA}{L} = E$$

The analysis was carried out for $k = 0$, 1×10^3, 1×10^4, 1×10^5 and 1×10^6 kN/m.

Figure 7.20(a) presents the analysis results in which the supporting member was attached to the structure before post-tensioning. Figure 7.20(b) presents the results of the analysis where the supporting member was attached after the post-tensioning.

Figure 7.21 shows the bending moments and deformation of the stress ribbon that is supported by a saddle. The surface of the

Figure 7.19 Stress ribbon at supports

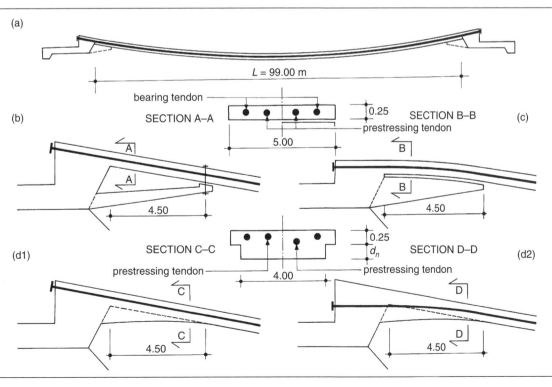

Figure 7.20 Bending moments in a structure with flexible supporting member

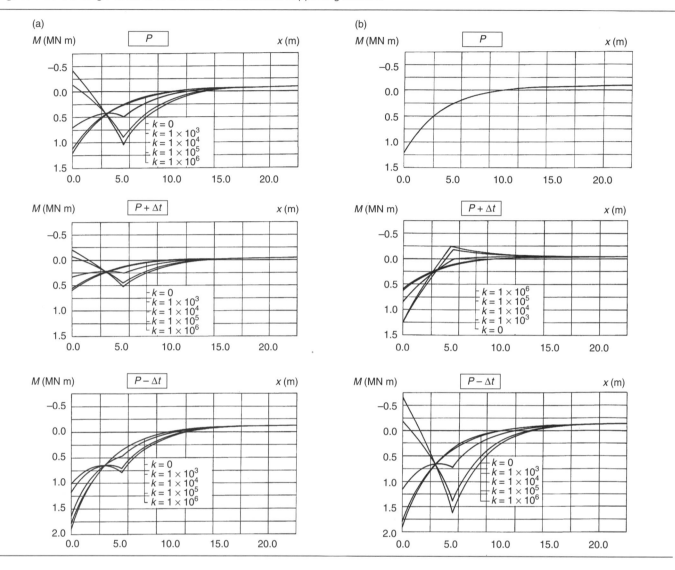

Figure 7.21 (a) Bending moments and (b) deformations in a structure with saddle

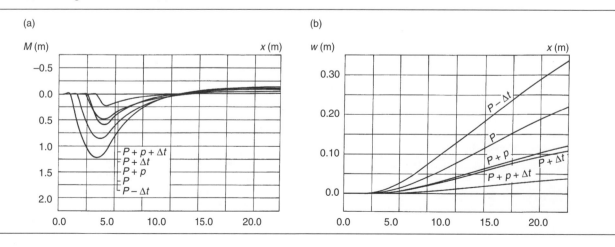

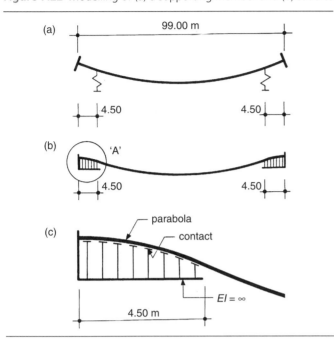

Figure 7.22 Modelling of (a) a supporting member and (b) saddles

saddle has the shape of a second-degree parabola whose tangent at the end of the saddle matches that of the stress ribbon. It was assumed that the saddle is very stiff. The saddle was therefore modelled by rigid members that were connected to the stress ribbon by *contact* members (Figures 7.22(b) and 7.19(c)). Since these members do not resist the tension, the ribbon can lift up.

Due to the post-tensioning the stress ribbon partially rises from the saddle. It is interesting to note that the stress ribbon never returns to the original position when it is loaded with temperature rise (Figure 7.21(b)).

Figure 7.23 shows the bending moments that occur in the stress ribbon when stiffened by short parabolic haunches. The haunches have a constant width of 4.00 m and variable depth d. The calculations were made for a depth $d_h = 0$, 0.25, 0.50, 0.75, 1.00 m. Figure 7.23(a) presents the results for a structure in which prestressing tendons are parallel to the surface of the deck; Figure 7.23(b) presents the results for a structure in which prestressing tendons follow the centroid of the haunch.

Although the described measures can significantly reduce the bending stresses, it is necessary to carefully design the stress ribbon in the vicinity of the supports. In designing the ribbon, the positive bending moments that cause tension stresses at bottom fibres are critical. Since the bearing and prestressing tendons are situated a sufficient distance away from the bottom fibres, it is possible to accept cracks there and design the ribbon as a partially post-tensioned member in which

crack width and fatigue stresses in the reinforcement are checked. If the ribbon above the saddle is assembled from precast members, it is necessary to guarantee compression in the joints. This can be provided with additional short tendons situated in the pier segments.

7.3. Piers and abutments

Typically, stress ribbon structures were designed for piers and abutments either with saddles or with short haunches. The typical arrangement can be seen in Figures 7.24 and 7.25. Procedures in different geographical regions will depend on the local conditions and chosen technology (Figures 7.26 and 7.27).

Figure 7.27(a) shows a cast-in-place deck supported by saddles. Figures 7.26(c) and 7.27(c) and 7.28 through 7.30 present a solution in which precast segments are supported by saddles. To accommodate the larger curvatures in the region, the length of pier segments is one-third of that of the typical segments. The segments are erected before the bearing tendons are placed and tensioned.

The segments directly above for pier columns are placed on cement mortar; the remaining segments are placed on neoprene strips. The stress ribbon is connected to the saddles using reinforcement between the segments situated above the pier columns.

Figures 7.26(b) and 7.27(b) and 7.31 through 7.34 depict an example of a cast-in-place haunch cast in a formwork, which was suspended on the already erected segments and the pier. In this case, the bearing tendons were supported by steel saddles anchored in the pier columns. The friction forces between the strands and the saddle are reduced using Teflon plates (Figure 7.26(e)). Until the deck itself provides a restraint, the piers can be stabilised using temporary struts.

Figures 7.26(d) and 7.27(d) and 7.35 through 7.37 show a solution in which the abutment and pier haunches were cast in place before the erection of the segments. The ducts, in which the bearing tendons are placed, have to allow a change of slope during the construction of the deck. A closure was therefore provided between the saddles and the segments.

The main advantage of the solution shown in Figures 7.26(b) and Figure 7.27(b) is that the shape of the saddle can be easily adjusted according to the geometry of the already erected deck; that is, it is not only dependent on the value of the initial stressing and actual weight of segments, but also on the temperature at the time of casting.

All other solutions require a very careful checking of the geometry of the already erected segments and an adjustment of the tendon forces before the joints are cast.

Figure 7.23 Bending moments in a structure with haunches

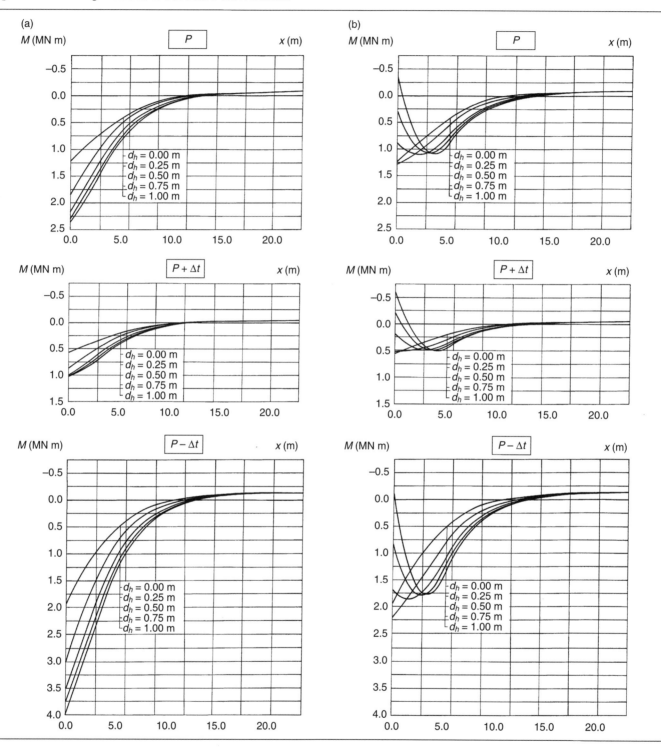

There is a large variety of possible structural and architectural arrangements of the piers and abutments. The saddles can be supported not only by piers, but they can be also suspended on cables.

Very careful consideration has to be given to the waterproof protection of the anchors of both the bearing and prestressing tendons at the abutments. A solution that complies with strict UK regulations is presented in Figure 7.38.

Figure 7.24 Abutments

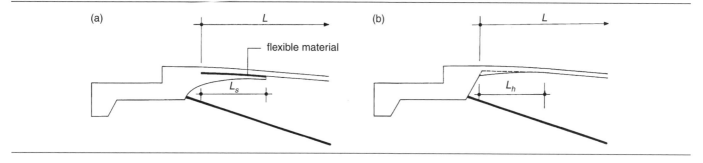

Figure 7.25 Intermediate piers

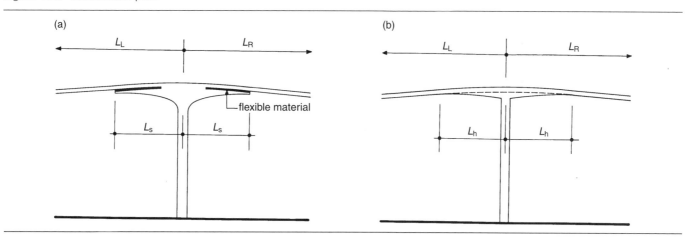

7.4. Transferring stress ribbon force to the soil

The stress ribbon is usually fixed at anchor blocks that are integral parts of the abutments. The abutments therefore need to transfer some very large horizontal forces into the soil by rock or ground anchors. Unfortunately, sound rock is close to grade only in exceptional cases and then the anchor block has the simple shape shown in Figures 7.39 and 7.40.

Figure 7.26 Erection of the stress ribbon at abutments

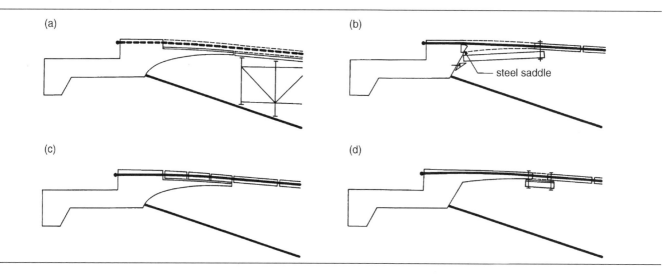

Figure 7.27 Erection of the stress ribbon at intermediate piers

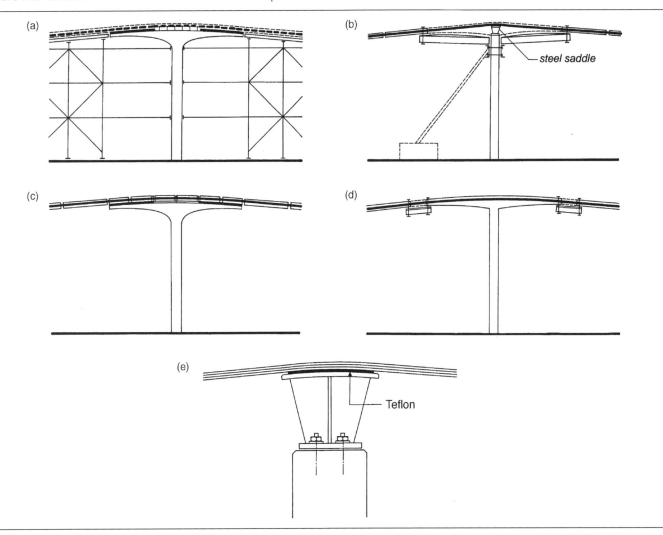

Figure 7.28 Grants Pass Bridge, Oregon, USA: intermediate pier

Figure 7.29 Grants Pass Bridge, Oregon, USA: intermediate pier

Figure 7.30 Grants Pass Bridge, Oregon, USA: intermediate pier during the erection

Figure 7.31 Prague-Troja Bridge, Czech Republic: intermediate pier

Figure 7.32 Prague-Troja Bridge, Czech Republic: intermediate pier

Figure 7.33 Prague-Troja Bridge, Czech Republic: intermediate pier during the erection

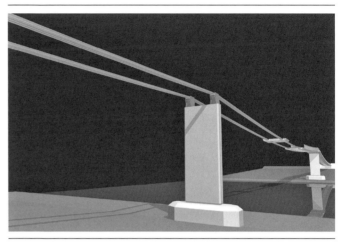

Figure 7.34 Prague-Troja Bridge, Czech Republic: formwork of the haunches

Figure 7.35 Maidstone Bridge, UK: intermediate pier

Figure 7.36 Maidstone Bridge, UK: intermediate pier

Figure 7.37 Maidstone Bridge, UK: intermediate pier during the erection

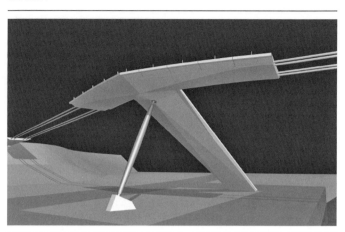

In most cases the competent soil is situated at a certain distance and the abutments have an arrangement as shown in Figure 7.41. In this case, it is necessary to check the soil pressure as well as the resistance against overturning and sliding not only for the maximum loading but for all stages of construction. It is important to realise that the rock anchors or ties have to be post-tensioned. That means they load the footing with an eccentric inclined force N_{an} that has vertical and horizontal components V_{an} and H_{an}.

During post-tensioning the capacity of the anchors is checked. By post-tensioning – or anchoring the footings to the soil – the variation of stresses at the anchor is eliminated and resistance against sliding is guaranteed.

Figure 7.38 Maidstone Bridge, UK: abutment (a) elevation and (b) cross-section

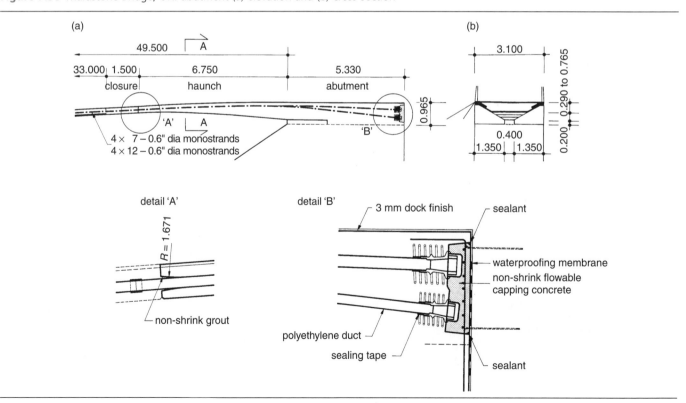

Figure 7.39 Rock anchors

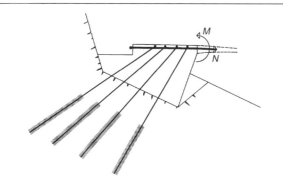

In Figure 7.41, the footing is provided with a shear key. Resistance against sliding therefore has to be checked at the bottom of the key. The safety against sliding is defined:

$$s = \frac{V \tan \phi + Ac}{H} = \frac{(V_q + V_{an}) \tan \phi + Ac}{H_q - H_{an}} \geq s_0 \quad (7.5)$$

where V_q is the vertical force from the dead and live load, H_q is the horizontal component of the force in the stress ribbon, ϕ is the angle of internal friction of the soil, A is the area of the footing at the level of the key and c is the cohesion.

The factor of safety against sliding s_0 depends on national codes and typically varies from 1.5 to 2.0.

The factor of safety has to be determined for all stages of the construction. In general, it is impossible to post-tension all rock anchors before erection of the deck since the horizontal force H_{an} is too large and can cause sliding of the footing in the direction of the post-tensioning. The post-tensioning is therefore usually done in two stages and the safety against sliding has to be checked for

1. post-tensioning of the first half of the rock anchors (Figure 7.41(b))
2. erection of the deck (Figure 7.41(c))
3. post-tensioning of the second half of the rock anchors (Figure 7.41(d))
4. service of the bridge, full live load and temperature drop (Figure 7.41(e)).

The use of rock or ground anchors requires a soil with adequate bearing capacity to not only resist the pressure from the stress ribbon, but also the pressure from the vertical component of the anchor force.

If there is insufficient capacity, drilled shafts can be used to resist both the vertical and horizontal components of the stress ribbon force (Figure 7.42). Although the drilled shafts

Figure 7.40 Redding Bridge, California, USA: rock anchors

have a relatively large capacity for resisting horizontal forces, their horizontal deformations are significant and it is necessary to consider them in the analysis. Elastic horizontal deformations can be eliminated with an erection process in which the structure is pre-loaded before casting of the closure joints. Nevertheless, with time, plastic deformations can cause considerable horizontal displacements of the abutments that consequently cause an increase of the sag of the stress ribbon.

Figure 7.43 shows a solution in which the drilled shafts are combined with ground anchors. In this type of design, it is necessary to take into account the progressive tensioning of the anchors and the possible horizontal movement of the footings. During service, the footing can move from the position shown in Figure 7.43(d) to that shown in Figure 7.43(e). The structural arrangement and the static analysis also have to take into account this phenomenon. The corresponding variation of stresses in the anchors also has to be considered.

An elegant solution is presented in Figure 7.44 in which the stress ribbon is supported by battered micropiles. They transfer the load from the stress ribbon into the soil because of their tension and compression capacity. The maximum tension in the piles occurs in the last row where the tension force from the horizontal force $^{H}T_i$ is increased by a tension force from the uplift (V_i) created by the vertical force V and bending moment M:

$$V_i = -\frac{V}{n} \pm \frac{M}{I} z_i$$
$$I = \sum z_i^2 \quad (7.6)$$

where n is the number of micropiles and z_i is the distance of the micropile from the centre of gravity of the micropile group.

Figure 7.41 Rock anchors

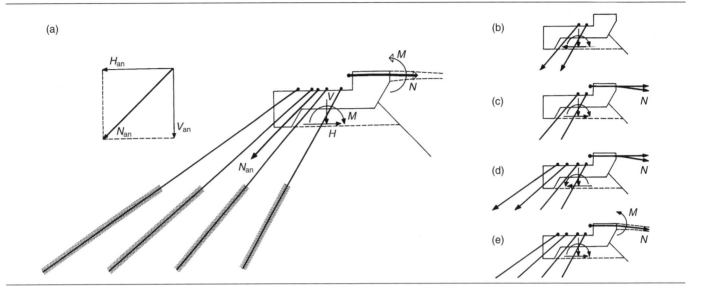

7.5. Erection of the deck

One of the main advantages of the stress ribbon structure is that the erection of the deck can be done independently of the conditions of the terrain under the bridge since the formwork of precast segments can be suspended from the bearing tendons.

Although there are some differences given by the arrangement of the stress ribbon above supports, there are two basic erection possibilities that are distinguished by the arrangement of the bearing tendons.

7.5.1 Erection possibility A

If the bearing tendons are situated in the troughs, where they are protected by cast-in-place concrete, they can also be used as erection cables (Figures 7.45 and 7.46(a)). The general arrangement of the erection of segments is depicted in Figures 7.47 and 7.48. The construction process could be as follows.

(a) First the bearing tendons are drawn by a winch. The strands are wound off from coils and are slowed down at the abutment by a cable brake that also ensures equal length for all strands. An auxiliary rope can also be attached to the erected tendon which enables the back drawing of the hauling rope. After drawing each tendon is tensioned to the prescribe stress (Figure 7.49).

(b) The segments are erected in each span by means of a crane truck. The erected segment is first placed under the bearing tendons and lifted until the tendons are touching the bottom of the troughs (Figure 7.50). 'Hangers' are then placed in position and secured; the segment is attached to hauling and auxiliary ropes and, by pulling of the winch, the segment is shifted along the bearing tendon into the pre-determined position (Figure 7.51). Before the segment is attached to the previously erected segments, the tubes for coupling the ducts of prestressing tendons are placed (Figure 7.52). This process is repeated until all the segments are assembled.

(c) When all segments are erected (Figure 7.53) the formwork for the saddle is hung on neighbouring segments and piers and/or abutments (Figure 7.55). In structures with saddles

Figure 7.42 Drill shafts

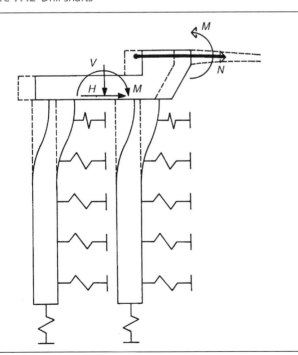

Figure 7.43 Drill shafts and ground anchors

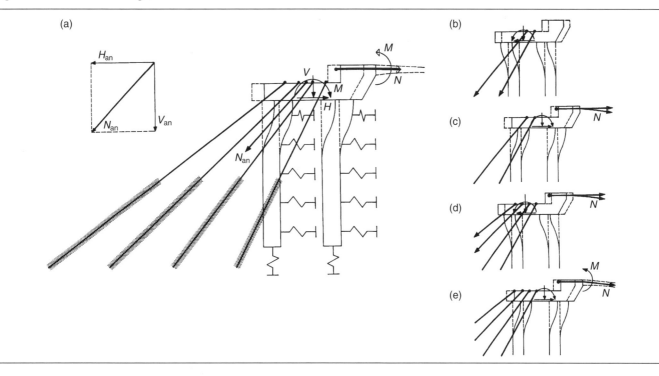

the formwork of the closures is suspended. The prestressing tendons (Figure 7.54) and rebar are then placed. After that, the joints, saddles and troughs or composite slab are all cast together (Figure 7.55). It is necessary to use a retarder in the concrete mix, which postpones the beginning of concrete setting until the concrete in all members is placed. In order to reduce the effects of shrinkage, temperature drop and accidental movement of pedestrians, it is recommended that the deck is partially prestressed as early as possible. When a minimum specified strength is attained, the stress ribbon is then prestressed to the full design stress.

7.5.2 Erection possibility B

If the bearing tendons are placed in the ducts, the process of construction has to be modified (Figures 7.46(b) and 7.56). The general procedure of the erection of segments is shown in Figure 7.56. The construction process could be as follows.

Figure 7.44 Battered micropiles

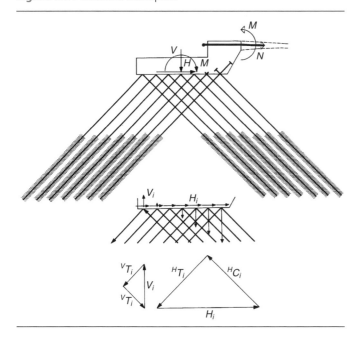

Figure 7.45 Construction sequences A

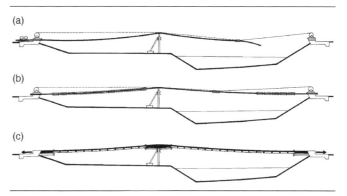

Figure 7.46 Erection of a segment: (a) erection A and (b) erection B

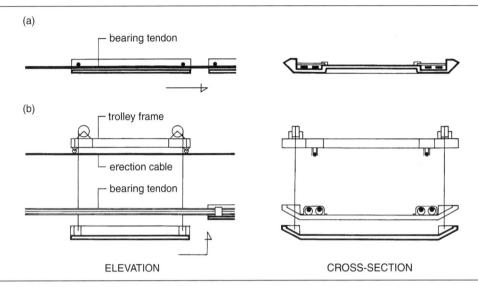

Figure 7.47 Prague-Troja Bridge, Czech Republic: erection scheme

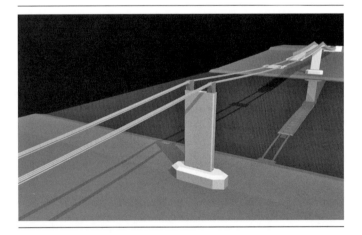

Figure 7.48 Grants Pass Bridge, Oregon, USA: erection scheme

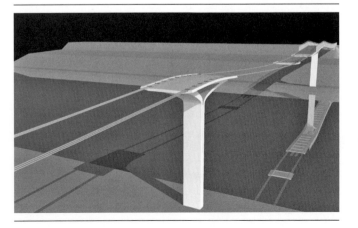

Figure 7.49 Grants Pass Bridge, Oregon, USA: bearing tendons

Figure 7.50 Prague-Troja Bridge, Czech Republic: erection of a segment

Figure 7.51 Prague-Troja Bridge, Czech Republic: shifting of a segment

Figure 7.52 Brno-Komin Bridge, Czech Republic: placing of tubes in joints

Figure 7.53 Grants Pass Bridge, Oregon, USA: deck after the erection of the segments

Figure 7.54 Grants Pass Bridge, Oregon, USA: bearing and prestressing tendons

(a) First an erection cable is erected and anchored at the abutments. The ducts are then progressively suspended, spliced and shifted along the erection cable into the design position. When the ducts are completed, the strands are pulled or pushed through the ducts and tensioned to the prescribed stress.

(b) The segments can also be erected by a crane truck. If the span across the obstacle is relatively short and the crane has sufficient reach it can erect all the segments (Figure 7.57). The erected segment is placed in a C-frame and slipped in under the bearing tendon. It is then lifted up so that it touches the bearing tendons. Thereafter, the 'hangers' are placed into position and secured. If the spans

Figure 7.55 Grants Pass Bridge, Oregon, USA: casting of the deck slab

Figure 7.56 Construction sequences B

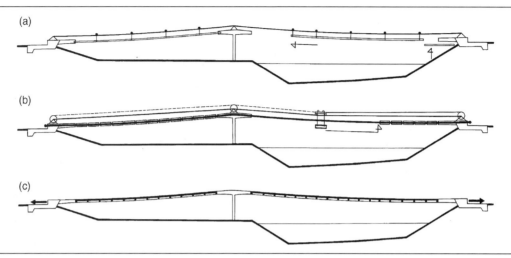

are longer, it is possible to apply an erection technique used for the erection of suspension structures (Figure 8.34). The segment is suspended on an erection frame supported on erection cables which is attached to hauling and auxiliary ropes. Using a winch, the segment is shifted along the erection cables into a pre-determined position where it is lifted until it touches the bearing tendons. The hangers are then placed and secured. This process is repeated until all segments are assembled.

(c) When all segments are assembled, the forms of the closures are hung and the prestressing tendons and rebars are placed. After that, the joints, saddles and troughs or a composite slab are cast (Figure 7.58) and post-tensioned. The remainder of this process is similar to the previously described erection method.

7.6. Static and dynamic analysis
7.6.1 Static function

As previously stated, the static behaviour of stress ribbon structures is given by their structural arrangement and by the process of construction. To help explain the static behaviour of a typical stress ribbon structure assembled of precast segments, two other systems will be discussed.

Figure 7.59 shows a structure in which bearing tendons support a deck of disconnected precast concrete panels of length of l_p (see also Figure 2.15). The weight g is composed of the weight of the bearing tendons ^Tg and the weight of the panels ^Sg. Let us consider the cable as a polygon which follows the funicular curve of equivalent loads placed at the vertices. At each joint i, the resultant force ^TR_i in bearing tendons balances the vertical

Figure 7.57 Maidstone Bridge, UK: erection of a segment

Figure 7.58 Maidstone Bridge, UK: casting of the deck slab

Figure 7.59 Stress ribbon structure stiffened by dead load

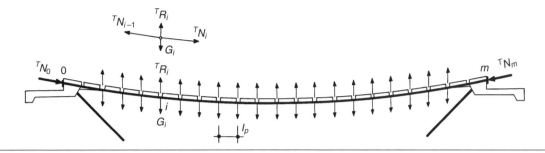

load $G_i = gl_p$. The anchor blocks are loaded by tension forces $^T N_0$ and $^T N_m$ that originate from the bearing tendons. During construction, the initial sag of the bearing tendons alone f_T increase to f_g when the segments are placed (Figure 7.60(a)).

Figure 7.60 Loading and deformation of the stress ribbon structure: (a) stiffened by dead load and (b) cast on the falsework

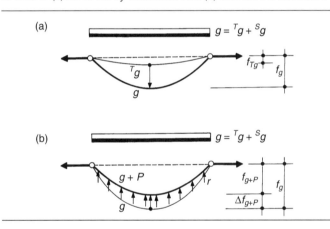

Figure 7.61 shows a structure that was cast on falsework. When the concrete reached a sufficient strength, the deck was prestressed. It is assumed that the deck has the shape of a polygon of straight members whose length corresponds to that of the precast panels. The weight g is also composed of the weight of the bearing tendons $^T g$ and the weight of the panels $^S g$. The prestressing lifts the stress ribbon up from the formwork. The force in the bearing tendons has the double function of partially carrying the dead load and partially prestressing the deck.

At each node i, the resultant internal force R_i is composed of a force $^T R_i$ that balances the vertical load G_i and of a force $^P R_i$ that loads the structure and consequently prestresses the deck. By prestressing, the initial sag of the stress ribbon f_g decreases to f_{g+p} (Figure 7.60(b)). The anchor blocks are loaded by tension forces N_0 and N_m that correspond to tension of a structure with sag f_{g+P}.

Figures 7.62 and 7.63 show a structure that was erected from precast segments (PS) with a cast-in-place composite slab (CS). The segments of length $l_s = l_p$ were suspended on

Figure 7.61 Cast-in-place stress ribbon structure cast on the falsework: (a) casting on the falsework and (b) prestressing

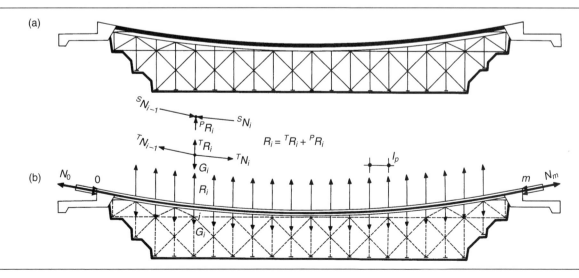

Figure 7.62 Precast stress ribbon structure: (a) after casting the joints and (b) prestressing

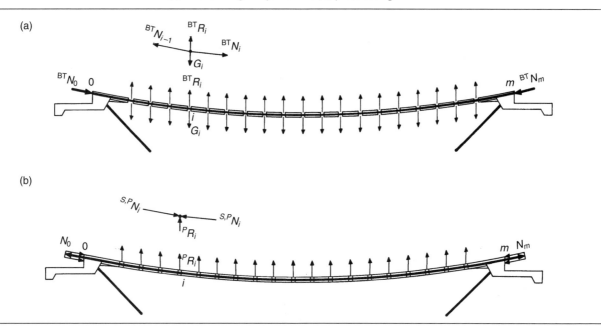

bearing tendons (BT). After casting the composite slab and joints, the structure was post-tensioned with prestressing tendons (PT). The precast segments have an area A_{PC}, composite slab A_{CS}, bearing tendons A_{BT} and the prestressing tendons A_{PT}. While the modulus of elasticity of the precast segments E_{PC} and the composite slab E_{CS} depend on their age, the modulus of elasticity of the tendons E_s is constant.

Weight g is composed of the weight of the bearing and prestressing tendons ^{T}g and the weight of the segments and cast-in-place slab ^{S}g.

It is interesting to follow the history of the forces that occur in the structure during the construction and during service. After casting the composite slab and joints, the bearing tendons carry all the dead load. They are stressed by a tension force $^{BT}N = N_g$. At each joint i, the resultant internal force $^{BT}R_i$ in the bearing tendons balances the vertical load $G_i = gl_s$ (Figures 7.62(a) and 7.63(a)). The anchor blocks are loaded by tension forces $^{BT}N_0$ and $^{BT}N_m$.

When the concrete in the joints reaches a sufficient strength, the deck is prestressed (Figures 7.62(b) and 7.63(b)). The structure has now become one continuous structural member with a composite section of modulus of elasticity E_{PC} and effective area $^{P}A_e$:

$$^{P}A_e = A_{PC} + \frac{E_{CS}}{E_{PS}} A_{CS} + \frac{E_S}{E_{PS}} A_{BT}.$$

By the stressing of the prestressing tendons, radial forces in the joints $^{P}R_i$ are created. The stress ribbon moves up as a consequence. Since the sag f_g has been reduced to f_{g+P}, the composite stress ribbon consisting of precast segments (PS), composite slab (CS) and bearing tendons (BT) is stressed not only by compression forces N_P but also by tension forces ΔN_{g+p} that correspond the sag reduction Δf_{g+p}:

$$\Delta N_{g+p} = N_{g+p} - N_g$$

$$\Delta f_{g+p} = f_{g+p} - f_g.$$

The normal force

$$^{S,P}N = -N_P + \Delta N_{g+p}$$

is distributed to the precast segments (PS), composite slab (CS) and bearing tendons (BT) in proportion to the modulus of elasticity and area of components.

The anchor blocks are loaded by tension forces N_0 and N_m that correspond to the tension of a structure of sag f_{g+P}.

After the prestressing tendons are grouted, they become a part of a composite section of area A_e and modulus of elasticity E_{PC}:

$$A_e = A_{PC} + \frac{E_{CS}}{E_{PS}} A_{CS} + \frac{E_s}{E_{PS}} A_{BT} + \frac{E_s}{E_{PC}} {^{PT}A_s}.$$

Figure 7.63 Loading, deformation and stresses of precast stress ribbon structure: (a) casting the joints, (b) prestressing, (c) service load and (d) effects of the creep and shrinkage of concrete

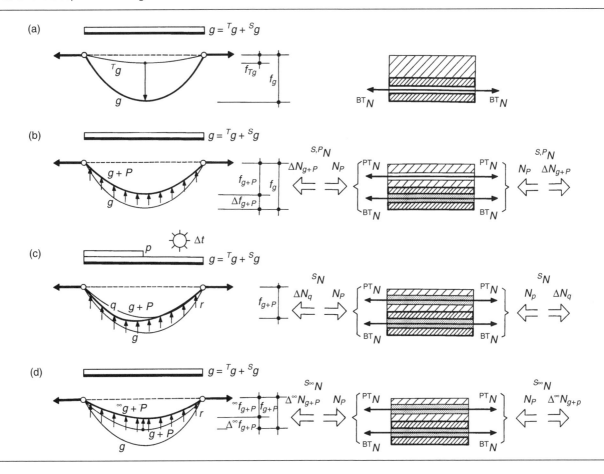

All service loads (Figure 7.63(c)) are resisted by this composite section that is stressed by the normal force

$$^S N = -N_P + \Delta N_q$$

where

$$\Delta N_q = N_q - N_g$$

and N_q is a normal force that originates in the stress ribbon loaded by dead load, live load, temperature changes and by ΔN_{g+p}.

$^S N$ is distributed proportionally to the modulus of elasticity and area into the precast segments (PS), composite slab (CS), bearing (BT) and prestressing (PT) tendons.

With time, the internal forces in the individual members of the composite section are redistributed because of the creep and shrinkage of concrete. Contrary to the redistribution of stresses that was described in Chapter 6, the situation here is more complex. Due to the shortening of the concrete, the stress ribbon moves up and the original sag f_{g+p} is reduced to f_{g+p}. Therefore the stress ribbon is additionally stressed by a tension force

$$^{S\infty} N = -N_P + \Delta^\infty N_{g+p}$$

where

$$\Delta^\infty N_{g+p} = N_{g+p} - N_g - N_{c+sh}$$
$$\Delta^\infty f_{g+p} = f_{g+p} - f_g - f_{c+sh}.$$

$^{S\infty} N$ is distributed into all members of the composite section. All service loads now act on a structure that has a reduced sag and therefore higher tension stresses.

It is therefore evident that the stresses in all structural members of the stress ribbon depend on the construction process, age of the concrete members and time at which the structure is loaded.

It is also clear that it necessary to distinguish between the behaviour of the structure during erection and during service.

During erection the structure acts as a cable (Figure 7.62(a)); during service it acts as a stress ribbon (Figure 7.62(b)) that is stressed not only by normal forces but also by bending moments. The shape and the stresses in the structure at the end of erection determine the stresses in the structure when in use. The change from cable to stress ribbon occurs when the concrete of joints starts to set.

All the design computations have to start from this basic stage. During the erection analysis, the structure is progressively unloaded down to the stage in which the bearing tendons are stressed. This is the way the required jacking force is obtained (Figures 7.64(a)–7.64(c)).

The designer sets the shape of the structure after prestressing (Figure 7.64(e)). Since this shape depends on the deformation of the structure due to prestress, the basic stage has to be estimated. The deformation of the structure due to prestress is computed and checked against the required final stage. This computation has to be repeated until reasonable agreement is obtained.

The basic stage is also the initial stage for the subsequent analysis of the structure for all service loads (Figures 7.64(d)–7.64(f)).

At the ultimate limit state there are no sufficient compression stresses in the joints between precast members; the joints therefore open (Figure 7.64(g)). The safety of the stress ribbon structure is then given by the tension strength of bearing and prestressing tendons. The stress ribbon structures should therefore be analysed as a cable at the ultimate limit state. Due to the non-linear behaviour of the cable, tension force in the bearing and prestressing tendons does not increase linearly with the increasing load. Stress ribbon structures therefore usually have a large ultimate capacity.

What follows is a description of a simplified analysis that the author used in his first structures. This approach can be used for preliminary analysis or for checking results obtained from modern analytical programs that are discussed further in Section 7.6.3.

7.6.2 Analysis of the structure as a cable
7.6.2.1 Erection stage
During the erection all loads are resisted by the bearing tendons that act as a cable. Since the tendons are not usually connected to the saddles they can slide freely according to the imposed load. This is true both for structures in which the bearing tendons are supported by steel or concrete saddles and for structures in which the bearing tendons pass through ducts in the support haunches.

Hence, the cables act as a continuous cable of m spans which crosses fixed supports (Figure 7.65). A change of any load causes friction in the saddles, the magnitude of which depends on the vertical reaction R and on a coefficient of friction μ. In all supports, friction forces ΔH act against the direction of the cable movement:

$$\Delta H = R\mu.$$

The stresses in the bearing tendons are also affected by their elongation in the anchorage blocks and by possible displacements of the end supports. The unknown horizontal force H_i is given by Equation (4.37), used for the analysis of the simple cable:

$$aH_i^3 + bH_i^2 + cH_i + d = 0. \tag{7.7}$$

Since both the length s and the elongation of the cable Δs are calculated for the whole length of continuous cable, we have:

$$s = \sum_{j=1}^{m} s_j = \sum_{j=1}^{m} \frac{l_j}{\cos \beta_j} + \frac{\cos \beta_j}{2H_j} D_j \tag{7.8}$$

$$\Delta s = \sum_{j=1}^{m} \Delta s_j = \sum_{j=1}^{m} \frac{2H_j}{EA \cos \beta} \left(s_j - \frac{l_j}{2\cos \beta_j} \right) \tag{7.9}$$

where

$$D_j = \int_0^{l_j} Q_j^2 \, dx.$$

The terms a, b, c and d have to be modified to:

$$a = \sum_{j=1}^{m} \frac{l_{j,i}}{EA \cos^2 \beta_{j,i}} + k,$$

$$b = Ln_i - \sum_{j=1}^{m} \frac{l_{j,i}}{\cos \beta_{j,i}},$$

$$c = \sum_{j=1}^{m} \frac{D_{i,j}}{EA}$$

and

$$d = \sum_{j=1}^{m} \frac{\cos \beta_{j,i}}{2} D_{j,i}.$$

The horizontal force H_j in span j is taken as the largest horizontal force acting in the most loaded span, reduced by the sum of the losses due to friction at the supports situated between span j and the most loaded span.

Figure 7.64 Static function

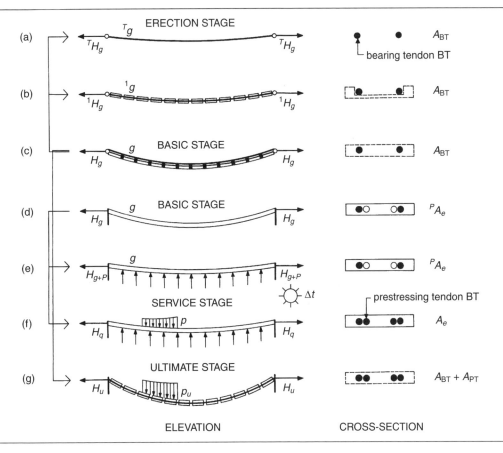

Since the terms a, b, c and d depend on the span lengths $l_{j,i}$ and vertical differences $h_{j,i}$ which in turn depend on horizontal force H_i, it is not possible to determine the unknown H_i directly by solving Equation (7.7). It is therefore necessary to determine H_i by iteration. First, the unknown H_i is calculated for zero deformation of supports and zero elongation of the cable at the anchor blocks. For this force the vertical reactions A_i, B_i and R_i, span length $l_{j,i}$, vertical difference $h_{j,i}$, members a, b, c and d and the new horizontal force H_i are calculated. This iteration is repeated until the difference between subsequent solutions is smaller than the required tolerance.

This analysis should be repeated for all erection stages. The goal of the analysis is not only to determine the jacking force for the bearing tendons, but also the deformations of the structure and the corresponding stresses that affect the substructure. As an example of an erection analysis, the history of the forces in bearing tendons of a two-span structure is presented in Figure 7.66.

Figure 7.65 Stage of erection: static function

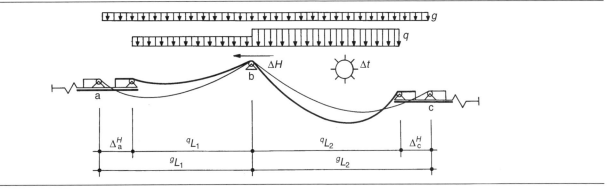

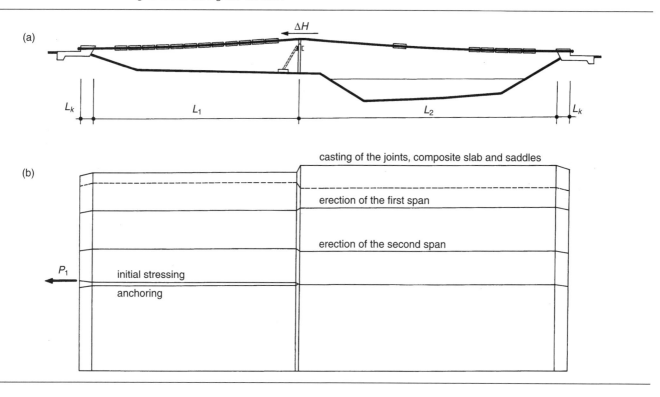

Figure 7.66 Forces in the bearing tendons during the erection

7.6.2.2 Service stage

Since the structure is very slender, local shear and bending stresses develop only under point loads and at the supports. Because these stresses are relatively small, they do not affect the global behaviour of the structure. This makes it possible to analyse the structure in two closely related steps.

In step 1, the stress ribbon is analysed as a perfectly flexible cable which is the supports (Figure 7.67(a)). The effect of prestressing is a shortening of the cable, which can be simulated as a temperature drop. The effect of creep and shrinkage can be analysed in a similar way. However, due to the redistribution of stresses between the individual components of the concrete section, an iterative approach has to be used. To facilitate this analysis, standard computer programs for continuous cables are used.

It is also possible to isolate and analyse the individual spans for the given loads and for different horizontal support movements (Figure 7.68). From the requirement that horizontal force be the same at each span, the horizontal force (H_i) is obtained. With this, the deformations of the supports of the single spans can be calculated.

In step 2, shear and bending stresses in single spans are calculated using the analysis of the bending of the simple cable (Section 4.2). The cable is analysed for the load $q(x)$ and for the horizontal force and deformations of the supports that were determined in step 1.

As an example of the described procedure, Figures 7.69 and 7.70 present results of the analysis completed for the Redding Bridge described in Section 11.1. Figure 7.69 shows the calculation model, applied loads and corresponding horizontal forces, deflections and rotation of the cable at the supports. The support bending moments were determined from these values.

Since the bending moments were relatively large, the support sections were analysed as partially prestressed members; a reduction of the bending stiffness caused by cracks was therefore considered in the analysis. Figure 7.70 shows the bending moments and normal stresses at the top fibres of the haunch. The analysis was performed for a stress ribbon with constant section, with haunch and with haunch of reduced stiffness.

7.6.3 Analysis of the structure as a geometrically non-linear structure

Modern structural programs allow us to follow the behaviour of stress ribbon structures both during erection and during service. These programs also need to capture the large deformation and the tension stiffening effects (Figure 2.23(a)). The structure can be modelled as a chain of parallel members that represent

Figure 7.67 Stage of service: static function: (a) analysis as a cable and (b) analysis as prestressed concrete band

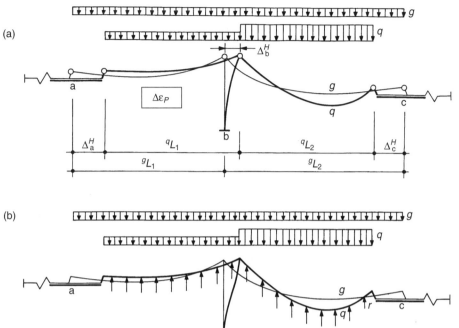

bearing tendons (BT), prestressing tendons (PT), precast segments (PS) and cast-in-place slab (CS) or trough (Figure 7.71). Bearing and prestressing tendons can be modelled as 'cable' members, for which the initial force or strain has to be determined. Precast segments and cast-in-place slabs can be modelled as 3D bars or as shell elements that have both bending and membrane capabilities.

Since the programs use so-called 'frozen members' it is possible to model a change of static system (from the cable into the stress ribbon) as well as the progressive erection of structure. The program systems also contain so-called 'contact' members that only resist compression forces. These members can be used for the modelling of saddles from where the stress ribbons can lift up.

In the analysis the initial stress in the tendons has to be determined. The initial forces are usually determined for the basic stage (Figures 7.64(c) and 7.64(d)) where the structure changes from cable to stress ribbon. The initial forces in the cable are determined using the cable analysis.

Some programs (for example LARSA) have 'cable members' that have zero stiffness (area and modulus of elasticity) in the initial stage. They are stressed by initial forces that exactly balance the external load. For all subsequent loads these elements are part of the structure, i.e. part of the global stiffness of the structure. This is the way to model the prestressing tendon.

Unfortunately, in some programs (for example ANSYS) the initial stage is modelled with an initial strain of tendons that also have actual stiffness (area and modulus elasticity), and therefore are part of the stiffness of the structure. Since a portion of the strain and corresponding stress is absorbed by their stiffness, it is necessary to artificially increase their initial strain such that the strain and corresponding stress in tendons exactly balances the load at the basic stage. That means that the initial stage has to be determined by iteration.

The analysis that starts from the basis stage can be used for both the analysis of the erection and service stages. The stresses in the structure during erection and the bearing tendons jacking forces are determined by simulating a progressive unloading of the structure. Since the superposition principle does not apply, the analysis of the service stage should be carried out according to the following flow chart.

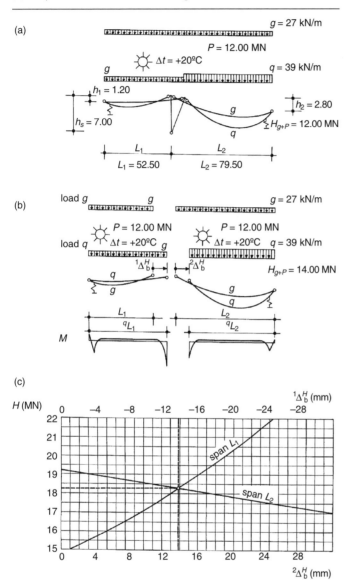

Figure 7.68 Analysis of a two-span structure: (a) continuous cable, (b) simple cable and (c) determining of the horizontal force

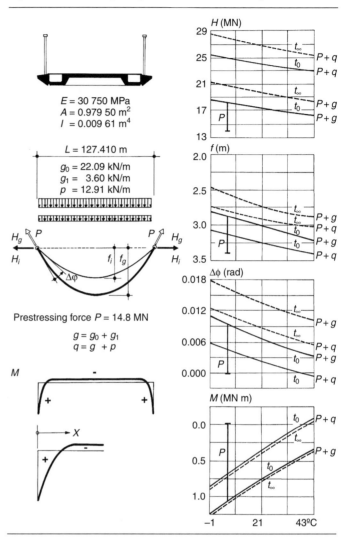

Figure 7.69 Redding Bridge, California, USA: analysis as a cable

Figure 7.71 shows (a) a shape and bending moment and (b) a calculation model of a one-span structure loaded by dead load, prestress and creep and shrinkage of concrete. It is evident that, due to creep and shrinkage, the sag is reduced and therefore all internal forces are higher at time t_∞.

Furthermore, since the area of the bearing and prestressing tendons is higher than in traditional concrete structures, a significant redistribution of stresses between steel and concrete occurs with time. In structures assembled from precast segments and cast-in-place slab, the redistribution of stresses between these members also has to be considered.

For the analysis of the creep and shrinkage it is necessary to perform a time-dependent analysis. It is not possible to analyse the structure in a single step for the initial strain caused by creep and shrinkage; this would cause significantly larger deformations and higher bending moments at the supports.

Unfortunately, common programs are not able to provide a time-dependent, geometrically non-linear structural analysis. The author therefore used the procedure combining time-dependent analysis with the finite-element software ANSYS (Chapter 6). The most advanced programs (for example RM2000, produced by the software firm TDV – Technische Datenverarbeitung – Graz, Austria) are able to provide a time-dependent geometrically non-linear structural analysis of composite stress ribbon structures.

Figure 7.70 Redding Bridge, California, USA: bending moments and stresses at a support haunch

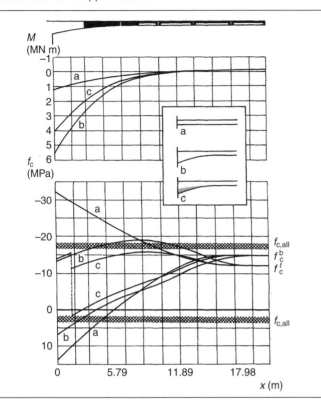

Figure 7.71 Stress ribbon structure: (a) deformation and bending moments and (b) modelling of the deck

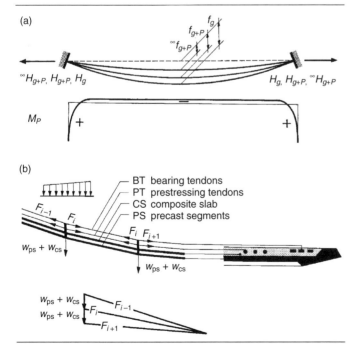

Figure 7.72 Bending moment at support haunches

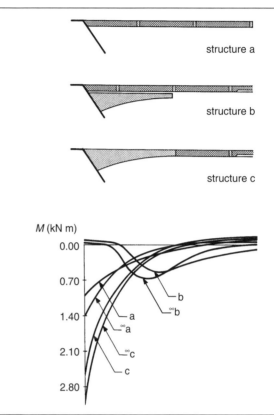

During the design of the Blue River Bridge (Section 11.1), a detailed time-dependent analysis was conducted of the structure. The structure is formed by precast segments with cast-in-place slab. The segments were suspended on bearing tendons and stressed by prestressing tendons. The actual bridge is formed by a stress ribbon that has cast-in-place haunches at supports.

To understand the problem, the analysis was done for three possible arrangements of the structure: (a) a structure (Figure 7.72) where the support region was detailed with a constant section, (b) with a 4.5 m saddle and (c) with 4.5 m parabolic haunch. The structure was analysed for the effects of prestress and creep and shrinkage of concrete using the CEB-FIP (MC 90) rheological functions.

Figure 7.72 shows the bending moments in the stress ribbon close to supports for time t_0 and time t_∞. Here it can be seen that the bending moments do not change significantly with time.

7.6.4 Dynamic analysis

As stated in Chapter 3, it is necessary to carefully check the dynamic response of stress ribbon structures for demands induced by people and wind. Response to earthquake loading

Figure 7.73 Typical natural modes

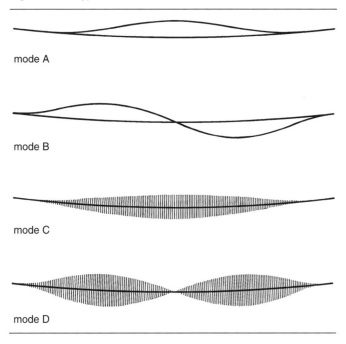

also has to be verified. Typically, the first step is to determine the natural modes and frequencies followed by a check of the dynamic response due to the moving load.

For preliminary calculations the vertical natural modes can be determined using the formulae for vibration of a simple cable given in Section 4.3. The dynamic test has proved their validity.

For final design, the dynamic analysis should be done with a calculation model that includes non-linear analysis. It is important to realise that the dynamic analysis is usually linear and that most programs are able to describe the special behaviour of the stress ribbon and cable-supported structures only by using the so-called tension stiffening effect.

A typical one-span stress ribbon structure is characterised by natural modes that are depicted in Figure 7.73. The vertical modes are denoted A and B, the first swing mode is denoted C and the first torsional mode is denoted D. Due to the vertical curvature of the prestressed band, a horizontal movement is always combined with torsion; it is therefore difficult to find a pure torsional mode.

Since the vibration following the first vertical mode (A) requires an elongation of the cable, the corresponding frequency is in some cases higher than the frequency of the second vertical mode (B).

When analysing multi-span structures, note that the bridge behaves as a continuous structure only when there is a horizontal displacement of the supports. For a small load, as caused by a group of pedestrians, the change of stresses is very small and the individual spans behave as isolated cables. When the structure is checked for motions that can cause unpleasant feeling, the dynamic analysis should therefore be done for the individual spans in addition to the overall structure.

7.6.5 Examples of the analysis

The results of calculation models and analysis of the Grants Pass and Maidstone Bridges (see Section 11.1) are presented in Figures 7.74–7.79. Both structures were modelled as 3D structures assembled from parallel 3D elements that modelled precast segments (PS), composite slab (CS), bearing (BT) and prestressing tendons (PT). The length of the elements corresponded to that of the segments.

For Grants Pass Bridge the saddles were modelled with 3D beam elements of 0.30 m length of varying depth. These elements were connected to the stress ribbon with contact members (Figure 7.75). In the case of the Maidstone Bridge, the saddles were modelled by 3D beam elements of 0.5 m length that had varying depth corresponding to that of the haunches. From the figures, it is evident that the calculation models can describe the actual arrangement of stress ribbon structures including their flexible connection to the soil.

The results of the dynamic analysis of the Grants Pass Bridge are presented in Figure 7.76. It is interesting to note that the first transverse frequencies lie in range 0.701–1.478 Hz. Although the bridge was heavily loaded with pedestrians during its opening, no unpleasant feelings by the users were reported.

Relatively flexible transverse flexible supports have significantly reduced the response of the structure to seismic demands, as can be seen from the response spectrum diagram.

Figure 7.78 presents the bending moment diagrams in the stress ribbon deck of the Maidstone Bridge. Due to the arrangement of the prestressing tendons at the abutments and pier haunches, the positive bending moments that usually appear at those locations were significantly reduced. From the natural modes and frequencies shown in Figure 7.79 it is evident that this complex structure vibrates in compound modes. This also demonstrates the good behaviour of a slender structure.

7.6.6 Designing structural members

Stress ribbon structures are designed as ordinary structural concrete. As such, it is reasonable to check all members as partially prestressed when crack width and fatigue stresses in tendons and reinforcing steel are being checked. Maximum compression stresses in the concrete should also be verified.

Figure 7.74 Grants Pass Bridge, Oregon, USA: calculation model of the structure

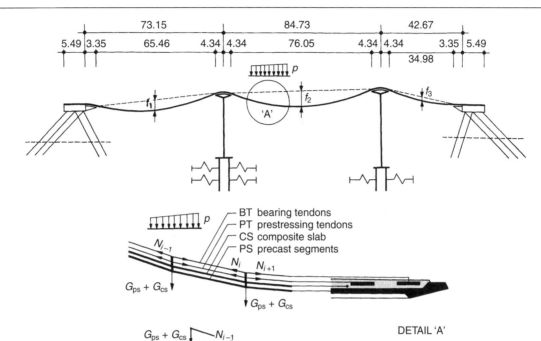

Since the stress range in the prestressed band is within the stress range of ordinary prestressed concrete structures, the stresses in the bearing and prestressing tendons should be treated as ordinary prestressing tendons in accordance with the appropriate national standard. For bonded and unbonded tendons, maximum service stresses should not usually exceed $0.7f_u$ and $0.6f_u$, respectively.

Figure 7.75 Grants Pass Bridge, Oregon, USA: calculation model of the pier

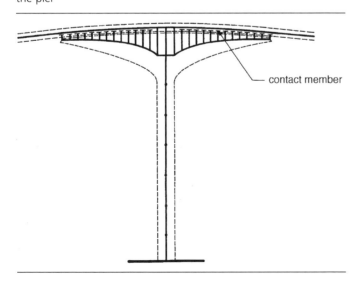

Since the joints and cracks in the concrete band open at increasing load, the stress ribbon behaves as a cable at ultimate loading (Figure 7.64(g)). The load is resisted both by the bearing and prestressing tendons. Since the additional load causes larger sag, the stresses in the tendons increase less than linearly with the load. This explains why it is possible to use relatively high allowable stresses in the tendons for service load.

7.7. Special arrangements

Ordinarily, common stress ribbon structures are composed of a straight deck with one or more spans that are supported by vertical piers. However, there are other possibilities that allow us to extend the field of applications of the step ribbon.

7.7.1 Stress ribbon supported by inclined struts

It is clear that a stress ribbon of two or more spans can also be supported by inclined struts. Figure 7.80 shows a possible arrangement of a two-span structure.

To eliminate bending in the strut due to dead load, it is necessary to balance the horizontal component of the axial force in the strut with additional tension in the deck. In an initial stage, the horizontal force in the shorter span H_1 has to balance the sum of the horizontal force in the longer span H_2 plus the horizontal component of the strut H_p, i.e.

$$H_1 = H_2 + H_p = \frac{gL_1^2}{8f_1} = \frac{gL_2^2}{8f_2} + H_p.$$

Figure 7.76 Grants Pass Bridge, Oregon, USA: natural modes and frequencies

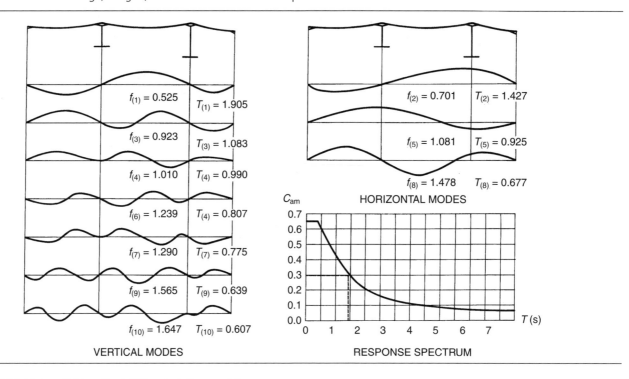

Figure 7.77 Maidstone Bridge, Kent, UK: calculation model

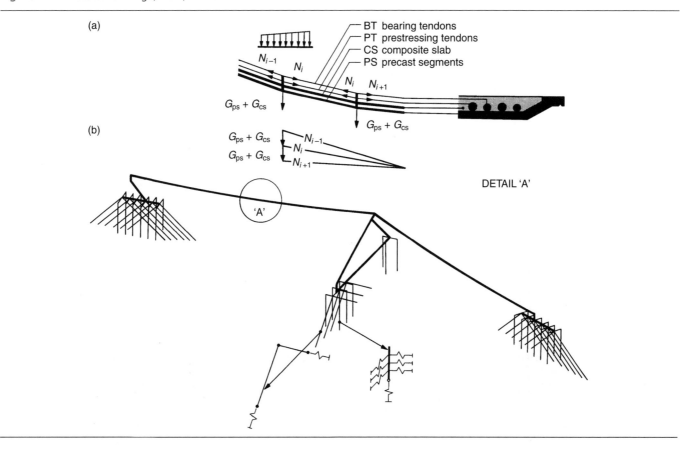

Figure 7.78 Maidstone Bridge, Kent, UK: bending moments in the deck

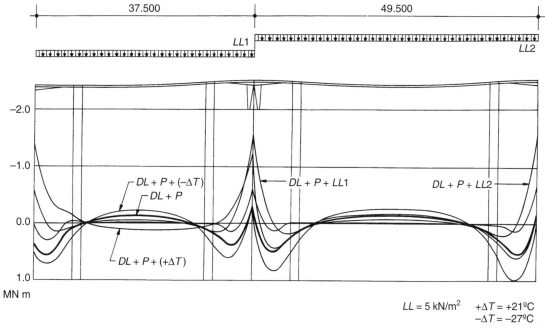

Figure 7.79 Maidstone Bridge, Kent, UK: natural modes and frequencies

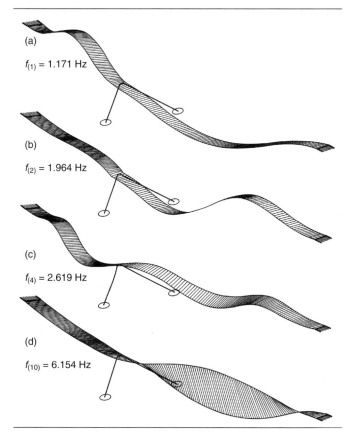

Tension forces due to live load and temperature changes are (in proportion to the inclination of the strut and its bending stiffness) larger in the shorter span. It is therefore necessary to increase the number of bearing tendons and level of pre-stressing in that span. This tension could also be resisted by external tendons anchored in the pier saddle and in the abutment.

When the structure has an arrangement similar to Figure 7.80, careful attention should be devoted to the motion of the shorter spans since the higher horizontal force can increase the vertical natural frequencies close to the critical value of 2 Hz.

Figure 7.80 Stress ribbon supported by inclined struts

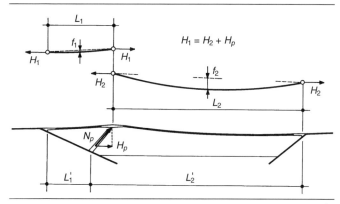

Figure 7.81 Stress ribbon suspended on stay cables

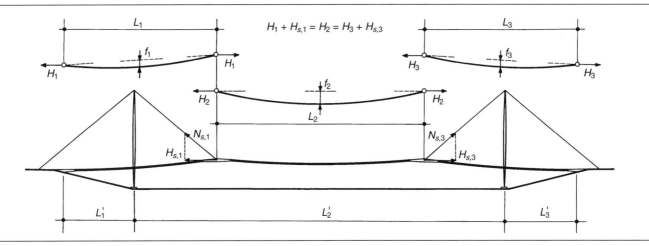

7.7.2 Stress ribbon suspended on stay cables

The stress ribbon can also be suspended on vertical or inclined stay cables anchored in towers and stress ribbon saddles. Figure 7.81 shows a possible arrangement for a three-span structure.

The geometry of the structure is also given by the initial stage, in which the equilibrium of forces has to be assured. The horizontal force in the main span is resisted by the sum of the side-span horizontal force and the horizontal component of the stay-cable force, i.e.

$$H_1 + H_{s,1} = H_2 = H_3 + H_{s,3} = \frac{gL_1^2}{8f_1} + H_{s,1}$$

$$= \frac{gL_2^2}{8f_2} = \frac{gL_3^2}{8f_3} + H_{s,3}.$$

7.7.3 Cranked alignment of the stress ribbon

Stress ribbon structures of two or more spans can also have a cranked alignment in plan. Although this solution carries several structural and static problems, it can solve particular problems of bridging a site and can improve the aesthetic impression of the structure.

Figure 7.82 shows a possible solution of a three-span structure in which cranks are done at both intermediate supports b and c. The piers that resist the corresponding transverse forces are situated at the bisectors of the angles α_b and α_c (Figure 7.82(a)). The transverse force can be resisted by a system of inclined members such as a compression strut and a tension tie (Figure 7.82(c)). It is also possible to design this with an inclined pier without a tie (Figure 7.82(d)). The inclination of the pier should be determined such that, for dead load, the transverse bending moment is zero. It should be noted that the vertical and horizontal forces V_p and H_p vary according to external loadings and changes of temperature, and that the piers have to be designed for all these loads.

For dead load, the piers should also have zero longitudinal moments. The geometry and forces in the individual spans should therefore be determined similarly to the case of straight stress ribbon structures, where horizontal forces in the individual spans should be the same.

In the initial state, the stress ribbon behaves as a continuous cable that carries the whole dead load. For all other loadings (including prestressing) it behaves as a stress ribbon that is acted upon by transverse and torsional bending moments and shear forces.

Only one structure of this type has been built and is located in Maidstone, UK (see Section 11.1). A similar structure was proposed by Cezary Bednarski for Hadrian Bridge, which is planned to be built in Scotland (Figure 7.83). In this design, the cranked alignment was combined with an inclined strut support in one span.

7.7.4 Star alignment of the stress ribbon

Stress ribbon structures can be also designed with a star alignment in plan. At the centre, individual spans can be connected to piers or to a tension ring.

Figure 7.84 shows a possible solution for a structure in which individual spans are connected at an intermediate support. Figure 7.85 shows a structure where the spans are connected in a middle platform. The structures are composed of three spans connected at a 60° angle. The geometry (sag and span) are determined in such a way that there is an equilibrium of forces in the horizontal plane at the connection point. It is possible to connect more spans and thereby create a large variety of structures.

Figure 7.82 Cranked alignment of the stress ribbon

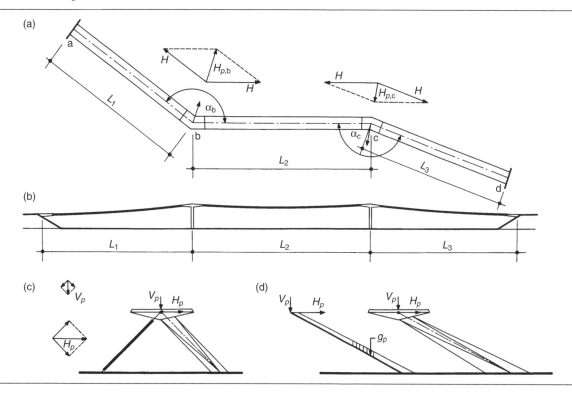

For structures connected at an intermediate pier, the span length and sags in the individual spans should comply with:

$$\frac{f_i}{L_i^2} = \frac{f_1}{L_1^2} = \frac{f_2}{L_2^2} = \frac{f_3}{L_3^2}.$$

Figure 7.83 Hadrian's Bridge, Scotland, UK (courtesy of Cezary Bednarski)

Figure 7.84 Star alignment of the stress ribbon with pier: (a) plan and (b) developed elevation

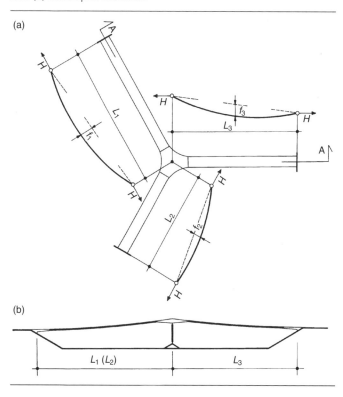

Figure 7.85 Star alignment of the stress ribbon with tension ring: (a) plan and (b) developed elevation

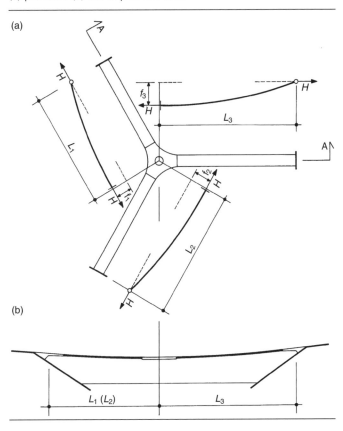

7.8. Stress ribbon supported by arch
7.8.1 Structural arrangement

The intermediate support of a multi-span stress ribbon can also have the shape of an arch (Figure 7.86). The arch serves as a saddle from which the stress ribbon can rise during post-tensioning and during temperature drop, and where the band can rest during a temperature rise.

In the initial stage, the stress ribbon behaves as a two-span cable supported by the saddle that is fixed to the end abutments (Figure 7.86(b)). The arch is loaded by its self weight, the weight of the saddle segments and the radial forces caused by the bearing tendons (Figure 7.86(c)). After post-tensioning the stress ribbon with the prestressing tendons, the stress ribbon and arch behave as one structure.

The shape and initial stresses in the stress ribbon and in the arch can be chosen such that the horizontal forces in the stress ribbon

Figure 7.86 Stress ribbon supported by arch: (a) stress ribbon and arch, (b) stress ribbon, (c) arch, (d) self-anchored system and (e) partially self-anchored system

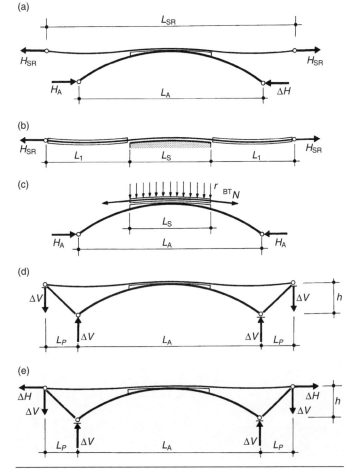

It is evident that the central platform, to which the individual spans are connected, should be horizontal. The initial forces in the bearing tendons therefore correspond to the forces of a cable of span $2L_i$:

$$H_i = \frac{g(2L_i)^2}{8f_i} = \frac{g4L_i^2}{8f_i}.$$

The span length and sags in individual spans should then comply with:

$$\frac{f_i}{4L_i^2} = \frac{f_1}{4L_1^2} = \frac{f_2}{4L_2^2} = \frac{f_3}{4L_3^2}.$$

In the initial state, the stress ribbon behaves like a continuous cable that carries the whole dead load. For all other loadings (including prestressing) it behaves as a stress ribbon that is also stressed by transverse and torsional bending moments and their corresponding shear forces.

At the time of writing, only one structure with a star layout has been built (located in Japan; see Section 11.1).

H_{SR} and in the arch H_A are same. It is then possible to connect the stress ribbon and arch footings with compression struts that balance the horizontal forces. The moment created by horizontal forces $H_{SR} h$ is then resisted by $\Delta V L_P$. In this way, a self-anchored system with only vertical reactions is created (Figure 7.86(d)). This self-anchored system eliminates the anchoring of horizontal forces in the upper soil layers.

In some cases, due to the slope limitations of the stress ribbon, the deck has to have a very small sag and the corresponding horizontal force becomes very large. A supporting arch that would balance this force would be extremely flat. If the topography requires an arch of higher rise, it is then possible to develop a partially self-anchored system.

The arch is designed for an optimum rise and its corresponding horizontal force. This horizontal force H_A is then transferred by the inclined props into the stress ribbon's anchor blocks. The anchor blocks only have to resist the difference, i.e.

$$\Delta H = H_S - H_A.$$

The moment created by horizontal forces $H_A h$ is then resisted by the couple $\Delta V L_P$.

It is possible to develop many partially self-anchored systems in which the arch helps to reduce the horizontal force of the stress ribbon. Figure 7.87 describes a static function of one possibility. In the initial stage, the arch is loaded only by its self weight and the weight of the saddle segments. In this case, the stress ribbon forms a one-span structure where the bearing tendons only carry the weight of the segments at either side of the saddle.

The horizontal force H_A is then transferred by means of the inclined struts to the anchor blocks of the stress ribbon that only now have to resist the difference:

$$\Delta H = H_{SR} - H_A.$$

The moment created by horizontal forces $H_A h$ is then resisted by a couple of vertical forces $\Delta V L_P$.

It is also obvious that the stress ribbon can be suspended from the arch. It is then possible to develop several fully or partially self-anchored systems. Figure 7.88 presents some concepts using such systems.

Figure 7.88(a) shows an arch fixed at the anchor blocks of the slender prestressed concrete deck. The arch is loaded not only by its own self weight and that of the stress ribbon, but also with the radial forces of the prestressing tendons.

Figure 7.88(b) shows a structure that has a similar static behaviour as the structure presented in Figure 7.86(d). The

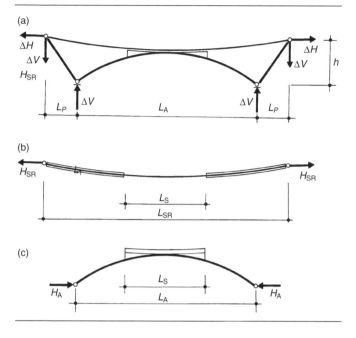

Figure 7.87 Stress ribbon supported by arch: (a) partially self-anchored system, (b) stress ribbon and (c) arch

two-span stress ribbon is suspended on an arch that serves as a 'saddle' on which the prestressed band changes curvature. In the initial stage the stress ribbon behaves as a two-span cable supported by the saddle (Figure 7.86(b)). The arch is loaded by its self weight, the weight of the saddle segments and the radial forces caused by the bearing tendons. When the stress ribbon is post-tensioned, the stress ribbon and arch behave as one structure.

To reduce the tension force at the stress ribbon anchor blocks, it is possible to connect the stress ribbon and arch footings by compression struts that fully or partially balance the stress ribbon horizontal forces.

Figure 7.88(c) shows a similar structure in which the slender prestressed concrete band has increased bending stiffness in the non-suspension portion of the structure not suspended from the arch.

Figure 7.88(d) presents a structure in which the change of curvature of the prestressed band is accomplished in a short saddle that is suspended from the arch. Since the arch is loaded by its self weight and by a point load from the stress ribbon, it should have the funicular shape corresponding to this load.

7.8.2 Model test

The author believes that a structural system formed by a stress ribbon supported by an arch increases the field of application of stress ribbon structures. Several analyses were undertaken to

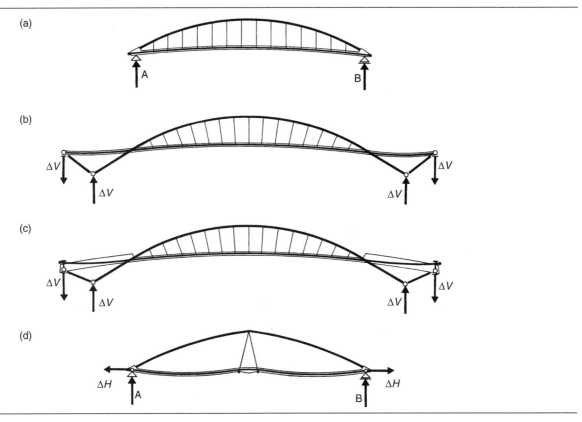

Figure 7.88 Stress ribbon suspended on arch: (a) tied arch, (b) tied arch with side spans, (c) tied arch with flexural stiff side spans and (d) two-span stress ribbon suspended on arch

verify this. The structures were checked not only with detailed static and dynamic analysis, but also on static and full aero-elastic models. The tests verified the design assumptions, behaviour of the structure under wind loading and determined the ultimate capacity of the full system.

The model tests were carried out for a proposed pedestrian bridge across the Radbuza River in Plzen. This structure was designed to combine a steel pipe arch with a span length of 77.00 m and 'boldness' of 973 m (ratio of square of the length L divided by the rise f) with a stress ribbon deck (Figure 7.89). Two steel pipes filled with concrete form the arch that supports the two-span stress ribbon deck.

The deck is a composite member consisting of precast concrete and cast-in-place parts. The deck width between railings is 4.00 m. The deck is fixed to the arch at mid-span. To keep the slope change within 8%, the deck is supported on steel plates in the crown location. At the ends, the stress ribbon is fixed to a diaphragm which is supported by two inclined cast-in-place concrete struts fixed to the foundations of the arch. Two tension pin piles support the diaphragm for uplift. Nine compression pin piles support the arch foundation. The structure thus forms a self-anchoring system, where the horizontal forces from the stress ribbon are transferred through the inclined concrete struts to the foundation where they are balanced against the horizontal component of the arch.

The behaviour of the structure was confirmed by detailed static and dynamic analysis performed with the program ANSYS. The calculation model (Figure 7.90) was able to describe the spatial behaviour of the structure as well as the construction process, including the non-linear behaviour of the structure.

7.8.3 Static model

The static physical model was done in a 1:10 scale. The shape and test set-up is shown in Figure 7.91. Dimensions of the model and cross-section, loads and prestressing forces were determined according to rules of similarity. The stress ribbon was assembled with precast segments of 18 mm depth and the cast-in-place haunches were anchored in anchor blocks made with steel channel sections. The arch consisted of two steel pipes and the end struts consisted of two steel boxes fabricated from channel sections. The saddle was made from two steel angles supported on longitudinal plates strengthened with vertical stiffeners.

Figure 7.89 Analysed bridge: (a) cross-section, (b) cross-section at mid-span and (c) elevation

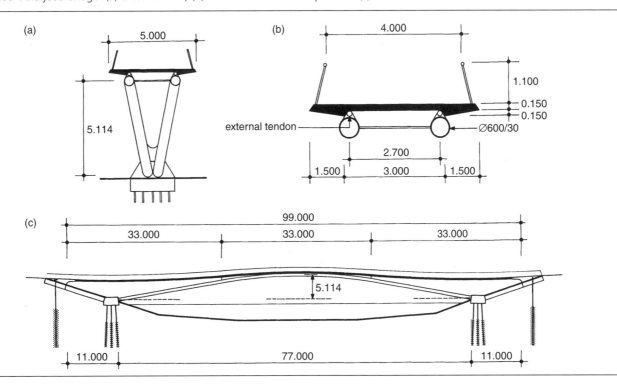

The footing common to the arch and inclined struts were assembled from steel boxes fabricated with two channel sections. They were supported by steel columns consisting of two 'I' sections. The end ties consisted of four rectangular tubes. The steel columns and the ties were supported by a longitudinal steel beam that was anchored to the test floor.

The precast segments were made from micro-concrete of 50 MPa characteristic strength. The stress ribbon was supported and post-tensioned by two monostrands situated outside the section. Their position was determined by two angles embedded in the segments. These angles were welded to transverse diaphragms situated outside the segments.

Figure 7.90 Calculation model

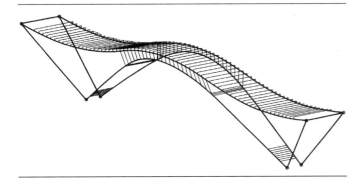

The loads, determined according to the rules of similarity, consisted of steel circular bars suspended on the transverse diaphragms and on the arch. The number of bars was modified according to desired load.

The erection of the model corresponded to the erection of the actual structure. After the assembly of the arch and end struts, the monostrands were stranded. The segments were then placed on the monostrands and the loads were applied. Next, the joints between the segments and the haunches were cast. When the concrete reached the minimum prescribed strength, the monostrands were tensioned to the design force.

Before the erection of the segments, strain gauges were attached to the steel members and the initial stresses in the structure were measured. The strain gauges at critical points of the stress ribbon were attached before the casting of the joints. During the erection of the segments, casting of the joints and post-tensioning of the structure, the deformations of the arch and the deck were carefully monitored. The forces in the monostrands were also measured by dynamometers placed at their anchors (Figure 7.92).

The model was tested for the five positions of live load shown in Figure 7.93(a). The tested structure was also analysed as a geometrically non-linear structure using the program ANSYS. For all erection stages and for the five load cases, the measurement results were compared to that of the analysis.

Figure 7.91 Model of the bridge: (a) cross-section, (b) elevation, (c) plan of the arch and (d) plan of the deck

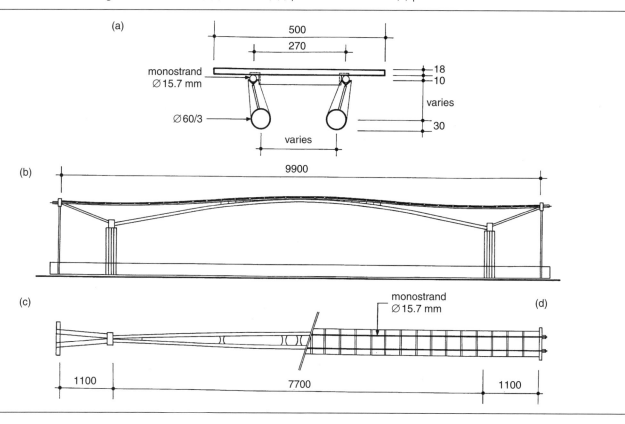

Figure 7.94 presents the calculated and measured normal stresses at several points along the stress ribbon and the calculated and measured deformations at several points along the ribbon and the arch, for one position of the live load. These results demonstrate a reasonable agreement between the analysis and the measurement.

Figure 7.92 Model of the bridge

Figure 7.93 Load: (a) service load and (b) ultimate load

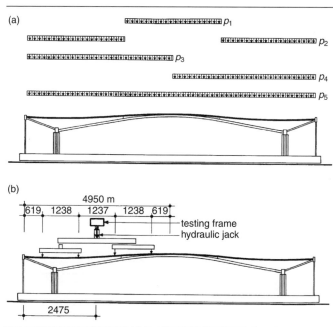

Figure 7.94 Load p3: (a) normal stresses at top fibres of the prestressed band, (b) normal stresses at bottom fibres of the prestressed band, (c) deflection of the prestressed band and (d) deflection of the arch

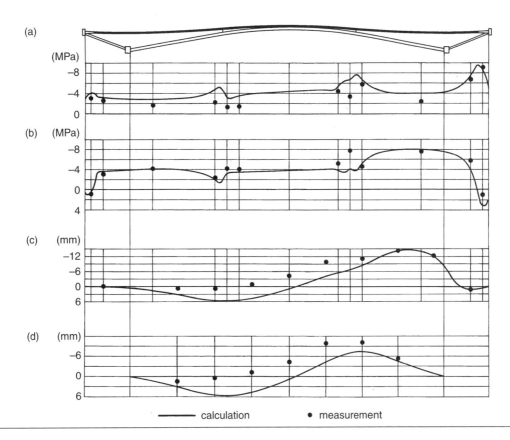

At the end of the tests the ultimate capacity of the overall structure was determined. It was clear that the capacity of the structure was not given by the capacity of the stress ribbon since, after the opening of the joints, the whole load would be resisted by the tension capacity of the monostrands.

Since the capacity of the structure would be given by the buckling strength of the arch, the model was tested for a load situated on one side of the structure (Figure 7.93(b)). The structure was tested for an increased dead load (1.3G) applied using the additional suspended steel rods, and then for a gradually increasing live load P applied with force control using a hydraulic jack reacting against a loading frame (Figure 7.95).

The structure failed by buckling of the arch (Figure 7.96) at a load 1.87 times higher that of the required ultimate load $Q_u = 1.3G + 2.2P$. The stress ribbon itself was damaged only locally by cracks that closed after the load was removed. The structure also proved to be very stiff in the transverse direction.

Figure 7.95 Ultimate loading: the structure before failure

Figure 7.96 Buckling of the arch

The buckling capacity of the structure was also calculated using a non-linear analysis in which the structure was analysed for a gradually increasing load. The failure of the structure was taken at the point when the analytic solution did not converge. Analysis was performed for the arch with and without fabrication imperfections. The imperfections were introduced as a sinus-shaped curve with nodes at arch springs and at the crown. Maximum agreement between the analytical solution and the model was achieved for the structure with a maximum value of imperfection of 10 mm. This value is very close to the fabrication tolerance.

The test has proven that the analytical model can accurately describe the static function of the structure both at service and at ultimate load.

7.8.4 Dynamic model

The dynamic behaviour of the proposed structure was also verified by dynamic and wind tunnel tests performed by M. Pirner (Institute of Theoretical and Applied Mechanics or ITAM, Academy of Sciences, Czech Republic). For these tests a geometrically and aerodynamically similar model in the scale 1:66 was produced (Figure 7.97). Both the deck and arch were cast from epoxy. To guarantee the similarity of forces, the arch springs were supported by adjustable bolts and the deck was supported and post-tensioned by wires in which designed forces were created. The anchor blocks and arch springs were supported by side steel boxes (abutments) mutually connected by a base beam. The whole model was supported as a simple beam.

The model was excited by an exciter situated mid-span on the base beam. The vibration of this beam was transferred into the structure through the abutments. The response of the structure was measured in 26 nodes uniformly distributed along the arch and deck. Table 7.1 presents the calculated and excited frequencies of the vertical modes shown in Figure 7.98. The scale of frequencies is in the range 12–13.

The results of the wind tunnel test (Figure 7.99) are shown in Figure 7.100, which depicts the relations between the model air-flow velocity and the vertical and horizontal root mean square (RMS) at nodes 1 and 2. The critical wind speed of the

Figure 7.97 Aerodynamic model of the structure: (a) cross-section and (b) elevation

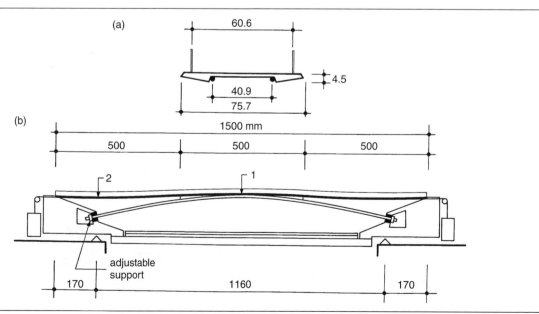

Table 7.1 Natural modes and frequencies

Mode	Structure calculation (Hz)	Model measurement (Hz)
V1	1.719	22.250
V2	1.799	32.500
V3	2.590	50.250

model was 11.07 m/s; the corresponding critical wind speed of the bridge is 90.03 m/s (Pirner and Fischer, 1999).

7.9. Structures stiffened by external tendons

Stress ribbon structures can be also stiffened by prestressing tendons placed outside the perimeter of the cross-section. They can be located parallel to the band, they can have the shape of suspension cables or they can be located under the deck. The latter two solutions reduce the amount of prestressing steel; however, they require additional structural members that have to be maintained. These solutions are therefore only suitable for special applications or for long spans.

To understand the behaviour of external tendons, a typical stress ribbon structure of 99 m span is compared to a structure of 198 m span stiffened by external cables.

The 99 m structure (Figure 7.101) is composed of 3.00 m long segments. The deck has a rectangular shape of 5.000 m width; the depth of the section is 250 or 125 mm. In the 250 mm thick structure, the bearing and prestressing tendons were internal; in the 125 mm thick structure they were situated externally. The modulus of elasticity of the concrete band was taken as $E_c = 36\,000$ MPa.

The amount of bearing and prestressing tendons was determined from the dead load, live load ($4\,kN/m^2$) and temperature change $\Delta t = \pm 20\,°C$ requirements. The analyses were performed for structures that after post-tensioning have a maximum slope at the abutments of $p = 5\%$ and $p = 8\%$. The corresponding drapes at mid-span are 1.2375 m and 1.980 m, respectively.

Figure 7.98 Natural modes

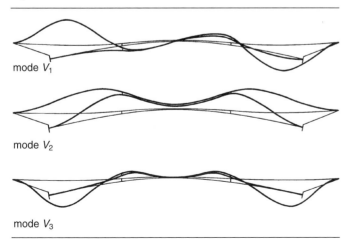

Figure 7.99 Aerodynamic model in the wind tunnel

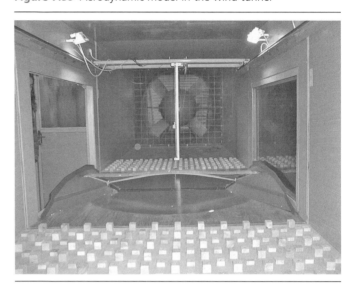

Figure 7.100 Vertical and horizontal response of the model

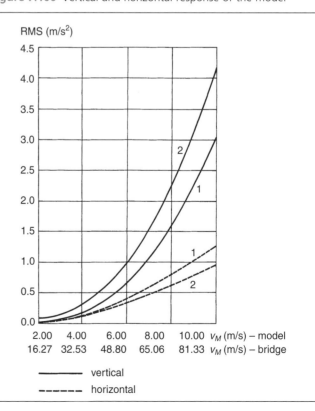

Figure 7.101 Stress ribbon $L = 99$ m: (a) cross-section, $d = 250$ mm, (b) cross-section, $d = 125$ mm and (c) elevation

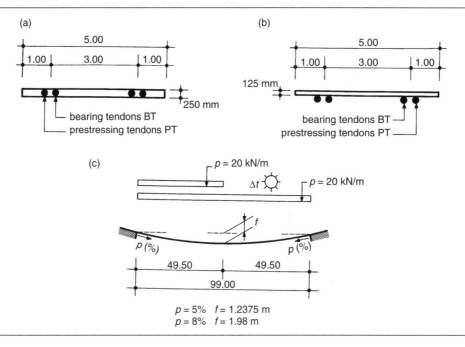

Figure 7.102 shows the deflections of the structure for (a) the live load placed over the whole length of the bridge; (b) the live load placed on one half of the deck; and (c) the temperature changes.

The basic natural modes (Figure 7.73) and frequencies are presented in Table 7.2. Since the first two vertical modes are very close, they are marked mode A and mode B and the table clarifies their order of appearance. Mode C represents the first swing mode that combines both transverse and torsional motion; mode D represents the torsional mode. The table also gives the values of the frequency ratio of mode D to mode B, which indicates a possibility of aerodynamic instability.

The structure of 198 m span had the same deck geometry as the 99 m bridge. The deck was suspended on bearing tendons situated inside the cross-section and post-tensioned by external tendons. Four different arrangements of external tendons were investigated (Figures 7.103 and 7.104) as follows.

- Structure A: tendons are situated under the deck. The tendons follow the geometry of the deck to which they are attached.
- Structure B: tendons (cables) are situated on both sides of the deck. In elevation the cables follow the shape of the stress ribbon, but in plan they have the shape of a second-degree parabola with a maximum drape of 5.00 m. At mid-span they are attached to the deck; along the length they are connected to the deck by stiff cross-beams extending from the segments.

Figure 7.102 Deformation of the stress ribbon $L = 99$ m: (a) load on the whole length, (b) load on the one half of the length and (c) temperature changes

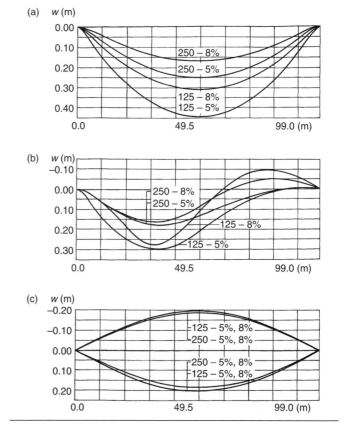

Table 7.2 Natural modes and frequencies

$L = 99$ m

Structure A	Mode A		Mode B		Mode C		Mode D		fD/fB
	Mode	Frequency (Hz)	Mode	Frequency (Hz)	Mode	Frequency (Hz)	Mode	Frequency (Hz)	
250 – 5%	1st	0.7698	2nd	1.0890	3rd	1.3460	7th	2.5400	2.3324
250 – 8%	2nd	0.9311	1st	0.8733	4th	1.4380	8th	2.5450	2.9142
125 – 5%	1st	0.7600	3rd	1.0460	2nd	0.9257	4th	1.5840	1.5143
125 – 8%	2nd	0.9215	1st	0.8300	3rd	1.0600	5th	1.5550	1.8735

$L = 198$ m

	Mode A		Mode B		Mode C		Mode D		fD/fB
	Mode	Frequency (Hz)	Mode	Frequency (Hz)	Mode	Frequency (Hz)	Mode	Frequency (Hz)	
Structure A									
250 – 5%	1st	0.4640	3rd	0.7310	2nd	0.6301	6th	1.2980	1.7756
250 – 8%	1st	0.3528	3rd	0.5784	2nd	0.5696	7th	1.2430	2.1490
125 – 5%	1st	0.4616	4th	0.7181	2nd	0.5096	5th	0.9086	1.2653
125 – 8%	2nd	0.5286	3rd	0.5686	1st	0.4899	7th	0.8404	1.4780
Structure B									
250 – 5%	1st	0.4639	2nd	0.7313	3rd	0.7863	7th	1.5450	2.1127
250 – 8%	1st	0.5282	2nd	0.5785	3rd	0.7065	7th	1.4690	2.5393
125 – 5%	1st	0.4613	2nd	0.7170	3rd	0.7729	6th	1.2820	1.7880
125 – 8%	1st	0.5281	2nd	0.5684	3rd	0.6892	7th	1.1940	2.1006
Structure C									
250 – 5%	1st	0.4343	2nd	0.6022	3rd	0.6388	7th	1.4350	2.3829
250 – 8%	2nd	0.4905	1st	0.4949	3rd	0.5996	8th	1.4300	2.8895
125 – 5%	1st	0.4341	2nd	0.5984	3rd	0.6273	6th	1.0830	1.8095
125 – 8%	2nd	0.4971	1st	0.4885	3rd	0.5749	7th	1.0960	2.2436
Structure D									
250 – 5%	1st	0.4406	3rd	0.6489	2nd	0.5822	6th	1.3640	2.1020
250 – 8%	2nd	0.5431	3rd	0.5544	1st	0.5392	7th	1.3290	2.3972
125 – 5%	2nd	0.4249	3rd	0.5298	1st	0.3651	6th	0.9178	1.7324
125 – 8%	3rd	0.4897	2nd	0.4045	1st	0.3963	7th	0.8984	2.2210

- Structure C: tendons have the shape of suspension cables. The inclined cables follow a second-degree parabola. The horizontal drape is 5.00 m and the vertical sag f is +5.00 m. At mid-span the cables are directly attached to the deck; elsewhere they are connected to the suspension cables at the joints by means of inclined hangers.
- Structure D: a tendon (a cable) is situated in the bridge axis below the deck. The cable has the shape of a second-degree parabola with a maximum sag f of +5.00 m. The cable is connected to the deck by means of vertical frames situated at the joints between the segments. In the transverse direction of the deck, the frames have a triangular shape.

The amount of strands in the bearing and prestressing tendons was determined from the dead load, live load ($4 \, kN/m^2$) and temperature change $\Delta t = \pm 20°C$ requirements. The analyses were performed for structures that after post-tensioning have a maximum slope at the abutments of $p = 5\%$ and $p = 8\%$. Corresponding drapes at mid-span are 2.475 m and 3.960 m, respectively.

Figure 7.103 Stress ribbon $L = 198$ m: (a) cross-section, $d = 250$ mm and (b) cross-section, $d = 125$ mm

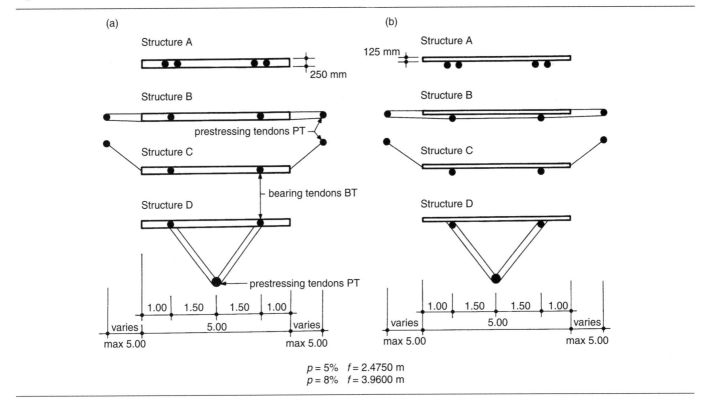

$p = 5\%$ $f = 2.4750$ m
$p = 8\%$ $f = 3.9600$ m

Figure 7.104 Stress ribbon $L = 198$ m: elevation and plan

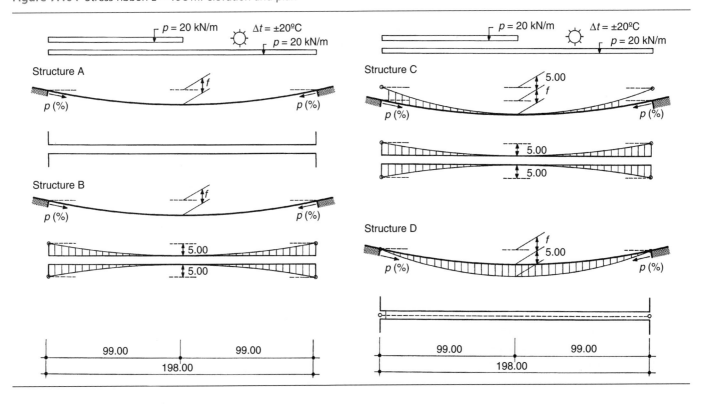

Table 7.3 Number of bearing and prestressing strands

	Structures A and B				Structures C and D			
	5%		8%		5%		8%	
Depth: mm	250	125	250	125	250	125	250	125
Bearing tendons	350	170	275	130	360	180	284	138
Prestressing tendons	350	170	275	130	120	60	120	60

The number of 15.5 mm prestressing strands given in Table 7.3 was determined from the zero-tension condition in the structure loaded with full live load. It is evident that the high level of initial compression stresses in the concrete requires the use of high-strength concrete.

Figure 7.105 shows the deflection of the structures for the case of (a) live load placed along the whole length of the deck; (b) live load placed on one half of the deck; and (c) temperature changes. Since structures A and B and structures C and D have the same vertical stiffness, their vertical deformations are the same and can be drawn in one figure.

Although the structures are stiffened by external cables, their basic natural modes are similar to the modes shown in Figure 7.73. The modes and corresponding frequencies are presented in Table 7.2. Similar to the 99.00 m structure, the first vertical modes are very close and are marked mode A and mode B. Table 7.2 clarifies their order of appearance. Due to the long span of the bridge mode C, which represents the first swing mode combining both horizontal and torsional motions, has low frequencies. Mode D represents the torsional mode which, if close to mode B, indicates the possibility of

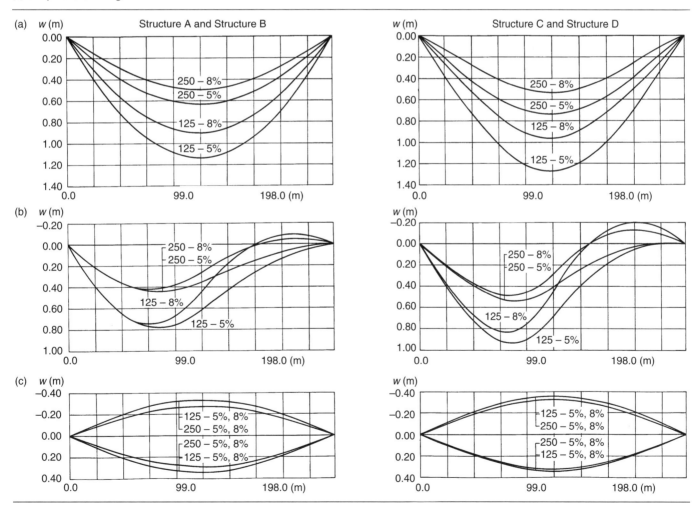

Figure 7.105 Deformation of the stress ribbon $L = 99$ m: (a) load on the whole length, (b) load on the one half of the length and (c) temperature changes

Figure 7.106 Structural arrangement of the stress ribbon: cross-sections

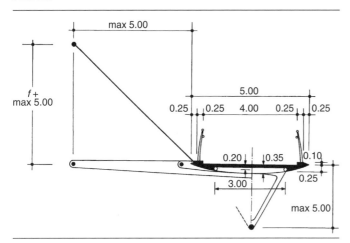

Figure 7.108 Structure B: cross-section

Figure 7.109 Structure B: view

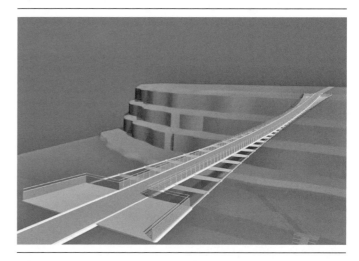

aerodynamic instability. The table also gives the ratio of torsional mode D to mode B.

Since the Structures C and D require a significantly smaller amount of prestressing, they are less stiff and therefore deform more than Structures A and B. The results show that the stiffness of the structure is mainly given by the tension stiffness of the prestressed concrete deck. The external tendons located on both sides of the deck also stiffen the structure in the transverse direction of the bridge.

The analysis was done for the simple rectangular section described above. The actual structural arrangement is shown in Figure 7.106 where all the studied options are included. Possible architectural arrangements are illustrated in Figures 7.107–7.113.

Figure 7.107 Structure A: cross-section

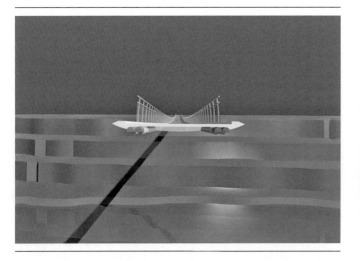

Figure 7.110 Structure C: cross-section

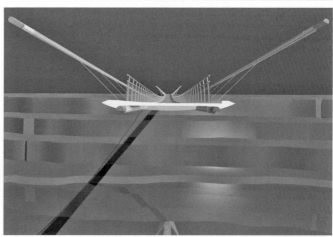

Figure 7.111 Structure C: elevation

Figure 7.112 Structure C: view

Figure 7.113 Structure D: cross-section

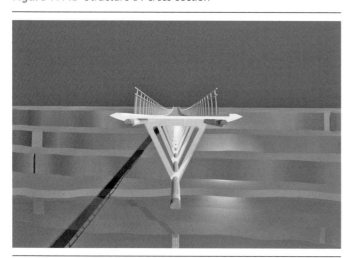

Although such arrangements were used in old rope structures, a solution with the prestressed concrete deck was used only recently. A structure similar to Structure D was built in Japan in 2001 (Section 11.1).

7.10. Static and dynamic loading tests

The design assumptions and quality of workmanship of the author's first stress ribbon structure DS-L and of the first stress ribbon bridge built in the US were checked by measuring the deformations of the superstructure at the time of prestressing and during loading tests. Dynamic tests were also performed on these structures. Only a few key results of a typical structure are given here. Since the shape of a stress ribbon structure is extremely sensitive to temperature change, the temperature of the bridge was carefully recorded at all times.

7.10.1 Static tests

The pedestrian bridge in Brno-Komin (see Section 11.1) was tested using vehicles weighing between 102 and 110 kN (Figure 7.114). The bridge was tested for two positions of vehicles. Initially, six trucks were placed symmetrically at the centre of the bridge, followed by four trucks placed on one side of the bridge. Figure 7.115 compares the measured and calculated deformations of the structure.

The pedestrian bridge in Prague-Troja (see Section 11.1) was tested by 38 vehicles weighing between 2.8 and 8.4 tons (Figure 7.116). The vehicles were first placed along the entire length of the structure and then they were placed on each span. During the test only the deformations in the middle of the spans and the horizontal displacements of all supports

Figure 7.114 Brno-Komin Bridge, Czech Republic: load test by heavy vehicles

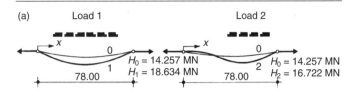

Figure 7.115 Brno-Komin Bridge, Czech Republic: load test (a) elevation, (b) deformation for load 1 and (c) deformation for load 2

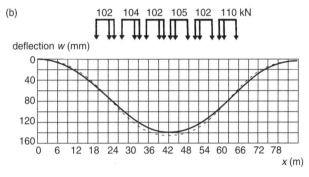

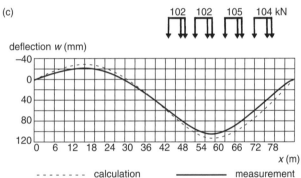

Table 7.4 Deflections at midspans of the pedestrian bridge in Prague-Troja

Loaded span		Span 1 (mm)	Span 2 (mm)	Span 3 (mm)
1, 2, 3	Calculation	40	200	56
	Measurement	40	186	57
1	Calculation	301	−124	−62
	Measurement	272	−92	−48
2	Calculation	−126	312	−78
	Measurement	−95	289	−50
3	Calculation	−38	−76	221
	Measurement	−25	−56	182

were measured (Table 7.4). As can be seen, the comparisons are very good.

The bridge across the Sacramento River in Redding was also tested using 24 cars with a total weight of 41.7 tons. Figures 7.117 and 7.118 present the results of the test for three positions of loading. As can be seen here, the comparison of results is also very good.

7.10.2 Dynamic tests

Stress ribbon structures DS-L built in Brno-Bystrc, Brno-Komín, Prerov and Prague-Troja were also subject to dynamic tests by M. Pirner (ITAM). In the course of the load tests, the agreement of excited natural frequencies with theoretical values was investigated. The structures were excited either by a human force (Figure 7.119), by a pulse rocket engine or by a mechanical rotation exciter (Figure 7.120). The bridge in Prague-Troja was dynamically tested

Figure 7.116 Prague-Troja Bridge, Czech Republic: load test by 38 vehicles

Figure 7.117 Redding Bridge, California, USA: load test by 24 cars

Figure 7.118 Redding Bridge, California, USA: deformation

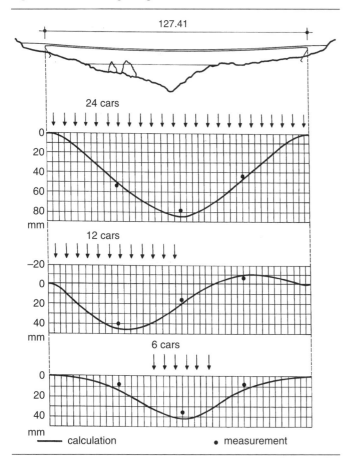

Figure 7.119 Brno-Komin Bridge, Czech Republic: dynamic test by running people

Figure 7.120 Brno-Komin Bridge, Czech Republic: dynamic test by mechanical exciter

again after 14 years of service. The second test proved that the dynamic response of the structure has not changed (Pirner *et al.*, 1998).

For illustration, Figure 7.121 shows the excited vertical modes. Table 7.5 presents calculated and measured frequencies together with corresponding logarithmic decrements of damping.

Figure 7.121 Prague-Troja Bridge, Czech Republic: exciter-induced modes

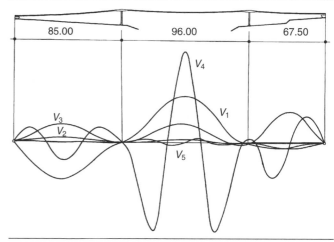

Table 7.5 Calculated and measured natural frequencies of the Prague-Troja Bridge, Czech Republic

Mode	Calculation (Hz)	Measurement (Hz)	Logarithmic decrement of damping
V1	0.514	0.525	0.033
V2	0.646	0.650	0.149
V3	1.204	0.925	0.034
V4	1.721	1.625	0.018
V5	2.416	2.275	

During the opening of the pedestrian bridges in Grants Pass (Oregon, USA) and Maidstone (Kent, UK), the bridges were loaded by many pedestrians (Figure 7.122). The bridges remained relatively stiff and pedestrians did not have a feeling of discomfort when walking across the bridge. No excessive vertical or horizontal motion or even so-called lock-in effect were reported.

The Sacramento River Bridge in Redding built on the trail is used not only by pedestrians and bicycles but also for horse riding (Figure 7.123). So far, no problems with excessive motion have been reported.

7.10.3 Wind tunnel tests

During the design of the author's first stress ribbon bridges DS-L, sectional models were tested in a wind tunnel by Studnickova (Klokner Institute, Prague, Czech Republic). The models that represented a 10.00 m long section of the deck were made in

Figure 7.122 Grants Pass Bridge, Oregon, USA: opening of the bridge

Figure 7.123 Redding Bridge, California, USA: horse riding

the scale 1:20. They were tested without and with suspended tubes that modelled suspended utilities (Figure 7.124).

Static lift (C_L), drag (C_D) and torsion moment (C_M) coefficients were measured. The deck was rotated with respect to the air flow, simulating wind inclination, in a range of angles of attack between $-12°$ and $8°$. The results of the measurements are presented in Figure 7.125.

The corresponding forces acting on the structure were:

Lift $\quad L = \frac{1}{2}\rho v^2 B C_L$

Drag $\quad D = \frac{1}{2}\rho v^2 B C_D$

Moment $\quad M = \frac{1}{2}\rho v^2 B C_M$

Figure 7.124 Bridges DS-L: sectional model

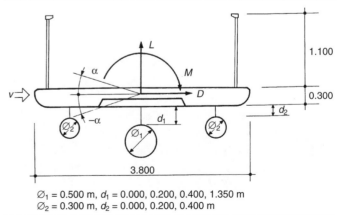

$\varnothing_1 = 0.500$ m, $d_1 = 0.000, 0.200, 0.400, 1.350$ m
$\varnothing_2 = 0.300$ m, $d_2 = 0.000, 0.200, 0.400$ m

Figure 7.125 Bridges DS-L: static lift (C_L), drag (C_D) and torsion moment (C_M) coefficients (eight figures)

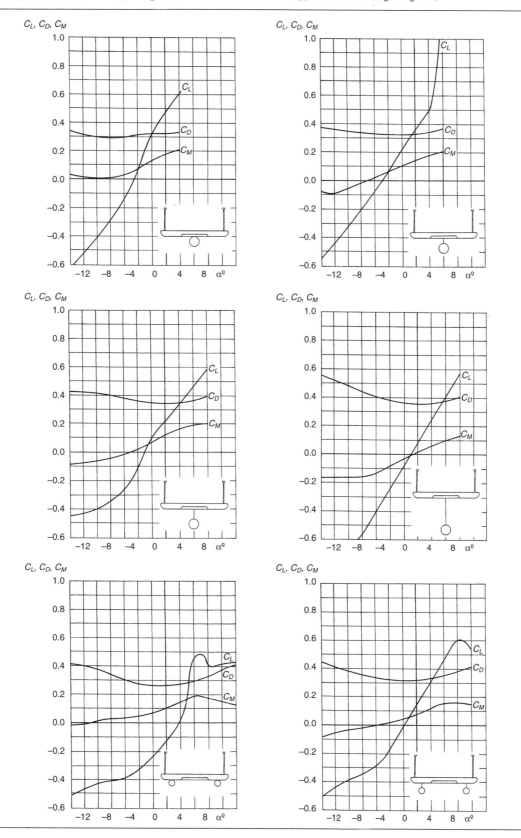

Figure 7.125 Continued

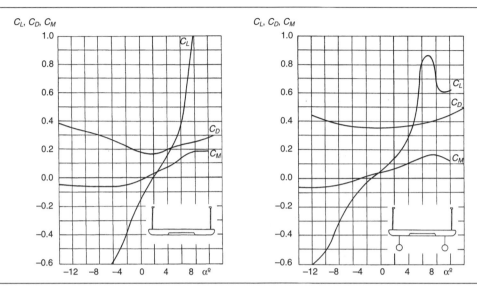

where ρ is the density of air (1.250 kg/m^3), v is the wind speed (m/s), B is the width of the deck (3.80 m) and C_L, C_D and C_M are dimensionless drag, lift and moment coefficients, respectively.

Similar model tests were carried out for the design of the Sacramento River Bridge in Redding. Figure 7.126 depicts the model and Figure 7.127 the results of the tests.

The design of the Grants Pass pedestrian bridge called for an observation platform situated at the mid-span of the second and third spans. To quantify the influence of this platform on the aerodynamic behaviour, wind tunnel tests of the one-span stress ribbon structure of the span of 99.0 m were carried out by M. Pirner (ITAM).

For these tests the stress ribbon structure supported by the arch described in Section 7.8 was utilised. When wind tunnel tests of this structure were completed, the arch was removed and external wires were tensioned to the design load. In this way, a new geometrically and aerodynamically similar model of the structure of span 99.00 m and width 5.00 m was obtained (Figures 7.128 and 7.129).

The structure was tested in a wind tunnel without and with a circular platform. The length of the platform (150 mm)

Figure 7.126 Redding Bridge, California, USA: sectional model

Figure 7.127 Redding Bridge, California, USA: static lift (C_L), drag (C_D) and torsion moment (C_M) coefficients

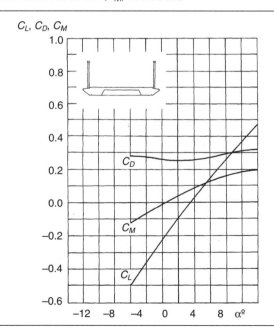

Stress ribbon structures

Figure 7.128 Aerodynamic model of the stress ribbon structure: (a) cross-section, (b) elevation and (c) plan

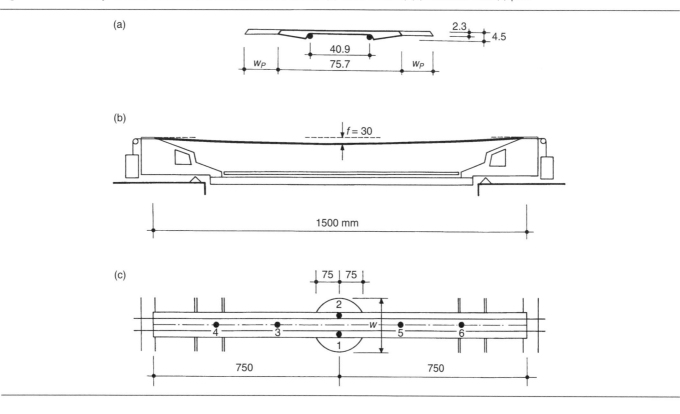

corresponded to the length of 9.90 m of the actual structure. The different widths w of the platform 105, 145 and 175 mm corresponded to the widths 6.930, 9.570 and 11.550 m of the actual structure.

The results of the wind tunnel tests of the structure without and with the platform of the maximum width are shown in Figure 7.130. The figure presents the relations between the model air-flow velocity and the vertical root mean square

Figure 7.129 Aerodynamic model in the wind tunnel

Figure 7.130 Vertical response of the model

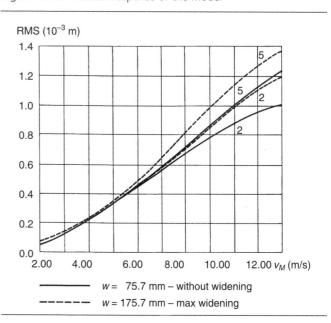

125

(RMS) of vertical deflections of nodes 2 and 5. It is evident that the platform does not influence the response substantially; the response increases by ≈15%. No symptoms of aerodynamic instability were observed.

REFERENCES

Pirner M and Fischer O (1999) Experimental analysis of aerodynamic stability of stress-ribbon footbridges. *Wind and Structures* **2(2)**: 95–104.

Pirner M, Fischer O and Urushadze S (1998) Diagnostic of the Troja footbridge by means of dynamic response. *Acta Techn*. CSAV 43.

Stress Ribbon and Cable-supported Pedestrian Bridges
ISBN 978-0-7277-4146-2

ICE Publishing: All rights reserved
doi: 10.1680/srcspb.41462.127

Chapter 8
Suspension structures

Suspension structures are described in many excellent books (Brown, 1996; Gimsing, 1998; Leonhardt, 1964; Pearce and Jobson, 2002; Wittfoht, 1972). Therefore only additional information about suspension structures with slender deck is discussed in this chapter.

8.1. Structural arrangement

As described in Chapter 2, a suspension structure is formed by a slender deck suspended or supported by a suspension cable (Figures 2.6(a) and 2.6(c)). The elevation of the deck can therefore have an optimum arrangement corresponding to local conditions. The structures can have one or more spans; the span length is practically unlimited (Figure 8.1).

The suspension cable can be anchored into the soil (Figures 2.11(a) and 8.2) and form a so-called earth-anchored system, or it can be anchored into the deck and create self-anchored systems (Figures 2.11(b) and 8.3). The suspension cables can be situated above the deck, under deck or above and under the deck.

The suspension cable has a funicular shape to the self-weight of the structure; it balances the effects of the self-weight and

Figure 8.1 Vranov Lake Bridge, Czech Republic

guarantees that the structural members are stressed by normal forces only. For service load, the suspension structure forms a complex system in which the deck distributes the load and all structural members contribute to the resistance of the structural system.

The advantage of the earth-anchored system is that the erection of the deck can be done independently of the terrain under the bridge (Figure 8.2). However, the suspension cables first have to be erected and the anchor blocks have to transfer a large tension force into the soil.

On the other hand, the self-anchored suspension bridges do not require an expensive anchor block and utilise the compression capacity of the concrete deck. However, the deck has to be erected first; the suspension cables can then be erected and tensioned (Figure 8.3). The fact that the erection of the deck requires a falsework and therefore depends on the terrain under the bridge prevents this system being used in many cases.

In several new applications, the erection of the structures is designed in such a way that anchor blocks are designed for erection loading only. When the erection is completed, part or all of the tension force is transferred from the anchor blocks into the deck. In this way, a partial or total self-anchored system is created (Figure 8.2(b)).

The earth suspension bridges are usually assembled of precast members (segments) that are suspended on suspension cables. Since the precast segments are mutually connected by pins, the suspension cables automatically have a funicular shape to the given load.

The self-anchored structures are usually cast in place on the falsework. The self-weight is transferred into the suspension cables by their post-tensioning, which can be done by jacking at their anchors or by lifting the tower. This operation requires that the camber of the deck and non-tension length of the cables be carefully determined.

Figure 8.2 Earth-anchored suspension structure: (a) erection, (b) service

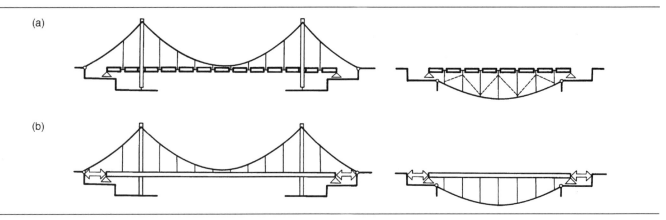

Figure 8.3 Self-anchored suspension structure: (a) erection, (b) service

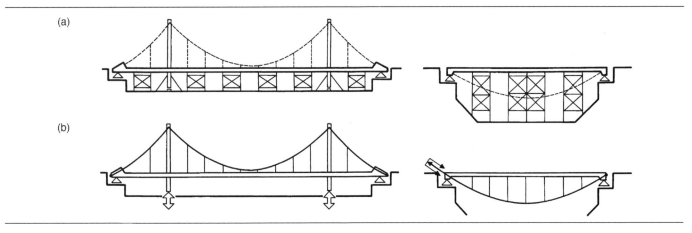

The classical suspension structures with cables situated above the deck are usually suspended on one or two towers and have one to three spans. The side spans can be completely or partially suspended on the suspension cables. The length of the suspended span is usually less than half of the length of the main span (Figure 8.4).

8.1.1 Geometry of the cable

The mutual distance between hangers of suspension structures with a slender deck is very short. For the preliminary determination of the geometry of the cables we can substitute the point load from hangers by a uniform load and assume that the dead load of the cables and suspenders g_C and the deck g_D are constant. Then the total dead load:

$$g = g_C + g_D = \text{constant}.$$

If the deck is suspended or supported along the whole length, the cables have the shape of a second-degree parabola (Figure 8.5). Since the horizontal force H_g in all spans has to be constant, the geometry of the cables in a span i is given by:

$$H_g = \frac{gL_i^2}{8f_i} = \frac{gL_{max}^2}{8f_{max}} = \frac{gL_1^2}{8f_1}$$

$$f_i = \frac{L_i^2}{L_{max}^2} f_{max}$$

where

$$y_1(x) = f_1(x) = 4f_1 \frac{x(L_1 - x)}{L_1^2}$$

and

$$y_2(x) = f_2(x) = 4f_2 \frac{x(L_2 - x)}{L_2^2}.$$

If the cables are anchored at the towers at different heights and the deck in side span is partially suspended on the cable, the

Figure 8.4 Typical arrangement of the earth-anchored suspension structure

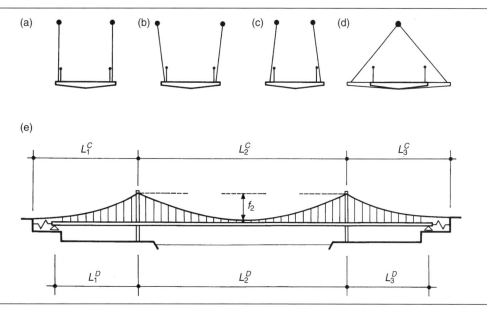

geometry of the cable can be determined from Equation (4.7) given in Chapter 4. If we set $f_2 = f_{max}$ the structure shown in Figure 8.6 is stressed by

$$H_g = \frac{gL_{max}^2}{8f_{max}} = \frac{gL_2^2}{8f_2}.$$

The geometry of the cable in span 2 is given by:

$$f_2(x) = 4f_2 \frac{x(l_2 - x)}{L_2^2}$$

$$y_2(x) = \frac{h_2}{L_2}x + f_2(x).$$

The geometry of the cables in span 1 can be derived from the course of bending moments in simple beam of span L_1^C loaded by load g_C and $g = g_C + g_D$:

$$A_1 = \frac{1}{L_1^C}\left[g_C L_a\left(L_1^D + \frac{L_a}{2}\right) + \frac{g(L_1^D)^2}{2}\right]$$

$x \leq L_a$

$$f_1(x) = A_1 x - \frac{g_C x^2}{2}$$

$x > L_a$

$$f_1(x) = A_1 x - g_C L_a\left(x - \frac{L_a}{2}\right) - \frac{g(x - L_a)^2}{2}$$

$$y_1(x) = h_1 - \frac{h_1}{L_1^C}x + f_1(x).$$

For span 3, we have

$$H_g = \frac{gL_2^2}{f_2} = \frac{g_C L_3^2}{f_3}$$

$$f_3 = \frac{g_C L_3^2}{gL_2^2}f_2$$

$$f_3(x) = 4f_3 \frac{x(L_3 - x)}{L_3^2}$$

$$y_3(x) = \frac{h_3}{L_3^C}x + f_3(x).$$

The possible arrangement of the cable-supported structures in the transverse direction is discussed in Section 2.2 (Figures 2.24 and 2.25).

Figure 8.5 Geometry of the cables: deck supported by cables

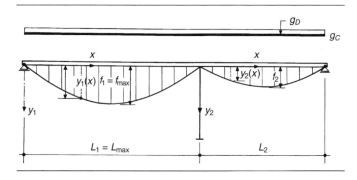

Figure 8.6 Geometry of the cables: deck suspended on cables

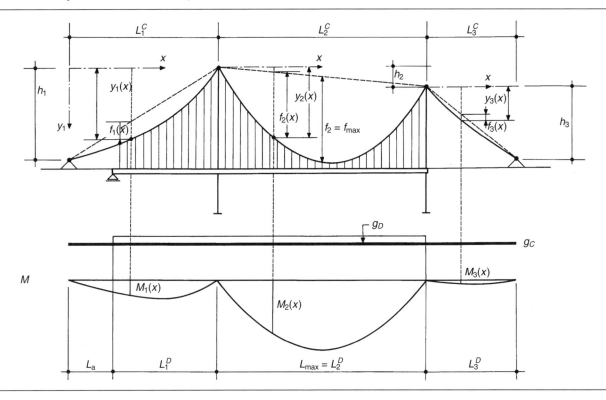

The suspension cables situated above the deck are usually suspended on cables developed by steel or prestressed concrete industry (Figure 2.31); the suspension cables situated under the deck are also formed by prestressed concrete band or by composite ties (Figures 2.29 and 2.40).

In many structures, the suspension cables are formed by spiral strands, locked coil strands or parallel wires (Figure 2.31(a)) and the hangers are formed by spiral strands or structural ropes (Figure 2.35). They are factory fitted with a combination of socket types to enable load transmittal between the structure and the cable (Figure 2.32) and they are delivered on site in a design length.

The suspension cables developed from the prestressing tendons are formed by parallel wires or prestressing strands (Figure 2.31(b)). They can be easily assembled on the site. To avoid a catwalk that is usually necessary for the erection of the cables on the site, the author used the stranded cables encased in steel tubes. Due to maintenance problems, hangers are preferred to bars. The tubes enable a simple connection of the hangers to the cables (Figure 8.7) and connection of the cables with the deck at mid-span (Figure 8.8).

8.1.2 Structures suspended on cables

A typical arrangement of the earth-anchored suspension structure is shown in Figure 8.4. The deck is suspended on suspension cables of three spans anchored at anchor blocks. The cables are deviated at the saddles (Figure 8.9) or they are anchored at anchor plates (Figure 8.10). The cables can also be anchored at anchor blocks that overlap at the tower (Figure 8.11).

The deck is usually suspended on two suspension cables that can be arranged in vertical (Figure 8.4(a)) or inclined planes. In

Figure 8.7 McKenzie River Bridge, Oregon, USA

Figure 8.8 McKenzie River Bridge, Oregon, USA

suspension structures of common spans, the concrete deck guarantees the transverse stiffness. As will be shown further, the outward inclination (Figure 8.4(b)) does not significantly increase the transverse stiffness of the structures. Also, an inward inclination (Figure 8.4(c)) does not considerably increase the torsion stiffness of the system. However, it creates a so-called 'gate effect' and a feeling of safety (Figure 8.12). Further, these structures do not vibrate in pure transverse modes; all transverse modes are accompanied by a distortion of the deck that contributes to the stiffness of the system.

The deck can be also suspended in the bridge axis on one central cable (monocable). This solution requires a torsionally stiff deck. With a slender deck it can be used with cross-inclined hangers (Figure 8.4(d)). To preserve the headroom, the cable

Figure 8.9 Weser River Bridge, Germany (courtesy of Schlaich, Bergermann and Partners)

Figure 8.10 Nordbahnhof Bridge, Stuttgart, Germany (courtesy of Schlaich Bergermann and Partners)

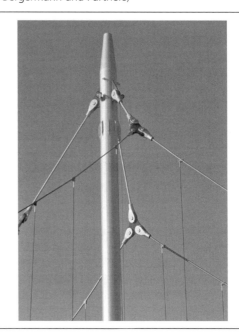

has to be sufficiently high above the deck and the deck slab has to be sufficiently widened beyond the deck slab.

In the longitudinal direction, the hangers can be vertical or inclined. The inclined hangers together with the suspension cables and the deck can form a stiff truss structural system. However, the structural details of connection of the hangers and suspension cables are not simple (Figure 2.27).

The anchor block can be formed not only by a concrete block but also by steel tension ties. Figures 8.13 and 8.14 depict steel ties used in the design of the Willamette River Bridge built in Oregon (see Section 11.2).

Figure 8.11 Willamette River Bridge, Oregon, USA

Figure 8.12 Vranov Lake Bridge, Czech Republic

Figure 8.14 Willamette River Bridge, Oregon, USA

The towers can have a different arrangement as discussed in Section 2.2 (Figure 2.25). They can be supported by bearings (Figure 8.15) or they can be fixed into the footing (Figure 8.16). However, to balance the horizontal force during the erection it is necessary to support them by hinges. The fixing is provided to resist the service load only.

As it was pointed out in Section 2.1, the restriction of the horizontal movement of the deck significantly reduces the deflection and corresponding bending stresses of the suspension deck. However, due to the effects of temperature changes and creep and shrinkage of concrete, it is difficult to make a fixed connection between the deck and the abutments. For this reason, the author used a partial fixing of the deck.

In the case of the Vranov Bridge (see Section 11.24), the arched deck is suspended on the cables and is flexibly connected to the

Figure 8.13 Willamette River Bridge, Oregon, USA

abutments. These are in turn mutually connected to the anchor blocks by prestressed concrete tie rods (Figure 8.17). The flexible members are formed by tartan plates that were pressed against the already erected structure, with an erection segment and jacks. After compression, the space between the abutment and the erection segment was filled with concrete and the segment was connected to the abutment. The value of compression was determined in such a way that, under the maximum shortening of the deck due to creep and shrinkage of concrete and temperature drop, a minimum compression of 0.5 MPa remains in the joint. In this way the tension from the suspension cables partially post-tensions the deck and creates a system where compression stresses in the deck stiffen the whole structure.

For a live load, temperature changes and effects of wind, the structure forms a closed system where the load is resisted both

Figure 8.15 Nordbahnhof Bridge, Stuttgart, Germany (courtesy of Schlaich Bergermann and Partners)

Figure 8.16 Willamette River Bridge, Oregon, USA

by the compression capacity of the concrete deck and by the tension capacity of the suspension cables. Since the expansion tartan joints behave non-linearly, the portion of the load resisted by the deck and the cables depends on the temperature and the age of the structure.

In the case of the McKenzie Bridge, neoprene pads were placed between the abutment edge walls and end segments (Figure 8.18(b)). The compression between the pads and abutments is provided by adjustable bolts of length dependent on the shortening of the deck due to the creep and shrinkage of concrete. In this way, a partially self-anchored system was created.

Due to local conditions, the precast deck of the side spans of the Willamette River Bridge is fix connected with curved rams and stairs. The expansion joints were therefore designed to be between the segments near the towers. The decks of the main and side spans are connected by stoppers (shock transition units) that prevent movement of the deck due to sudden load but allow movement of the deck due to temperature changes and creep and shrinkage of concrete.

The cable sag f_{max} at the mid-span of the main span of length L_{max} is usually from 0.12 to $0.08 L_{max}$. The deck can only be suspended on suspension cables or it can be also fix connected to them at mid-span (Figure 8.8). As was discussed, the horizontal movement of the deck can be allowed or restricted.

To quantify the above parameters, extensive studies were done. The influence of the deck stiffness and restriction of the horizontal movement of the deck were discussed in Section 2.1 (Figure 2.19). The influence of the different values of the sag, connection of the cable to the deck and restriction of the horizontal movement are presented in Section 8.3.

Figure 8.17 Vranov Lake Bridge, Czech Republic: a partial fixing of the deck

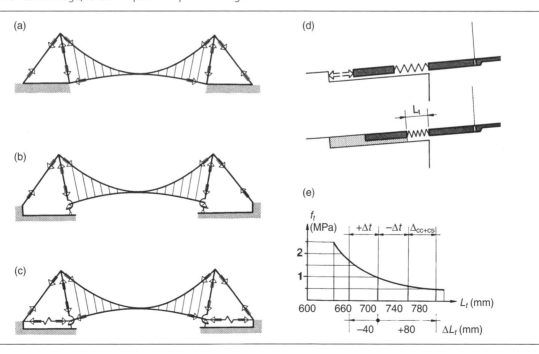

Figure 8.18 McKenzie River Bridge, Oregon, USA: a partial fixing of the deck

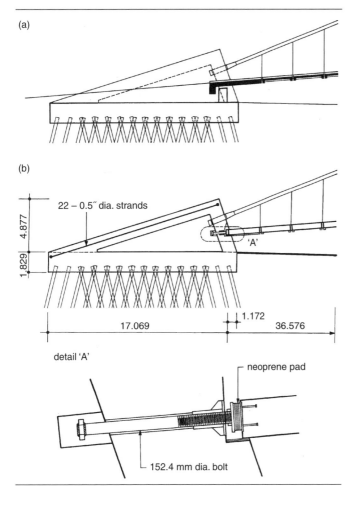

Figure 8.19 Max-Eyth-See Bridge, Stuttgart, Germany: construction sequences

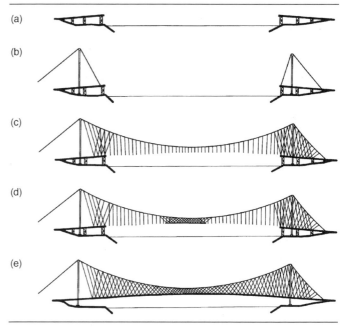

on pontoons (Figure 8.20) or using the technology developed for stress ribbon bridges. This technology will be described further in detail.

The erection of a bridge comprises four stages (Figure 8.21)

1. erection of towers
2. erection of suspension cables
3. erection of segments
4. grouting of cables and post-tensioning of the deck.

Figure 8.20 Max-Eyth-See Bridge, Stuttgart, Germany: erection of the deck (courtesy of Schlaich Bergermann and Partners)

8.2. Erection of the structures

As already discussed, suspension structures can be earth- or self-anchored. The advantage of the earth-anchored system is that the erection of the deck can be done independently of the terrain under the bridge.

8.2.1 Earth-anchored suspension structures

8.2.1.1 Structures suspended on cables

The construction usually starts with construction of the anchor block, abutments and footing. The towers and suspension cables with hangers are then erected. The suspension cables formed by spiral strands, locked coil strands or parallel wires are delivered onsite in designed lengths and the cables formed by prestressing strands are usually assembled onsite.

After the suspension cables are erected, the deck is assembled (Figure 8.19). The segments can be erected by cranes situated

Figure 8.21 McKenzie River Bridge, Oregon, USA: construction sequences

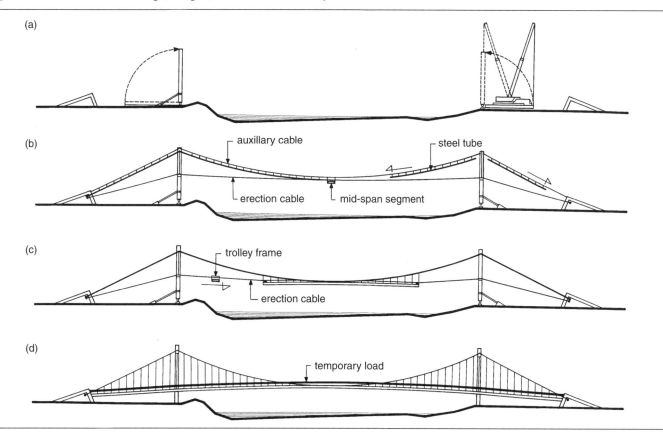

8.2.1.1.1 ERECTION OF THE PYLON

Concrete towers are usually cast in a horizontal position with hinges that allow their rotation into the design position (Figure 8.22). Steel towers are also assembled close to their design position. The towers are then lifted by cranes (Figures 8.23 and 8.24) and their position is secured by temporary erection struts. Since their tops are moved horizontally during the erection of the deck, they are stabilised in an inclined position.

Figure 8.22 McKenzie River Bridge, Oregon, USA: casting of the tower

Figure 8.23 McKenzie River Bridge, Oregon, USA: lifting of the tower

Figure 8.24 Willamette River Bridge, Oregon, USA: erection of the tower

Figure 8.26 Vranov Lake Pedestrian Bridge, Czech Republic: partial lifting of the tower

The 30 m high towers of the Vranov Bridge were also cast in a horizontal position. Since there was no room for erection cranes, they were lifted into the design position by the tension of the cables anchored in the opposite towers. The towers were first raised to a partially upright position by the tension pull of the short vertical cables anchored at the tops of the towers and temporary supports (Figures 8.25(a) and 8.26). The cables were pulled by hydraulic jacks supported by steel anchor members situated on the steel girders, transferring the load to the supports (Figure 8.27).

In the second stage, the towers were raised into their final design position by the tension pull of the cables anchored at their tops (Figures 8.25(b) and 8.25(c)). The erection cables spanning the bay were pulled by hydraulic jacks. The cables were first tensioned in such a way that tension in the cables balanced the dead load of the partially upright towers. The temporary supports were then removed and the first tower was raised into the final position by the tension pull caused by the jacks

Figure 8.25 Vranov Lake Pedestrian Bridge, Czech Republic: construction sequences of the erection of the towers

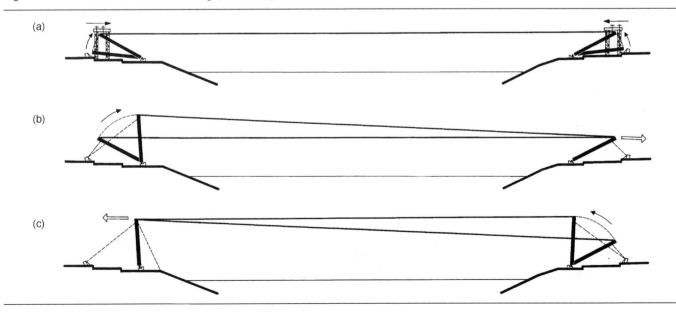

Figure 8.27 Vranov Lake Pedestrian Bridge, Czech Republic: partial lifting of the tower by stressing jacks

situated at the second tower (Figure 8.28). When the position of the first tower was secured by the temporary cables, the second tower was raised by the tension pull caused by the jacks situated at the first tower.

8.2.1.1.2 ERECTION OF THE SUSPENSION CABLES

The erection of the suspension cables begins with the erection of the mid-span segment (Figure 8.21(b)). The segment is suspended on an erection frame that moves on erection cables formed by strands. These cables are anchored at the end-anchor blocks and are supported by temporary saddles suspended on post-tensioned bars anchored at the pylon's

Figure 8.28 Vranov Lake Pedestrian Bridge, Czech Republic: lifting of the tower

Figure 8.29 McKenzie River Bridge, Oregon, USA: erection cables

heads (Figure 8.29). The segment is shifted along these cables into the design position. In this way, the central erection platform allowing the erection of the suspension cables is prepared.

After the erection of the mid-span segment, the main suspension cables are installed. Steel tubes are progressively suspended on erection strands and shifted into the designed position (Figure 8.30). The strands forming the main suspension cables are

Figure 8.30 McKenzie River Bridge, Oregon, USA: erection of the main suspension cables

Figure 8.31 Willamette River Bridge, Oregon, USA: erection of the island segments

Figure 8.33 Willamette River Bridge, Oregon, USA: island segments suspended on the main suspension cables

then pulled and tensioned according to their designed length. In this way, the self-weight of the island segment is transferred from the erection cables into the main suspension cables.

In the case of the Willamette River Bridge where the design included an observation platform at mid-span, the erection of the deck was slightly modified. Due to the platform, the suspension cables had to be threaded through the central wider (island) deck segments. The central platform was not formed by one central segment, but by eight island segments. After their erection (Figures 8.31 and 8.32) the suspension cables were installed and strands were tensioned. In this way, the weight of island segments was transferred from the erection cables to the suspension cables (Figure 8.33).

8.2.1.1.3 ERECTION OF SEGMENTS

Before the erection of segments, the temporary erection struts at the towers are released. The bearing cables that were used for erection of the central segment are also used as erection cables for the erection of span segments. The segments are shifted along the cables to the designed position (Figures 8.21(c), 8.34 and 8.35). At first the front end of the segment was pin connected with a previously erected segment (Figure 8.36); the rear end was then suspended on the main suspension cables (Figure 8.37). During the erection, the structure was progressively changing in shape from concave to convex (in accordance with the length of the suspenders and load; Figures 8.38–8.40).

8.2.1.1.4 GROUTING OF CABLES AND POST-TENSIONING OF THE DECK

After the erection of all the segments, the joints between the segments are cast and post-tensioned.

Figure 8.32 Willamette River Bridge, Oregon, USA: island segments suspended on the erection cables

Figure 8.34 Willamette River Bridge, Oregon, USA: erection of a typical segment

Figure 8.35 McKenzie River Bridge, Oregon, USA: lifting of a typical segment

To eliminate tension stresses in the cement mortar of the suspension cables, the deck is temporarily loaded before the cables are grouted (Figure 2.41). The load was created by plastic tubes filled with water. When the mortar reached a sufficient strength, the loading was released. In the case of the Vranov Bridge, the temporary load was created by the tensioning of external tendons that loaded the structure by radial forces.

8.2.1.2 Structures supported by cables

Figure 8.2(a) depicts an earth-anchored suspension structure supported by cables situated under the deck. The suspension tension chord is usually formed by a stress ribbon assembled from precast segments suspended on bearing tendons. The

Figure 8.36 McKenzie River Bridge, Oregon, USA: hangers: (a) cross section, (b) partial elevation

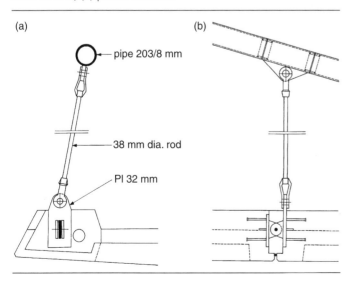

Figure 8.37 McKenzie River Bridge, Oregon, USA: suspension of a typical segment

stress ribbon forms an erection platform for the erection of the struts and deck.

The stress ribbon is assembled as for classical stress ribbon structures. The segments with vertical or inclined struts are shifted into the design position along the bearing tendons (Figure 8.41). The deck is progressively assembled from precast segments. The segments are shifted along the erection

Figure 8.38 Vranov Lake Pedestrian Bridge, Czech Republic: geometry of the structure during the erection

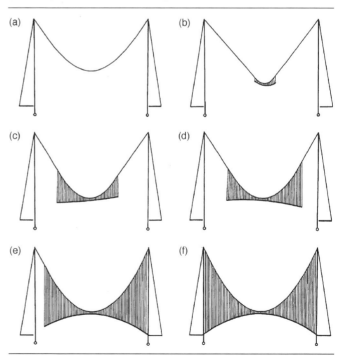

Figure 8.39 Vranov Lake Pedestrian Bridge, Czech Republic: structure during the erection of segments

Figure 8.40 Vranov Lake Pedestrian Bridge, Czech Republic: structure during the erection of segments

Figure 8.41 Ishikawa Zoo Bridge, Japan: erection of a stress ribbon segment (courtesy of Sumitomo Mitsui Construction Co., Ltd)

Figure 8.42 Ishikawa Zoo Bridge, Japan: erection of a deck segment (courtesy of Sumitomo Mitsui Construction Co., Ltd)

rails supported by struts (Figure 8.42) or are transported into the design position by an erection truck (Figure 8.43). After the erection of all segments the joints between all precast members are cast and post-tensioned.

8.2.2 Self-anchored suspension structures

Note that self-anchored suspension structures are actually classical concrete structures prestressed by external tendons situated outside the perimeter of the cross-section of the deck. The goal of the design and erection is to create a structure that – for the self-weight – is stressed by normal forces only.

The construction of the self-anchored structure begins by the casting of footings and abutments. The towers are then temporarily erected and their position is secured. After that, the deck is assembled from precast or cast-in-place segments (Figure 8.44).

Figure 8.43 Nozomi Bridge, Japan: erection of a deck segment (courtesy of Oriental Construction Co., Ltd)

Figure 8.44 Tobu Bridge: falsework (courtesy of Oriental Construction Co., Ltd)

Figure 8.45 Johnson Creek Bridge: construction sequences

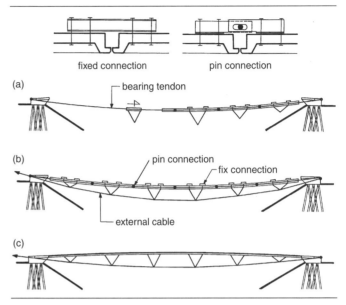

Subsequently, the suspension cables, hangers or struts are placed and cables are post-tensioned. The post-tensioning can be done in many ways, for example by post-tensioning the cables by hydraulic jacks or by casting the deck in a camber. These operations have to be carefully designed and executed to guarantee the required shape of the structure and state of stresses in all structural members.

It is evident that the self-anchored bridge can be assembled as a classical suspension structure temporarily anchored in anchor blocks. In the case of Ganmon Bridge which was built in Japan (see Section 11.2), the bearing tendons of the stress ribbon were temporarily coupled with tendons anchored at the abutments. When the erection of the deck was completed, the couplers were released. In this way, the released force from the stress ribbon prestressed the structure.

In a preliminary design of the Johnson Creek Bridge on Springwater Trail, Oregon (Section 11.2), the author developed a partially self-anchored system in which the deck was erected independently of the terrain under the bridge and the abutments resist a relatively small horizontal force. The deck is formed by precast segments and a composite deck slab which is post-tensioned by external cables anchored into the end abutments.

The erection procedure was developed from the erection of stress ribbon structures. The erection bearing tendons that are situated within a composite portion of the deck are erected and tensioned first. The segments that are erected with steel struts are then suspended on the tendons and shifted along them into the design position (Figure 8.45(a)). In the design position, the segments are mutually connected by two types of steel members that guarantee their fix or pin connection.

The external cables are then erected and tensioned (Figure 8.45(b)). The structure moves into the design position by tensioning of the cables (Figures 8.45(c) and 8.46). The joints between the segments are cast and the deck is then post-tensioned by external tendons. In this way, a self-anchored structure is created.

8.3. Static and dynamic analysis

Suspension structures should be analysed as geometrically non-linear structures. Modern program systems enable the function of the suspension structures to be expressed during both the erection and service. However, it is necessary to use programs which analyse the structure with large deformations and that use so-called tension stiffening. A suspension structure should be modelled as a space 3D structure formed by suspension cables, hangers and a deck.

The suspension cables and hangers should be modelled by cable elements. The deck can be modelled as a beam assembled from 3D bars that are connected to the hangers by transverse members (Figure 8.47(a)), or it can be modelled by shell

Figure 8.46 Johnson Creek Bridge: model of the erection

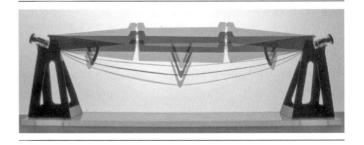

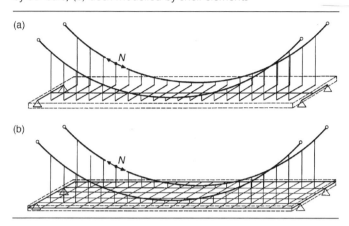

Figure 8.47 Modelling the suspension structure: (a) deck modelled by 3D bars, (b) deck modelled by shell elements

elements that have both bending and membrane capabilities (Figure 8.47(b)).

If the deck is composed of precast segments and a cast-in-place deck slab post-tensioned by prestressing tendons, it can be modelled as a deck of a stress ribbon structure, i.e. as a chain of mutually connected parallel members that represent prestressing tendons (PT), precast segments (PS) and a composite slab (CS). Prestressing tendons should be modelled as 'cable' members, in which initial force or strain has to be determined. Precast segments and a composite slab can be modelled as 3D bars or as shell elements.

Since the programs use so-called 'frozen members' it is possible to express the change of the static system, i.e. the change from hinge connection into the fix connection of the deck segments and/or the change from a cable to a stress ribbon. Initial stresses or strains in suspension cables have to be determined in the analysis. The initial forces are usually determined for the structure at the basic stage.

As discussed in Section 7.6, it is necessary to realise that some programs have 'cable members' that have zero stiffness (area and modulus of elasticity) in the initial stage. They are stressed by initial forces that correspond exactly to a load given in input data. For all other loads, these elements are part of the structure, i.e. part of the global stiffness of the structure. Unfortunately, in some programs the initial stage is specified for cables of actual stiffness. Since a portion of the strain and corresponding stress is absorbed by their stiffness, it is necessary to increase their initial strain in such a way that the strain and corresponding stress in tendons exactly balances the load at the basic stage. This means that the initial stage has to be determined by iteration.

For earth-anchored structures, the initial stage corresponds when the joints between precast members are cast and the deck becomes a structurally stiff member that contributes to the resistance to the load. For self-anchored structures, the initial stage corresponds to when suspension cables are post-tensioned and falsework is removed.

8.3.1 Initial stage of earth-anchored suspension structure

The initial stage of the suspension structure is – similar to the stress ribbon structure – a state in which the suspension cable carries the self-weight of the structure and the structure has a required geometry.

During erection, the whole load is usually resisted by suspension cables that act as a perfectly flexible cable which is able to resist the normal force only. Under this assumption, the cable curve will coincide with the funicular curve of the load applied to the cable and to the chosen value of the horizontal force H or sag. The erected segments, which are usually mutually pin connected, do not contribute to the resistance of the load.

8.3.1.1 Initial stage of a structure suspended on cables situated in vertical planes

Determination of the initial stage of the structure suspended on vertical cables is relatively simple, and is derived from the analysis of single cables. According to Equation (4.7) the geometry of the cable is given by the following equations:

$$p^0(x) = \frac{Q(x)}{H}$$

$$p(x) = y'(x) = p^0(x) + \frac{h}{l} = p^0(x) + \tan\beta$$

$$f(x) = \frac{M(x)}{H}$$

$$y(x) = \frac{M(x)}{H} + \frac{h}{l}x = f(x) + x\tan\beta$$

where $Q(x)$ and $M(x)$ are shear force and bending moment on a simple beam of span l.

Figure 8.48 shows a suspension structure formed by a cable of horizontal span length L, suspended at hinges A and B. The difference in height of hinges A and B is h. The cable is stressed by the self-weight and by point loads that represent the weight of the hangers and weight of the suspended segments. Since the segments are mutually pin connected, the hangers carry the load corresponding to one-half of the weight of the adjacent segments. The geometry of the cable is given by a requirement that at node i the cable has a sag f_i.

The exact shape is determined iteratively. At first the geometry is determined for load

$$G_i = \frac{g_C}{2}(\Delta_C^L + \Delta_C^R) + g_H\Delta_H + \frac{g_D}{2}(\Delta_S^L + \Delta_S^R) \quad (8.1)$$

Figure 8.48 Initial state: vertical cable

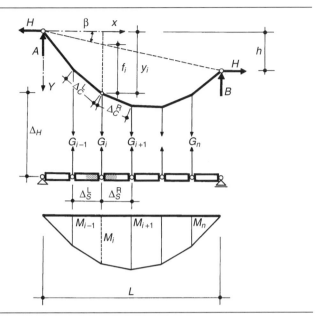

for $i = 1, \ldots, n$ where g_C, g_H and g_D are weights of the unit length of the cable, hangers and deck segments. The definition of notation Δ_C^L, Δ_C^R, Δ_H, Δ_S^L, Δ_S^R is evident from Figure 8.48.

As a first step, it is assumed that

$$\Delta_C^L = \Delta_S^L$$
$$\Delta_C^R = \Delta_S^R$$
$$\Delta_H = 0.$$

For the load G_i, the bending moment diagram is determined. The unknown horizontal force

$$H = \frac{M_i}{f_i} \quad (8.2)$$

is then determined. The shape of the cable at any point x is then given by

$$f(x) = \frac{M(x)}{H}$$
$$y(x) = \frac{M(x)}{H} + \frac{h}{L}x = f(x) + x \tan\beta. \quad (8.3)$$

For each node i, the distances Δ_C^L, Δ_C^R and Δ_H and new load G_i are determined. The new horizontal force and geometry of the cable are then calculated. The calculation is finished when a sufficient accuracy is obtained. Since the segments are relatively heavy (compared to the weight of the cables and hangers), only a few iteration steps are required.

The determination of the geometry of the cables formed by strands grouted in steel tubes is more complex. Since during the erection the strands are not grouted in the steel tubes, the vertical force from hangers has to be distributed between two components: tangent and radial to the cable. The tangent force is resisted by a compression capacity of the steel tube acting as an arch fixed at mid-span; the radial force is resisted by a tension capacity of the strands. The geometry of the suspension cable is therefore close to the circle. Figure 8.49 compares the function of the traditional suspension cable and the cable formed by strands in steel tubes. When the strands are grouted in tubes, the load is resisted by the composite action of the strands, tubes and cement mortar. The initial stage has to be determined iteratively.

8.3.1.2 Initial stage of a structure suspended on inclined cables

Analysis of a structure suspended on inclined cables is more complex. Figure 8.50 depicts a symmetrical suspension structure whose deck has a variable width and is suspended on two suspension cables supported at points a, and b_L and b_R. The difference in height between hinges is h. The position of the cables in the transverse direction is given by coordinates z_i'.

The cables are stressed by the self-weight and by point loads that represent the weight of the hangers and weight of the suspended segments. Since the segments that form the deck are mutually pin connected, the hangers carry the load corresponding to one-half of the weight of the adjacent segments. The geometry of the cables is given by a requirement that at node i the cables have a sag f_i.

The unknown horizontal force H in both cables is first determined. From the horizontal force, the vertical coordinates y_i in all nodes i of the cable can be determined using the procedure described in the previous chapter. From Figure 8.50 it is evident that

$$y_i' = y_a - y_i = y_a - \frac{h}{L}x_i - f_i \quad (8.4)$$

The force in the inclined hanger

$$N_i = \sqrt{G_i^2 + F_i^2} \quad (8.5)$$

depends on the known vertical component G_i and on the unknown horizontal component F_i. This value depends on the unknown inclination of hangers given by coordinate z_i.

To determine this value it is necessary to express the forces acting on the cable using these unknown coordinates. According to Figure 8.50(a) the value of the forces F_i can be

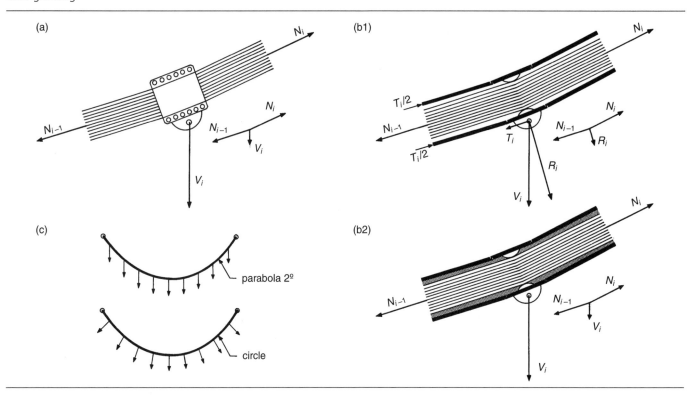

Figure 8.49 Static function: (a) classical cable, (b) cable formed by strands in steel tubes and (c) funicular shape, 1: before grouting and 2: after grouting

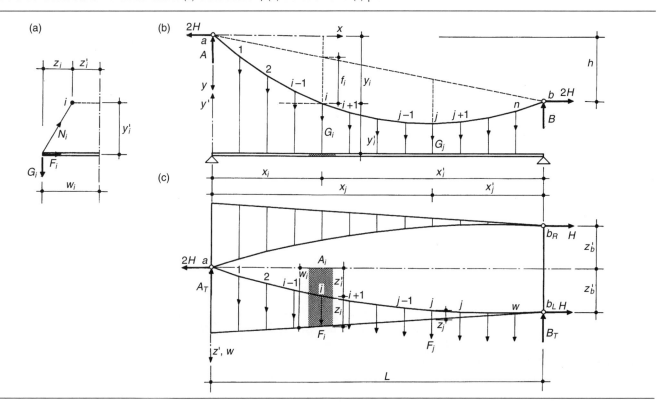

Figure 8.50 Initial state – inclined cable: (a) cross section, (b) elevation and (c) plan

Figure 8.51 Earth-anchored suspension structure (deck suspended on cables): static function

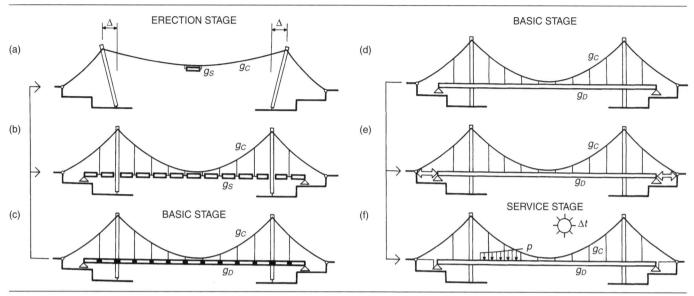

written:

$$\frac{G_i}{F_i} = \frac{y'_i}{z_i}$$

$$F_i = G_i \frac{z_i}{y'_i} = k_i z_i,$$

where

$$k_i = \frac{G_i}{y'_i}. \tag{8.6}$$

The transverse reaction A_T at hinge a is given by

$$\begin{aligned} A_T &= H\frac{z'_b}{L} + \sum_{j=1}^{n} \frac{F_j x'_j}{L} \\ &= H\frac{z'_b}{L} + \sum_{j=1}^{n} \frac{k_j x'_j}{L} z_j. \end{aligned} \tag{8.7}$$

The transverse bending moment at node i is defined:

$$\begin{aligned} M_i &= A_T x_i - H z'_i - \sum_{j=1}^{i} F_j(x_i - x_j) \\ &= H\frac{z'_b \cdot x_i}{L} + x_i \sum_{j=1}^{n} \frac{k_j x'_j}{L} z_j - H z'_i - \sum_{j=1}^{n} k_j z_j (x_i - x_j). \end{aligned} \tag{8.8}$$

It is obvious that at each node of the cable the bending moment

$$M_i = 0, \tag{8.9}$$

where $i = 1, \ldots, n$. In this way we have n equations for n unknown coordinates z_i. By solving this system, we can obtain the coordinates z_i. The forces F_i and N_i are then determined.

8.3.2 Analysis of the suspension structures

8.3.2.1 Earth-anchored suspension structures

Analysis of the structure for both the erection and the service load starts from the initial basic stage. During the erection analysis the structure is progressively unloaded until the stage at which the structure is formed by a cable only (Figure 8.20(c)) or by a cable with island segment (Figure 8.22(b)). An initial inclination of the towers is also determined (Figure 8.51).

Analysis of the structure for service load also starts from the same initial basic stage. However, after casting and post-tensioning of the joints between the segments, the deck has an actual stiffness and contributes to the resistance of the structure to service load.

Analysis of suspension structures supported by cables is similar. If the structures are supported by a stress ribbon, not only the deck and struts are removed but also the segments forming the ribbon (Figure 8.52). Analysis of the structure for service load starts by post-tensioning of the stress ribbon.

The same analytical model is used for analysis for the ultimate load.

Note that the stresses in the deck redistribute with time due to the creep and shrinkage of concrete. After erection, the deck

Figure 8.52 Earth-anchored suspension structure (deck supported by cables): static function

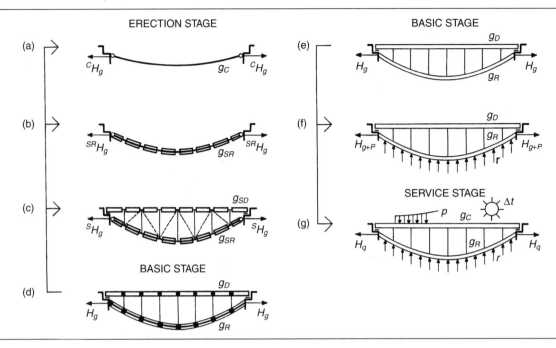

functions as a chain of simple beams. After casting and post-tensioning of joints, a free rotation of joint is restricted not only for the effects of the service load but also for effects of creep and shrinkage. This restriction means that the moments transfer from span into the supports and the course of bending moments are similar to a continuous beam (Figure 8.53). The redistribution depends on the age of the concrete at the time of post-tensioning (see Chapter 6, Figures 6.8 and 6.10). However, since the bending moments due to the dead load are, compared to the bending moment due to the live load, very small, this redistribution is not critical for the design.

8.3.2.2 Self-anchored suspension structures

Analysis of self-anchored structures also starts from the basic stage for which optimum forces in suspension cables were determined. The shape of the structure during the assembly or casting is obtained by unloading the structure by removal of the prestressing force in cables. Analysis of the structure for service load is the same as for the earth-anchored structure.

8.3.3 Parametric study

To understand the behaviour of a typical suspension structure of span $L = 99$ m, extensive parametric studies were carried out. At first, the influence of the sag, connection of the deck with suspension cables at mid-span and restriction of the horizontal movement at support were studied on a 2D suspension structure. Space behaviour of a 3D suspension structure was then analysed.

8.3.3.1 2D suspension structure

The structure of the span $L = 99.00$ m is formed by a concrete deck suspended on a suspension cable of the mid-span sag f. The structure was analysed for two arrangements of the suspension cables called A and B (Figure 8.54). In option A, there is a 2 m gap between the cable and deck; in option B the cable is connected with the deck at mid-span.

Figure 8.53 Redistribution of bending moment in the deck

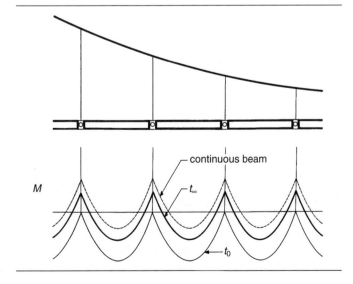

Figure 8.54 Parametric study of 2D structure: (a) loads, (b) studied structures and (c) characteristic deformations

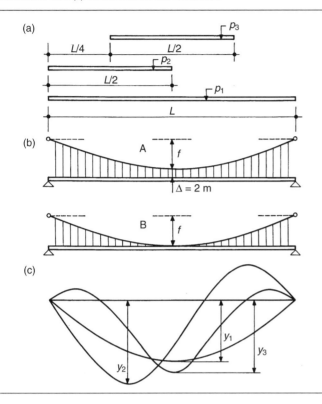

The concrete deck of the area $A_D = 1.500 \, \text{m}^2$ and modulus of elasticity $E_D = 36 \, \text{GPa}$ was analysed for four values of bending stiffness characterised by moments of inertia $I_D = 0.005, 0.050, 0.500$ and $5.000 \, \text{m}^4$.

The initial state was determined for three values of the mid-span sag f with corresponding areas A_C and modulus of elasticity $E_C = 190 \, \text{GPa}$ of the suspension cable formed by strands.

$f = L/8 = 12.375 \, \text{m}; \quad A_C = 8.490 \times 10^{-3} \, \text{m}^2$

$f = L/10 = 9.000 \, \text{m}; \quad A_C = 9.905 \times 10^{-3} \, \text{m}^2$

$f = L/12 = 8.250 \, \text{m}; \quad A_C = 11.886 \times 10^{-3} \, \text{m}^2$

The hangers situated at a distance of 3.00 m are formed by bars of area $A_H = 1.608 \times 10^{-3} \, \text{m}^2$ and modulus of elasticity $E_H = 210 \, \text{GPa}$.

The structure was analysed for three loads.

- Load 1: dead load plus live load 16 kN/m situated along the whole deck length.
- Load 2: dead load plus live load 16 kN/m situated on one-half of the deck length.
- Load 3: dead load plus live load 16 kN/m situated in the central portion of the deck length.

The structure was analysed as a geometrically non-linear structure by the program ANSYS. In the initial state the forces in the cables exactly balanced the effects of the dead load. The deformation of the structure and internal forces and moments in all structural members were obtained. The characteristic deformations are presented in Figure 8.54 where maximum values of the deformations y_1, y_2 and y_3 caused by loads 1, 2 and 3 are marked. It is obvious that deck bending moments are proportional to the deck deformation.

Figure 8.55 depicts only the maximum values of the deformations y_1, y_2 and y_3 for loadings 1, 2 and 3 that originate in the structure of the moment of inertia $I_D = 0.005 \, \text{m}^4$ and $I_D = 0.050 \, \text{m}^4$.

From the results it is evident that the maximum reduction of the deflection is caused by restriction of the horizontal movement at supports. The maximum reduction of the deflection is achieved by a combination of horizontal fixing and connection of the cable to the deck. The different sag of suspension cables mainly influences the amount of steel in the cables.

8.3.3.2 3D suspension structure

Space behaviour was verified on a structure of span $L = 99.00 \, \text{m}$ (Figures 8.56 and 8.57). The concrete deck was suspended on two suspension cables of sag:

$f = 0.1; \quad L = 9.90 \, \text{m}.$

The structure was analysed for two arrangements of the suspension cables referred to as A and B (Figure 8.56). Analysis was also carried out for a deck situated in a vertical parabolic curve with maximum rise at mid-span $r = 1.98 \, \text{m}$, marked C. In this option the suspension cables were connected to the deck at mid-span.

At the abutments, the deck was supported by movable bearings (M) or by fixed bearings (F). The suspension cables are vertical (V) or inclined inwards (In) or outwards (Out).

For all above-mentioned arrangements, natural modes and frequencies were determined. The behaviour of the suspension structure is characterised by following modes (Figure 8.58):

Mode L	horizontal vibration of the structure supported by movable bearings;
Mode V_1	vertical vibration in the first vertical mode;
Mode V_2	vertical vibration in the second vertical mode;
Mode S	swing vibration – transverse vibration combined with torsion; and
Mode T	torsional vibration.

Structures suspended on vertical cables vibrate in pure transverse modes; structures suspended on inclined cables vibrate

Figure 8.55 Parametric study of 2D structure – deformation of the deck: (a) $I_D = 0.005\,\text{m}^4$ and (b) $I_D = 0.050\,\text{m}^4$

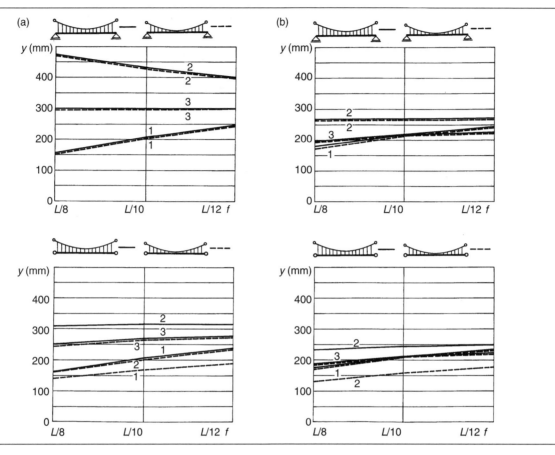

in more complex swing modes that combine transverse and torsional movement.

Tables 8.1–8.3 present the results of the analysis. Since the ratio of torsional frequency f_T to the first bending frequency f_{V1}

$$r = f_T/f_{V1}$$

indicates the possibility of the aerodynamic instability, its value is also given in the tables. The value r should be greater than $r_{cr} = 2.5$, which is considered as the critical value.

From the results it is evident that the fixed connection of the deck to the cable increases the stiffness of the structures. For the given span length, the inclination of the suspension

Figure 8.56 Parametric study of 3D structure: studied structures

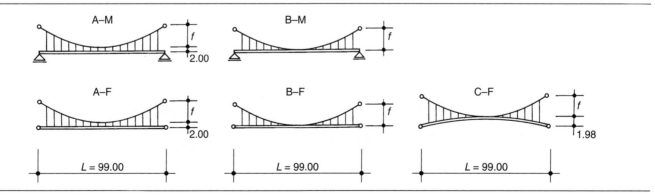

Figure 8.57 Parametric study of 3D structure: arrangement of the main suspension cables

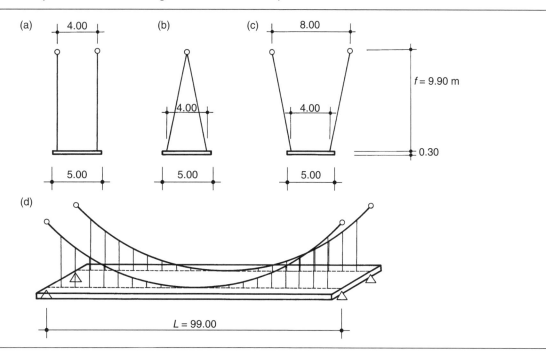

Figure 8.58 Natural modes

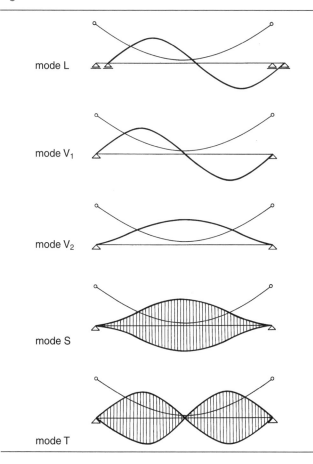

cables does not increase the transverse and torsional stiffness. However, note that the dynamic analysis is done linearly and therefore the values in the tables do not cover the reduction of deformation caused by the fixing of the deck (Figure 2.22).

8.3.4 Examples of analysis

The described modelling and analyses are illustrated on two structure examples. The static function of the Vranov Lake Bridge (see Section 11.2) is evident from Figure 8.59. The structure was modelled as a space structure. The towers were modelled as a 3D frame, the suspension cables and hangers as cable member and the deck as a 3D grid formed by two longitudinal bar members and transverse bar members. The longitudinal members were situated at edge girders and the transverse members at the diaphragms. The structure was stiffened by external tendons that were modelled by cable members and situated parallel to the edge girders.

Analyses were carried out for both the erection and service stage on one calculation model. Figure 8.60 presents the first vertical, first swing and first torsional modes and frequencies. Since the first 12 natural frequencies corresponding to vertical modes are less than the walking frequency of 2 Hz, the bridge proved to be insensitive to pedestrian loading. Even the loading imposed by vandals (a group of people trying to make the bridge vibrate in an eigenmode) caused negligible effects with an amplitude of several millimetres.

Table 8.1 Natural frequencies

	Mode L f_H: Hz	Mode V_1 f_A: Hz	Mode V_2 f_B: Hz	Mode S f_C: Hz	Mode T f_D: Hz	f_T/f_{V1}
V-A-M	0.24	0.44	0.64	0.88	2.93	6.66
V-A-F	–	0.43	0.64	1.84	2.93	6.81
V-B-M	0.40	0.64	0.88	0.88	3.11	4.86
V-B-F	–	0.64	0.87	1.85	3.12	4.86
V-C-F	–	0.66	0.99	1.69	3.18	4.82

Table 8.2 Natural frequencies

	Mode L f_H: Hz	Mode V_1 f_A: Hz	Mode V_2 f_B: Hz	Mode S f_C: Hz	Mode T f_D: Hz	f_T/f_{V1}
In-A-M	0.24	0.43	0.64	0.89	2.93	6.81
In-A-F	–	0.43	0.64	1.78	2.93	6.81
In-B-M	0.39	0.64	0.90	0.89	3.09	4.83
In-B-F	–	0.64	0.86	1.78	3.11	4.86
In-C-F	–	0.66	0.99	1.73	3.19	4.83

Table 8.3 Natural frequencies

	Mode L f_H: Hz	Mode V_1 f_A: Hz	Mode V_2 f_B: Hz	Mode S f_C: Hz	Mode T f_D: Hz	f_T/f_{V1}
Out-A-M	0.24	0.43	0.64	0.89	2.94	6.66
Out-A-F	–	0.42	0.64	1.79	2.94	6.81
Out-B-M	0.39	0.64	0.90	0.89	3.12	4.86
Out-B-F	–	0.64	0.86	1.80	3.13	4.86
Out-C-F	–	0.66	1.08	1.66	3.16	4.82

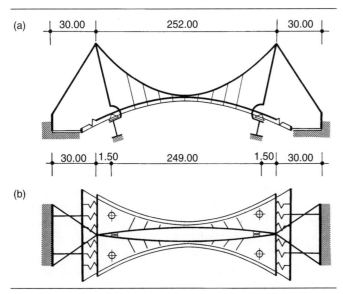

Figure 8.59 Vranov Lake Pedestrian Bridge, Czech Republic: calculation model: (a) elevation and (b) plan

The Willamette River Bridge was analysed as a geometrically non-linear structure by the program ANSYS. The global function of the structure was checked on the model depicted in Figure 8.61 which expressed the actual boundary condition and space function of the bridge. This model was used to determine the internal forces for the effects of the live, wind and seismic load. In all earthquake analyses performed, the stoppers were considered for both functioning and failing and the design took the worst case into consideration.

This calculation model was also used for the analysis of all erection stages including the stage in which the self-weight of island segments is transferred from erection cables into the main suspension cables. The initial stage for both analyses was the structure loaded by self-weight at the end of the erection of segments.

Figure 8.62 shows the first two vertical, swing and torsional modes and frequencies. The response of the structure to pedestrians was checked for a pulsating point load $F = 180\,\text{kN}$

Figure 8.60 Vranov Lake Pedestrian Bridge, Czech Republic: natural modes and frequencies

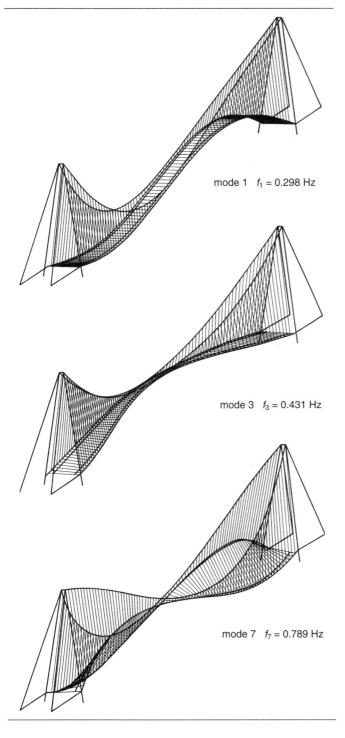

Figure 8.61 Willamette River Bridge, Oregon, USA: calculation model: (a) detail of the tower and (b) bridge

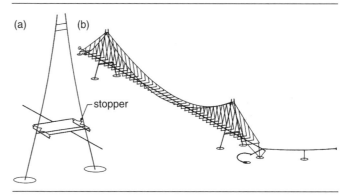

Special consideration was given to the design of the ties due to their primary static function and fatigue loading and a very detailed analysis was therefore performed. The ties were modelled by a space structure assembled from shell elements.

8.3.5 Static and dynamic loading tests

The quality of the workmanship and static assumptions were also verified by static and dynamic loading tests. The Vranov Lake Bridge was tested by 24 trucks parked along the whole length of the bridge and by ten trucks parked on one side and in the middle of the span (Figure 8.63). The deformation of the structure due to the load of 24 trucks, which created 50% of design load, was 0.250 m. The difference between the calculated and measured values was less than 5% of the total deflection.

The aerodynamic stability of the Vranov Lake Bridge was tested on a geometrically and aerodynamically similar model of the

Figure 8.62 Willamette River Bridge, Oregon, USA: natural modes and frequencies

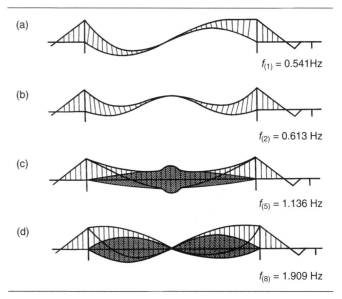

moving along the main span. The maximum calculated acceleration is much less than the limit presented in Section 3.3. The bridge is stiff and the users did not have a feeling of discomfort when walking or standing on the bridge.

Figure 8.63 Vranov Lake Pedestrian Bridge, Czech Republic: loading test

structure built to a scale of 1:130 (Figure 8.64). The tests were carried out by M. Pirner (ITAM). The deck was assembled from segments cast from micro-concrete and the suspension cables were formed by piano wires. The bridge is aerodynamically stable for all wind speeds from 0 to 55 m/s (0 to 198 km/hour) (Pearce and Jobson, 2002).

Even though the ratio r of the pure torsion frequency $f(8) = 1.909$ to the first bending frequency $f(1) = 0.541$ of the Willamette River Bridge is $r = 3.529$ and acceptable behaviour under aerodynamic excitation of the bridge was observed, it was still felt to be necessary to check the response of the structure in the wind tunnel (Figure 8.65). The main reasons were the overall slenderness of the structure and the existence of the mid-span observation platform.

A geometrically and aerodynamically similar model of the structure built to a scale of 1:68.7 was tested by M. Pirner (ITAM). The deck was assembled from segments cast from epoxy and the suspension cables were formed by piano wires.

Figure 8.64 Vranov Lake Pedestrian Bridge, Czech Republic: wind tunnel test

Figure 8.65 Willamette River Bridge, Oregon, USA: wind tunnel test

The results of the tests confirmed an excellent function of the structure. Figure 8.66 shows the arrangement of the model and excited natural modes and frequencies. It was not possible to excite pure torsional modes during the dynamic test of the model. The bridge is aerodynamically stable for all wind speeds from 0 to 65 m/s (0 to 234 km/hour).

Figure 8.66 Willamette River Bridge, Oregon, USA: aerodynamic model: exited mode and frequencies

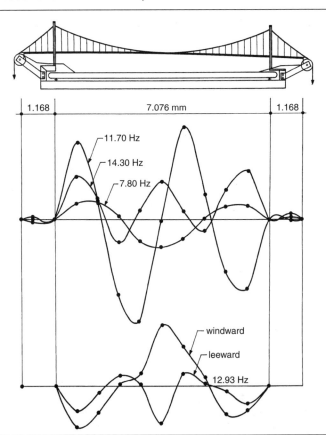

Figure 8.67 Johnson Creek Bridge on Springwater Trail, Oregon, USA: static model: (a) cross-section and (b) elevation

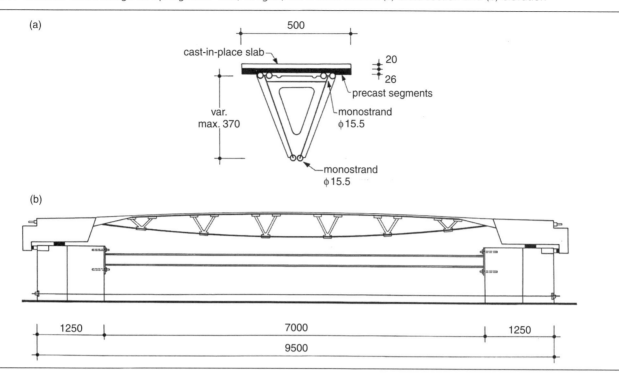

The structural behaviour of the Johnson Creek Bridge on Springwater Trail, Oregon (see Section 11.2) was verified on a static model built to a scale of 1:8 (Figure 8.67). The model was assembled as a stress ribbon deck supported and prestressed by external cables. Both the deck and cables are anchored to the end-anchor blocks. These anchor blocks were supported by concrete pedestals that were mutually connected by steel members and post-tensioned bars. The stress ribbon deck was assembled from precast segments 26 mm deep and a composite slab of thickness 20 mm. The transition between the composite stress ribbon and anchor blocks was formed by cast-in-place haunches. The connection of external cables (two 15.5 mm diameter monostrands) to the deck was made by steel struts.

The model was erected as the proposed structure. After casting, the anchor blocks were temporarily fixed to the pedestals and the bearing cables with steel struts and precast segment were erected. The main suspension cables were then pulled through ducts situated at the steel struts. By tensioning of the suspension cables, the structure was lifted into the design position. The formwork of the haunches was then suspended on the segments and the anchor block, joints between the segments, composite slab and haunches were cast. When the concrete reach sufficient strength, the fixing of the anchor blocks to the pedestals was released and the suspension cables were tensioned up.

To guarantee a model similarity, steel bars representing the self-weight of the structure were suspended on the deck. The live load was represented by additional concrete cylinders and sand bags placed on the stress ribbon deck.

The construction of the model was carefully monitored. The structure was tested for three positions of the live load situated along the whole length, in the middle (Figure 8.68) and on one-half of the deck length. The measured deformations and strains were in good agreement with the results of the static analysis (Table 8.4). The ultimate load test was carried out for the position of the live load situated on one-half of the length

Figure 8.68 Johnson Creek Bridge, Oregon, USA: static model: load situated in the middle of the span

Table 8.4 Johnson Creek Bridge, Oregon, USA static model: deformation of the deck at mid-span

	Load 1: mm	Load 2: mm	Load 3: mm
Measurement	29	28	19
Calculation	28	28	18

Figure 8.69 Johnson Creek Bridge, Oregon, USA: static model: ultimate load

(Figure 8.69). The structure failed by shear at the longitudinal joint between the precast segments and cast-in-place slab at loads higher than the design ultimate load.

REFERENCES

Brown DJ (1996) *Bridges. Three Thousand Years of Defying Nature*. Reed International Books Ltd., London.

Gimsing NJ (1998) *Cable Supported Bridges: Concept & Design*. John Wiley & Sons, Chichester.

Leonhardt F (1964) *Prestressed Concrete: Design and Construction*. Wilhelm Ernst & Sons, Berlin.

Pearce M and Jobson R (2002) *Bridge Builders*. Wiley-Academy, John Wiley & Sons, Chichester, UK.

Schlaich J and Bergermann R (1992) *Fußgängerbrücken. Ausstellung und Katalog*, ETH Zürich.

Wittfoht H (1972) *Triumph der Spannweiten*. Beton-Verlag GmbH, Düsseldorf.

Chapter 9
Cable-stayed structures

Cable-stayed structures have been described in several excellent books (Gimsing, 1998; Mathivat, 1983; Menn, 1990; Podolny and Scalzi, 1976; Walther *et al.*, 1998); only additional information about structures formed by a slender deck is presented in this chapter.

9.1. Structural arrangement

As described in Chapter 2, a cable-stayed structure is formed by a slender deck that is suspended on stay cables anchored in the tower and deck (Figures 2.6(b) and 2.6(d)). The elevation of the deck can therefore have an optimum arrangement corresponding to local conditions. The structures can have one or more spans.

It is possible to anchor back stays in the soil and create a totally or partially earth-cable-stayed structure. For economical reasons, this solution makes sense only in special cases or for structures of long spans. The prevailing portion of cable-stayed structures is formed by self-anchored systems that stress the footings by vertical reactions only (Figure 9.1).

Classic cable-stayed structures have two or three spans that are suspended on one or two towers (Figure 9.2). The towers can be vertical or inclined. Multi-span cable-stayed structures have also been built. Special attention has to be paid to the deflection of the deck and to the analysis of the dynamic response.

It is possible to suspend only the main span and anchor the back stays at short side spans or anchor blocks (Figure 9.3(a)). The optimum span length of the side spans is from 0.40 to 0.45 of the length of the main span. The length depends on the arrangement of the stay cables and side supports. If the deck has an expansion joint between the side and approach spans, the structure requires a shorter side span and that backstays be anchored (Figure 9.3(b)).

If the structure is continuous, the side spans can be longer and stay cables symmetrical to the tower can be extended into the approach spans (Figure 9.3(c)). The cable-stayed structure can be significantly stiffened by supports situated in side spans (Figure 9.3(d)).

Stay cables can have different arrangements as shown in Figure 9.4. While a statically superior radial arrangement (Figure 9.4(a)) brings structural difficulties in relation to the anchoring

Figure 9.1 Neckar River Bridge, Germany

Figure 9.2 Arrangement of the cable-stayed structure

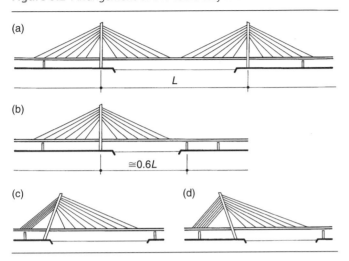

Figure 9.4 Arrangement of the stay cables

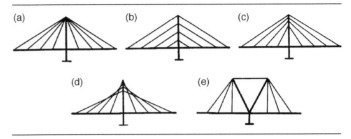

arrangements can be combined to create the most suitable solution (Figure 9.6).

To reduce the bending moments in the deck it is necessary to anchor stay cables at a relatively short distance (3–6 m). The bending moments due to dead load are then very low and the deck can be very slender (from 0.30 m).

of all stays at one point (Figure 2.39), a statically less advantageous parallel arrangement (Figure 9.4(b)) simplifies structural details. A semi-radial arrangement (Figure 9.4(c)) represents a reasonable compromise both from a static and structural point of view.

The relatively light load of pedestrian bridges justifies the solution presented in Figures 9.4(d) and 9.5. The basic

If the towers have sufficient height ($0.2L$), the stiffness of the cable-stayed structure is given by the stiffness of the system formed by compressed masts and deck and by a tension stiffness of the stay cables (Figure 9.7). If low towers are designed, the structure requires sufficient bending stiffness of the deck.

Cable-stayed structures do not usually utilise increasing stiffness of the structural system by anchoring the deck at the abutments (Figure 2.20). However, the advantage of restricting the horizontal movement at the abutments was demonstrated by Menn in the design of the Sunniberg Bridge. The bridge, which has five spans of length from 59 to 134 m, is curved in plan with a curvature radius $R = 503$ m. The slender deck running at 60 m above grade is suspended on pylons protruding 15 m above the deck.

Figure 9.3 Cable-stayed structure – length of the side spans

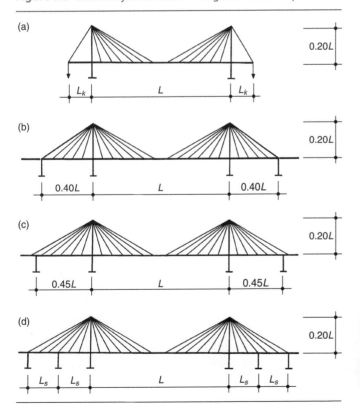

Figure 9.5 Badhomburg Bridge, Germany (courtesy of Schlaich, Bergermann and Partners)

Figure 9.6 Židlochovice Bridge, Czech Republic

Figure 9.8 Sunniberg Bridge, Switzerland, tower: (a) cross-section, (b) longitudinal section, (c) transverse bending moments and (d) longitudinal bending moments

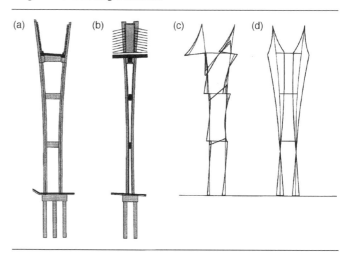

The stiffness of the structure comes from the plan curvature of the deck, which is fixed at the abutments (Figure 9.9(a)). While vertical deflection of the deck of traditional multi-span cable-supported structures has to be controlled by intermediate anchor piers or by bending stiffness of the deck; in this bridge the vertical deformation of the deck is controlled by the transverse stiffness of the curved deck. Any vertical load causes the horizontal movement of the deck that acts in the horizontal plane as an arch (Figure 9.9). The transverse movement of the deck creates transverse moments in the piers forming transverse frames (Figure 9.8(c)).

A similar increase in the stiffness is used in the author's design of the Bohumin Pedestrian Bridge (Section 11.3).

The towers can be made from concrete or steel. They can be formed by individual columns, or they can be H-, V- or A-shaped. The possible arrangement of the cable-supported structures in the transverse direction is discussed in Section 2.2 (Figures 2.24 and 2.25).

Figure 9.7 Cable-stayed structure: (a) classical and (b) extradosed

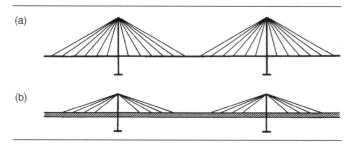

The deck can be totally suspended on the stay cables or can be supported at towers. The deck can be also frame connected with tower legs. At the towers, the cables can be deviated at the saddles (Figure 2.37(a)), anchored at anchor plates (Figure 2.37(b)) or they can overlap and be anchored at blocks (Figures 2.37(c) and 9.10).

The deck is usually suspended on two planes of cables: vertical or inclined. Suspension in the bridge axis requires a torsionally stiff girder. Suspension in two inclined planes creates a feeling of safety (Figure 9.11), while suspension in one central plane can naturally divide cyclists from pedestrians (Figure 9.12).

Figure 9.9 Sunniberg Bridge, Switzerland: (a) calculation model and (b) cross-section of the deck

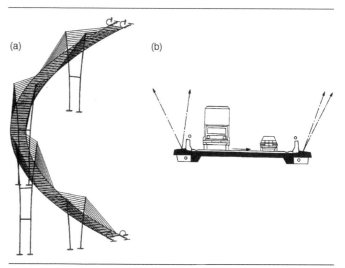

Figure 9.10 Židlochovice Bridge, Czech Republic

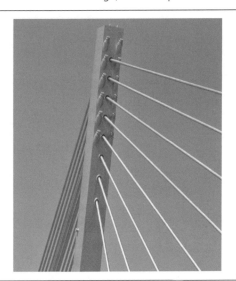

Figure 9.12 Bohumin Bridge, Czech Republic: suspension on one central plane of stay cables

In cable-stayed structures of common spans, the concrete deck guarantees the transverse stiffness. Although the outward inclination of the stay cables increases the transverse stiffness and inward inclination increases the torsion stiffness of the system, the increase is not significant. However, these structures do not vibrate in pure transverse modes; all transverse modes are accompanied with distortion of the deck that contributes to the stiffness of the system.

The possible arrangements of the structure supported by stay cables is presented in Figure 9.13. However, an elegant solution with one upright (Figure 9.14) requires a relatively stiff deck. A slender deck with several uprights shown in Figure 9.13(b) is, compared with self-anchored suspension structures, too complicated.

Figure 9.11 Bohumin Bridge, Czech Republic: suspension on two inclined planes of stay cables

Figure 9.13 Cable-stayed structure supported by stay cables: (a) one stay and (b) multiple stays

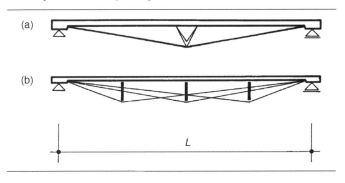

Figure 9.14 Osormort Bridge, Catalonia, Spain (courtesy of Carlos Fernandez Casado, SL, Madrid)

Figure 9.15 Pin connection of the stay cable

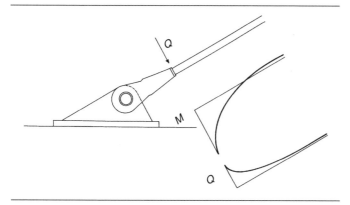

The stay cables situated above the deck are usually suspended on cables developed by steel or prestressed concrete industry (Figure 2.31). In the first application, the stay cables were also formed by prestressed concrete ties or walls (Figure 2.40). A similar arrangement can be used for cables situated under the deck.

The stay cables can be formed by spiral strands, locked coil strands or parallel wires (Figure 2.31(a)). They are factory fitted with a combination of socket types to enable load transmittal between the structure and the cable (Figure 2.32) and are delivered on site in design length. The socket also enables a pin connection of the stays with the deck and tower (Figure 9.15).

The stay cables developed from the prestressing tendons are formed by individual bars, parallel bars, parallel wires or prestressing strands (Figure 2.31(b)). They can easily be assembled on the site. Typical arrangement of the cable is shown in Figure 9.16.

Due to large fatigue stresses that originate in stay cables, all details are designed in such a way that the forces in the cables can be adjusted and it is possible to progressively exchange the cables (AASHTO, 1997).

Due to the deformation of the deck and towers, relatively large bending moments originate at the anchors of the stay. It is reasonable to reduce these local moments and transfer the critical moment from anchors.

Many designers assume that the pin connection of the stay cable eliminates bending stresses in the stay cable. Pin connection can eliminate local bending caused by an erection misalignment, but cannot eliminate bending stresses caused by service load. The rotation of the pin is caused by a shear force and corresponding bending moment that stresses the cable. Rotation occurs only when rotation overcomes pin friction. The shear force that rotates the pin stresses the cable. It is therefore necessary to design the cable for corresponding local moment and/or design a measure that reduces the bending.

9.2. Erection of the structures

The process of cable-stayed structure erection has to guarantee that the forces in the stay cables, together with post-tensioning of the deck, balance the effects of the dead load.

Figure 9.16 Typical arrangement of the stay cable

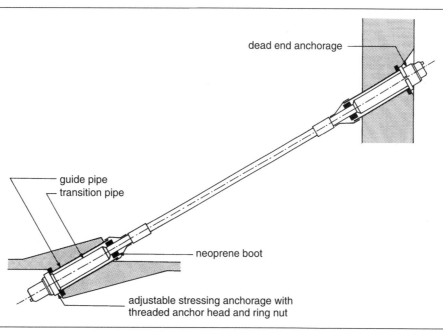

Figure 9.17 Erection of the deck on a falsework

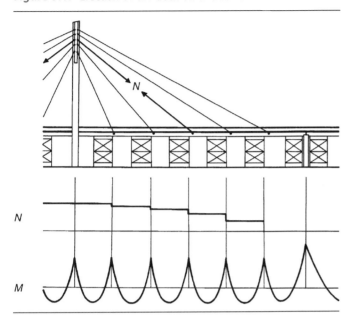

Figure 9.18 Erection of the deck in a cantilever

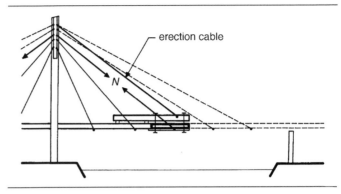

tensioned and in a design shape. The bars were anchored in temporary erection struts that also guarantee the design sag (Section 11.3). This procedure guarantees that the stays were stressed by design forces.

The main advantage of cable-stayed structures is that they can be progressively erected in self-anchored cantilevers independently of the terrain under the bridge (Figure 9.18). They can be assembled from precast segments or cast in movable travellers. Precast segments can be lifted by a winch supported by a beam anchored to the deck. However, the beam or traveller can cause relatively large bending moments in a slender deck. It is therefore suitable to suspend the beam or traveller on an erection stay cable and transfer the load directly to the tower (Figure 9.18). The connection between the traveller and already cast deck has to transfer the horizontal component of the cable force.

Figure 9.19 shows a traveller that was used in the construction of Diepoldsau Bridge, built across the Rhone River in

Cable-supported structures can be assembled from precast members or can be cast-in-place on the falsework. However, the structural arrangement of a cable-stayed structure calls for the erection in free cantilevers in which the effects of the dead load are resisted by stay cables (Mathivat, 1983; Podolny and Scalzi, 1976; Podolny and Muller, 1982).

If the structure is cast in place, the cables are usually installed after the deck is completed. Since the tensioning of one cable (Figure 9.17) influences the tension in all other cables which have already been tensioned, it is necessary to design an adjustment of their tension. In the construction of Hungerford Bridge (London, UK) the stay cables, formed from bars, were installed

Figure 9.19 Erection of the Diepoldsau Bridge, Switzerland: (a) elevation, (b) partial elevation and (c) cross-section

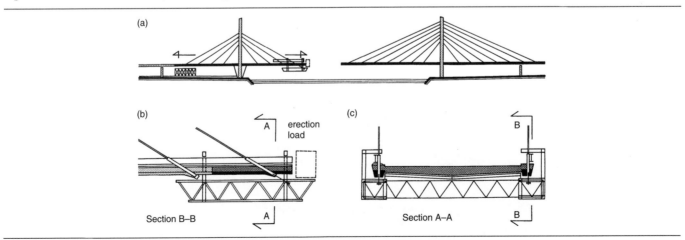

Figure 9.20 Židlochovice Bridge, Czech Republic: (a) elevation, (b) connection of the longitudinal and transverse members and (c) longitudinal normal stresses in the deck

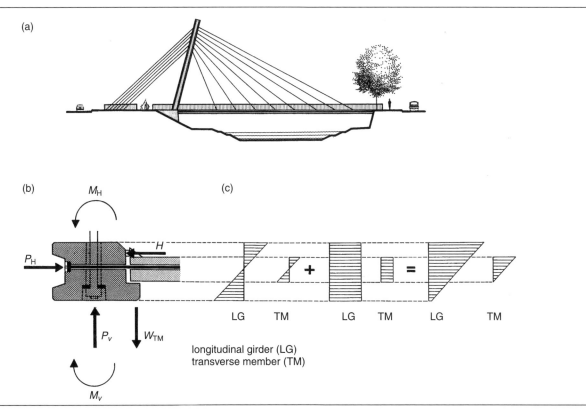

Switzerland (Walther et al., 1998). The bridge of maximum span 97.00 m was formed by a solid slab of an average thickness of 0.45 m; the deck was cast in segments of length 6.00 m. The traveller was suspended on the final stay cables that also served as erection cables. The horizontal component of the cable force was resisted by precast edge girders in which the stay cables were also anchored. To guarantee the linear behaviour of the cable, it was necessary to load the traveller by a temporary load that was progressively unloaded during the casting of the segment.

The deck can also be erected from two longitudinal precast edge girders and transverse solid slab members connected by longitudinal and transverse post-tensioning. Figure 9.20 shows a bridge built across the Svratka River in the Czech Republic which is suspended on an inclined tower.

The design of the bridge has addressed two special problems: elimination of torsion of the longitudinal girders during the erection of the structure and determination of the level of the post-tensioning of the longitudinal girders and transverse members in such a way that the redistribution of stresses between them is a minimum (Strasky et al., 2002).

Erection eccentric transverse post-tensioning of the joints between longitudinal and transverse members solved the first problem (Figure 9.20(b)). The transverse members were provided with steel brackets with nuts and screws situated on the surface close to their ends. After a transverse member was erected, the screws were drawn until their heads touched the longitudinal girders. The post-tensioning bars were then partially tensioned. The couple of the forces acting on the girder (under the screw head and bar anchor) created a moment that balanced the torsion.

The second problem was solved by the design process of erection of the deck that was designed in such a way that the prestress of the longitudinal girders is larger then the prestress of the transverse members. Since the joints between the transverse members were cast after the erection of the deck, the main compression caused by stay cables loads only the edge girders. The deck was successively erected in 5 m long parts. The process of the assembly was as follows (Figure 9.21).

1. The edge longitudinal girders were erected; each new girder was post-tensioned to the same value as the previously erected girders (Figure 9.21(a)).

Figure 9.21 Židlochovice Bridge, Czech Republic: progressive erection of the deck

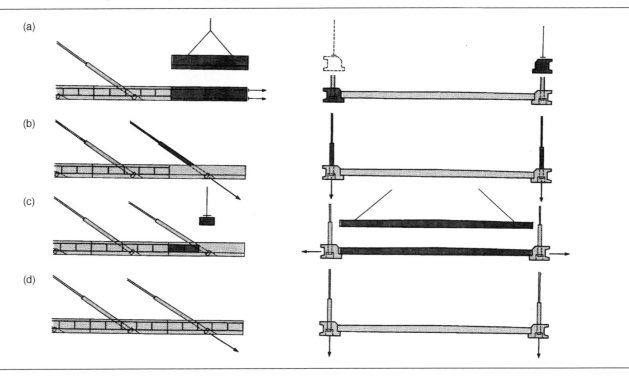

2. Stay cables were installed and tensioned to the prescribed level (Figure 9.21(b)).
3. Transverse members were erected and the joints between the transverse members and the longitudinal girders were post-tensioned (Figure 9.21(c)).
4. The forces in the stay cables were adjusted (Figure 9.21(d)).

For small pedestrian bridges, progressive casting of the deck in movable travellers or erecting deck from precast match-cast segments is too expensive. To simplify construction, a very simple technology has recently been developed. This technology was utilised in the construction of two pedestrian bridges built in Eugene, Oregon (see Chapter 11). The deck of both bridges was erected in cantilevers above busy highways (Figure 9.22).

To simplify production, the segments were not match-cast. To allow cantilever erection, the segments were cast with projecting structural tubes that mated with adjacent segments (Figures 9.23 and 9.24). During the erection, the anchor plates of the stays were bolted to erected segments. To guarantee the stability of the erected cantilever, the anchor plates of the stays are also bolted to the adjacent, previously erected segments. To allow adjustment of the erected structure and allow erection tolerances, there are elliptical slots in the plates above adjacent segments. These bolts were tightened after adjustment of the geometry. The bending moments were then resisted by a couple of forces originating in tubes and anchor plates.

The precast deck of the I-5 Gateway Pedestrian Bridge was erected in balanced cantilevers from the central tower (Figures 9.22 and 9.25(a)). The closures and composite slab was also placed in balanced cantilevers directed away from the tower (Figure 9.25(c)). After that, the suspended spans together with

Figure 9.22 I-5 Gateway Bridge, Oregon, USA: segment erection

Figure 9.23 I-5 Gateway Bridge, Oregon, USA: segment connection: (a) bridge elevation, (b) connection detail and (c) section A–A

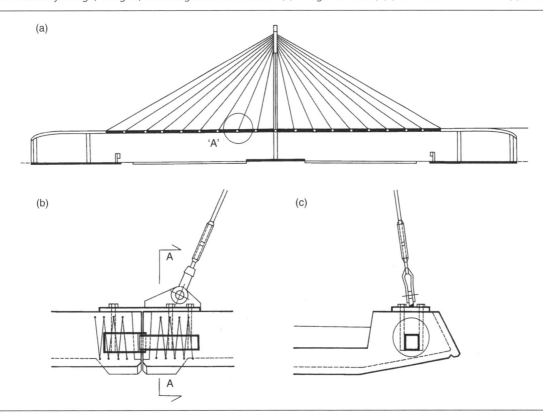

curved spans were post-tensioned. The precast deck of the Delta Ponds Bridge was erected similarly; the precast segments were erected in the cantilever that was directed from the already-cast backspans.

Figure 9.24 Delta Ponds Bridge, Oregon, USA: connection detail

Figure 9.25 I-5 Gateway Bridge, Oregon, USA: construction sequences: (a) erection of segments, (b) suspension of the closure formwork and (c) casting of the joints, closures, composite deck slab and subsequent deck prestressing

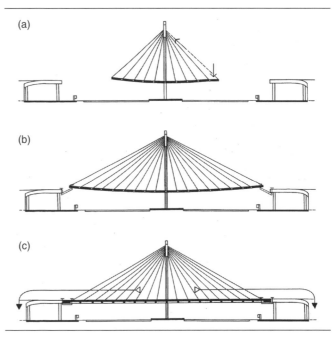

Since the erected cantilevers form a stable system the individual cantilevers can also be cast on the falsework, subsequently suspended on the towers and then rotated into the design position.

9.3. Static and dynamic analysis

Cable-stayed structures can be analysed similarly to suspension structures as geometrically non-linear structures. The stay cables can be modelled as suspension cables and the towers and deck by 3D bars or shell elements (Figure 9.26). However, the larger stiffness of cable-stayed structures given by the height of the towers and by an initial tension in the stay cables allows the stay cables to be substituted by bars. If the tension stiffness of the deck is not utilised (Figure 2.20) the cable-stayed structure can be analysed linearly and it is possible to use a superposition of the static effects. The structure can then be analysed in two steps. In the first step, the stresses due to the dead load and prestress are determined. In the second step, the structure is analysed as common structures by linear programs. Since the cable-stayed structures form progressively erected hybrid systems, the time-dependent analysis is mandatory.

9.3.1 Stay cables

The stay cables that are anchored at towers and the deck behave as cables that are loaded by their own weight and by deflection of their supports. Figure 9.27 shows deformation of anchor points of the longest stay cable of the Elbe River Bridge (Figure 2.10). It is evident that the new position of the cable is influenced mainly by a horizontal deformation of the tower and vertical deflection of the deck. Due to these deformations the length of the cable, and consequently the stress in the cable, have changed.

To understand the function of the cable the following study was done. Figure 9.28(a) depicts cables of length $l_0/\cos\beta$ where l_0 is 60, 120 and 180 m. The cable was assembled of 0.6″ strands of

Figure 9.26 Modelling of the cable-stayed structure: (a) deck modelled by 3D bars and (b) deck modelled by shell elements

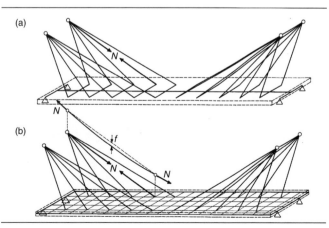

Figure 9.27 Static function of the stay cable

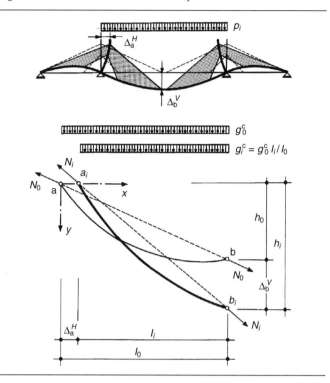

Figure 9.28 Normal stresses in stay cables

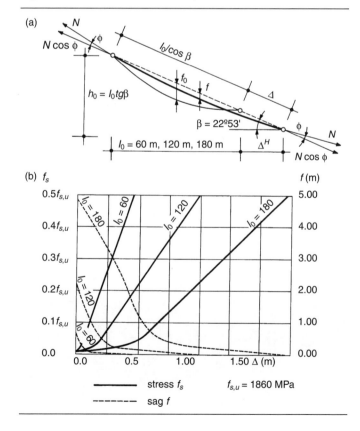

modulus of elasticity $E_s = 190$ GPa. The cables were stressed by an initial stress:

$$f_{s,i} = 0.005 f_{s,u} = 0.005 \times 1860 = 93.0 \text{ MPa}.$$

The cables were then loaded by deflection Δ that was increased in stepped increments of 0.05 m. The results of the analyses are presented in Figure 9.28(b) where stresses in the cables f_s and corresponding sags f are plotted.

It is evident that the studied cables behave almost linearly (the change of stress is linearly proportional to the change of the deflection) for the stresses $f_s > 0.1 f_{s,u}$. This means that for these stresses the stays can be modelled as pin-connected bars (Figure 9.30). The common structures can be solved by linear programs. However, it is necessary to check that the stresses in the cables lie within the range that guarantees their linear behaviour. The same is true for structures in which the stay cables are situated under the deck.

For longer stay cables, the initial stresses have to be higher. The non-linear behaviour is usually taken into account by using the so-called Ernst modulus E_i (Figure 9.29) (Ernst, 1965).

$$E_i = \frac{E_s}{1 + \frac{(\gamma l)^2 E_s}{12 f_s^3}}$$

where E_s is the modulus of elasticity of steel, f_s is the stress in the cable, γ is the density of the cable and l is the horizontal span of the cable.

The forces in the stay cables vary according to the position and intensity of the load. Figure 9.30 shows the two positions of the live load which cause maximum stresses in the longest back stay cable of the Elbe River Bridge. By linear analysis we can observe that, due to the load situated in the main span, the cable is stressed by tension due to the load situated in side spans by compression.

It is evident that the initial cable tension has to be designed in such a way that the maximum tension stresses are less than permissible stresses and minimum tension has to be greater than the tension that guarantees the linear behaviour. The design of the initial tension is also influenced by the redistribution of stresses due to the creep and shrinkage of concrete. This will be discussed further in Section 9.3.2.

Figure 9.30 Stresses in stay cables: (a) dead load, (b) max tension, (c) min tension and (d) allowable range of stresses

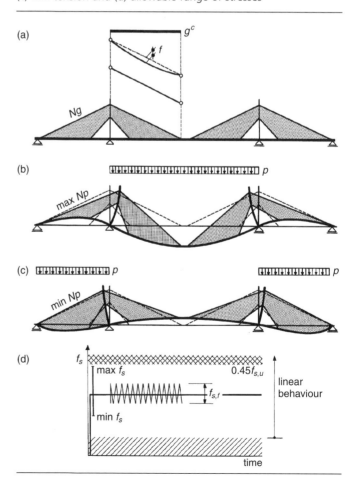

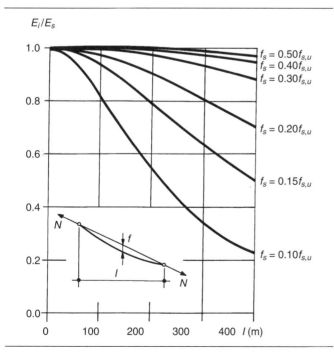

Figure 9.29 Ernst modulus

Figure 9.31 Bending moments and shear forces in stay cables

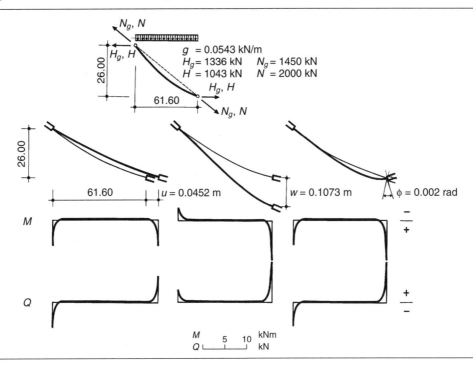

Stay cables are usually fixed to the pylons and deck. Due to their deformations, significant bending and shear stresses originate at the stay anchors. Figure 9.31 shows the deformations and corresponding shear forces and bending moments that originate due to the design live load in the longest stay cable of the Elbe Bridge (Figures 2.10, 4.11 and 4.12). Since corresponding stresses significantly influence the design of stay cables, it is necessary to develop details that reduce them.

Figure 9.32(a) shows a course of the bending moment and bending stresses that originates in the vicinity of their anchors in the stays that were loaded by rotation of the deck $\phi = 0.002$ radians (Figure 9.31). The figure shows the bending moments and stresses for (1) the stay cables of the constant cross-section and (2) the stay cable that is strengthened at anchors (Figure 4.12). It is evident that local strengthening causes the bending moment to increase. The resultant bending stresses are smaller, however.

Figure 9.32(b) shows courses of the bending moment in the stay of the constant section (1) that is loaded by the same rotation. In this case, the stay is supported by a spring at a distance of 1.7 m from the anchor. The spring has a different stiffness that varies from 0 to 500 MPa. It is evident that the bending moment can be significantly reduced and the peak of bending moment can be transferred from the anchor to the spring.

This is the reason why modern stay cables have the arrangement shown in Figure 9.16. The cables are usually strengthened at the anchor area and are guided by strong pipe which ends with a neoprene ring. The neoprene ring, together with the pipe, creates a flexible support (a spring) which reduces the local bending moments.

As already discussed, the bending moments also originate at the anchors that are pin connected (Figure 9.15).

From the above it is evident that the maximum stresses that originate in the cable are substantially influenced by the bending of the cable. The designer should also check the fatigue stresses that can originate in the stays (Figure 9.30(d)). Although the national standards do not usually specify the fatigue load, the designer should use an engineering judgement and consider a reasonable value of fatigue stresses when designing flexible structures.

9.3.2 Redistribution of stresses due to creep and shrinkage

The importance of the accurate determination of the stay force was discussed in Chapter 6. The redistribution of the stresses was demonstrated on the example of a simple beam that, after 14 days of curing, was suspended at mid-span on a vertical cable (Figure 6.8). It was shown that the initial force in the

Figure 9.32 Bending of the cable: (a) influence of local strengthening and (b) influence of support by a spring

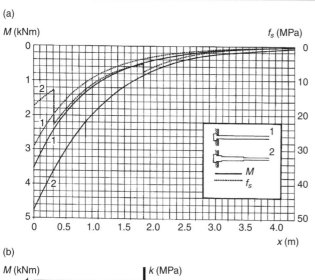

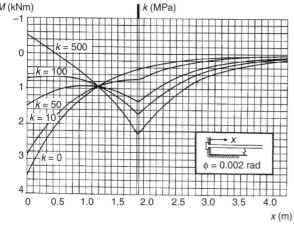

stay cable has to correspond to the reaction of a two-span continuous beam and there is no redistribution of forces in time. However, the cable was vertical and therefore the beam was not compressed.

Since the stay cables are inclined, they stress the beam by normal forces. The normal stresses create horizontal deformations that, due to the creep of concrete, increase in time. Shrinkage of concrete also causes horizontal deformations of the deck. The stay anchor horizontally moves and its initial force is reduced. The change of the stay force causes a redistribution of stresses in the structure.

To quantify this phenomenon, extensive studies were completed; only a few results are presented here. Figure 9.33(a) depicts the same beam that was, after 14 days of curing, suspended at mid-span on a very stiff inclined stay cable ($E_s A_s = \infty$). Before suspending, a force

$N = R/\cos \beta$

was created at the cable. R is a reaction at the intermediate support of a continuous beam of two spans of length 2×6 m.

A time-dependent analysis was performed using the CEB-FIP (MC 90) creep function. In time, the bending moments are redistributed from the anchor point to the span; at time $t_\infty = 100$ years the moment at the anchor point is approximately 60% of the initial moment.

Another analysis was also performed on the assumption that there is no shrinkage in the beam. In this case, the redistribution is very small (Figure 9.33(b)).

A time-dependent analysis was also completed for the structure in which the initial force was increased by 10% (Figure 9.33(c)). The value of the redistribution of bending moments was similar to the redistribution of moments shown in Figure 9.33(a), but the final value was closer to the bending moment that corresponds to the force $N = R/\cos \beta$.

Figure 9.33(d) presents the results of the analysis in which the force in the stay cable was adjusted after one year. The redistribution of moments was significantly reduced; at time $t_\infty = 100$ years the moment at the anchor point is approximately 80% of the initial moment.

It is evident that the redistribution of moments can reach significant values. However, note that in the structures with a slender deck which is suspended on multiple stay cables, the bending moments due to the dead load are (compared to the bending moment due to the live load and temperature changes) very small.

To quantify the influence of the redistribution of stresses in actual cable-supported structures, a cable-stayed structure as depicted in Figure 9.34 was studied. The goal of the study was to determine optimum forces in the stay cables of progressively erected cable-stayed structures. Analyses were completed for a non-composite and a composite deck of a different stiffness. Only the results of the analysis carried out for a slender deck are presented here.

A parametric study was completed for a symmetrical cable-stayed structure with radial arrangement of the stay cables. To eliminate the influence of the tower's stiffness, the towers were modelled as movable bearings that can rotate and move in the horizontal direction of the bridge. Since the stay cables have a symmetrical arrangement, the side spans are stiffened by additional supports.

Figure 9.33 Redistribution of bending moments in a two-span beam

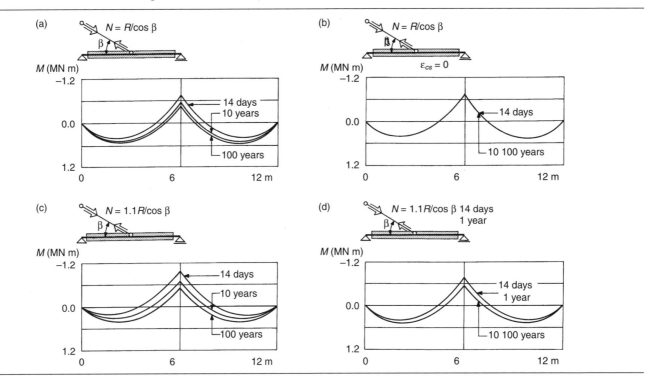

Figure 9.34 Redistribution of bending moments in a cable-stayed structure

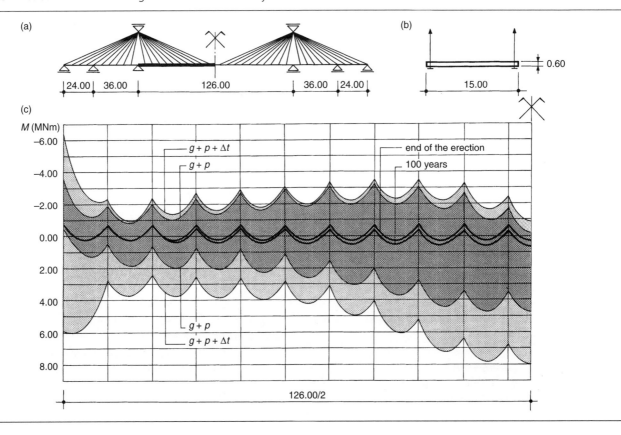

The structure was analysed for the effects of the dead and live loads given by the Czech standards and for temperature changes $\Delta t = \pm 20°\text{C}$. The area of the cables was determined from the conditions of max $f_s \leqslant 0.45 f_{s,u}$.

The deck was progressively cast in 6 m long sections. It was assumed that the traveller has a similar arrangement and weight as the traveller used in the construction of the Diepoldsau Bridge (Figure 9.19). The time-dependent analysis was done for a progressive erection in which one segment was completed (moving of the traveller, placing of the reinforcement, casting of the segment and post-tensioning of the stay cable) within seven days.

The forces in the stay cables were determined from several iterations. In the analysis, the structure was several times progressively demounted and erected in such a way that in the final stage shown in Figure 9.34 (after the erection of the symmetric cantilevers) the forces in all stay cables were $N_i \cong R_i / \cos \beta_i$. It is evident that after 100 years a significant redistribution of the bending moments has occurred. To quantify the value of the redistribution, bending moment envelopes determined for live load and temperature changes of $\Delta t = \pm 20°\text{C}$ are also shown. It is evident that the dead load moments are (compared to moments due to the live load) very small. It is therefore evident that the value of the redistribution is not (from the engineering point of view) critical and can be accepted. The redistribution can also be notably reduced if the forces are adjusted after one year.

REFERENCES

AASHTO (1997) Guide specification for design of pedestrian bridges. AASHTO.

Ernst HJ (1965) Der E-Modul von Seilen unter Berücksichtigung des Durchhanges. *Bauingenieur*, No. 2.

Gimsing NJ (1998) *Cable Supported Bridges: Concept & Design*. John Wiley & Sons, Chichester.

Mathivat J (1983) *The Cantilever Construction of Prestressed Concrete Bridges*. John Wiley & Sons, New York.

Menn C (1990) *Prestressed Concrete Bridges*. Birkhäuser Verlag, Basel.

Podolny W Jr and Scalzi JB (1976) *Construction and Design of Cable Stayed Bridges*. John Wiley & Sons, New York.

Podolny W and Muller J (1982) *Construction and Design of Prestressed Concrete Bridges*. John Wiley & Sons, New York.

Strasky J, Navratil J and Susky S (2001) Applications of time-dependent analysis in the design of hybrid bridge structures. *PCI Journal* **46(4)**: 56–74.

Walther R, Houriet B, Walmar I and Moïa P (1998) *Cable Stayed Bridges*. Thomas Telford Publishing, London.

Stress Ribbon and Cable-supported Pedestrian Bridges
ISBN 978-0-7277-4146-2

ICE Publishing: All rights reserved
doi: 10.1680/srcspb.41462.171

Chapter 10
Curved structures

Several curved cable-supported structures have been also built successfully. They utilise a space arrangement of the cables and prestressing tendons to minimise the bending and shear stresses in the deck.

The curved deck can be suspended on both its edges or on an outer edge or an inner edge. The arrangement depends on the local conditions, radius of the curvature and on the required span length of the crossing.

10.1. Suspension of the deck on both edges

If the deck can be suspended on both edges, it is possible to apply an arrangement that was developed in the design of the Ruck a Chucky Bridge by TY Lin International (Lin and Burns, 1981). Although this bridge has not been built, its design clearly shows how all internal forces can be balanced by the arrangement of stay cables.

Figure 10.1 Ruck a Chucky Bridge, California (courtesy of TY Lin International)

The bridge with span 396.24 m crosses the reservoir in the plan curvature of 628.00 m (Figures 10.1 and 10.2). The deck is suspended on the stay cables arranged in hyperbolic paraboloid formation to create an array of tensile forces, which produce pure axial compression in the curved deck. The vertical-force components of the cables balance the weight of the deck (Figure 10.3(a)). The resultant forces of the horizontal components act in the direction of the curved axis and are designed to reduce the horizontal bending moments at critical points to zero (Figure 10.3(b)). The design demonstrates how a pure engineering approach can create a structure of great beauty and elegance.

While the Ruck a Chucky Bridge was suspended on the cables anchored at the hills, another bridge proposed by TY Lin crosses the river (Figure 10.4). The deck is suspended on two towers, each formed by two single columns with backstays anchored at the banks.

If the radius curvature is sufficiently large, the deck is usually suspended on V-shaped towers as shown in Figures 9.8 and 9.9. However, the forces in the stays have to balance both vertical and transverse effects of the dead load.

Figure 10.2 Ruck a Chucky Bridge, California: plan (courtesy of TY Lin International)

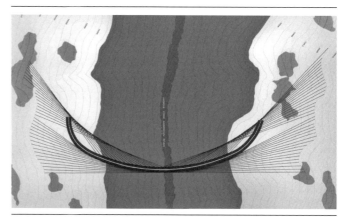

Figure 10.3 Ruck a Chucky Bridge, California: (a) balancing vertical forces and (b) balancing transverse forces

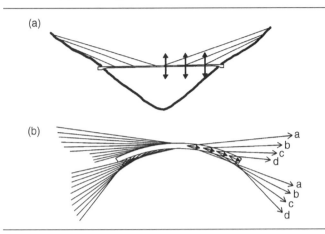

Figure 10.4 Model of the curved cable-stayed bridge (courtesy of TY Lin International)

Figure 10.5 Suspension of the curved deck: (a) outer edge and (b) inner edge

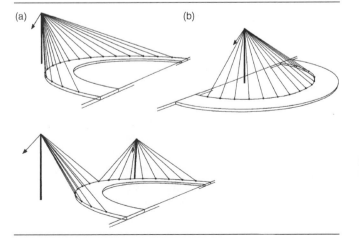

Figure 10.6 Rosewood Golf Club Bridge, Japan (courtesy of Shimizu Corporation)

10.2. Suspension of the deck on an outer edge

Figure 10.5 shows the curved deck that is suspended only on the (a) outer or (b) inner edge. From the first view, the suspension of the deck on the outer edge gives a feeling of safety (Figure 10.6). Although the vertical component of the cable forces can reduce the torsional moment, the horizontal components of the cable forces create significant transverse moments in the deck. It is difficult to find an arrangement of the cables that balances the effects of dead load. This solution is suitable for suspension of the two curved ramps that merge into one deck, as shown in Figure 10.7 (see also Section 11.3.8).

Figure 10.7 Glorias Catalanas Bridge, Barcelona, Spain (courtesy of Carlos Fernandez Casado, SL, Madrid)

Figure 10.8 Kelheim Bridge, Germany (courtesy of Schlaich, Bergermann and Partners)

Figure 10.10 Malecon Bridge, Madrid, Spain (courtesy of Carlos Fernandez Casado, SL, Madrid)

10.3. Suspension of the deck on an inner edge

On the other hand, it is relatively simple to find an arrangement of the cables anchored in the inner deck edge that balances the effects of the dead load. This arrangement was developed by Schlaich from Stuttgart for the Kelheim Bridge (Germany) which was completed in 1987 (Figures 10.8 and 10.9; Section 11.2.8) (Schlaich and Seidel, 1998). Another excellent example of this method is Malecon Bridge in Madrid which was designed by Carlos Fernadez Casado, LS (Figures 10.10 and 10.11) (Troyano and Mantreola, 1995).

Development of the suspended system is evident from Figures 10.12–10.15. Figure 10.12(a) shows a circular deck that is supported by single columns situated in the bridge axis and fixed to abutments. Due to the self-weight, the deck is stressed by bending moments M_y, M_x and M_z. Their courses are depicted in Figure 10.13.

Figure 10.12(b) shows the same deck in the bridge axis that is suspended on the radial cables anchored at the single mast.

Figure 10.11 Malecon Bridge, Madrid, Spain: inclined stay cables (courtesy of Carlos Fernandez Casado, SL, Madrid)

Figure 10.9 Kelheim Bridge, Germany: inclined hangers (courtesy of Schlaich, Bergermann and Partners)

Figure 10.12 Curved cable-stayed structure: (a) continuous beam, (b) cable-stayed structure, (c) effects of the vertical components of the stays and (d) effects of the horizontal components of the stay

Figure 10.13 Internal forces in the curved deck

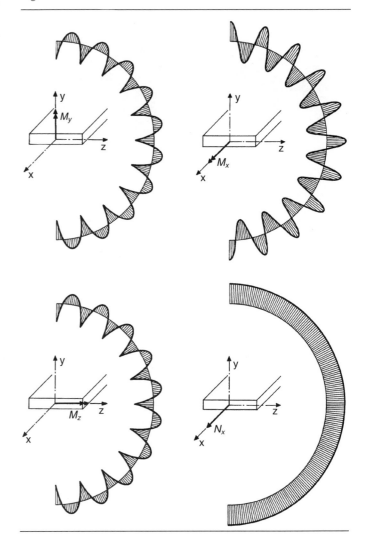

The cables are stressed by the force

$$N = G/\cos\beta$$

where G corresponds to the vertical reaction that originates at single supports of the continuous structure shown in Figure 10.12(a). It is evident that the vertical component of the cable force substitutes single supports (Figure 10.12(c)) and the horizontal component creates radial forces that stress the deck in the horizontal direction (Figure 10.12(d)). Since the deck is fixed at the abutments, it is stressed by uniform compression stresses.

The deck is therefore stressed by the same bending moments M_y, M_x and M_z as the continuous beam and by an additional normal force N_x (Figure 10.13). In the case that the distance between the stay cables is small (3–6 m) the bending and torsional stresses are also very small. This means that the stay cables balance the effects of the dead load and create compression in the deck.

Unfortunately, the stay cables anchored in the bridge axis prevent the use of one-half of the bridge deck. It is therefore necessary to anchor the stay cables at the inner edge and determine the solution which creates the same state of stresses. There are two possibilities as follows.

1. Anchoring the stay cables in a stiff member protruding above the deck. The anchor point of each cable is situated on the line in the direction of the centre of gravity of the deck slab (Figure 10.14(a)).
2. Anchoring the stay cable at the edge and add prestressing tendons into the deck slab. The slender deck has to be supplemented by an additional member situated at sufficient distance from the deck, or a deck of sufficient depth must be created. In this case, radial forces from the cable create a moment to the centre of gravity of the composite section that balances the forces.

Figure 10.14 Curved cable-stayed structure: (a) suspension above deck, (b) suspension at the deck, (c) effects of the stays and (d) effects of the post-tensioning

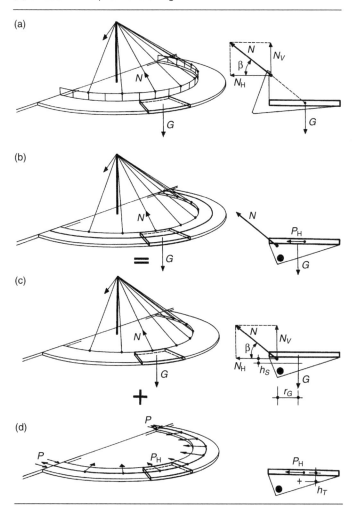

Figure 10.15 Curved suspension structure: (a), (b) suspension above deck and (c), (d) suspension at the deck

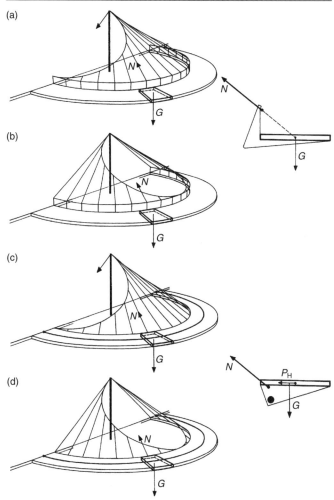

In the case depicted by Figure 10.14(c), a moment created by a couple of vertical forces

$$M_V = Gr_G = N_V r_G = (N \sin \beta) r_G$$

is balanced by a moment that is created by a horizontal component of the stay force and by radial forces due to post-tensioning, i.e.

$$M_H = N_H h_S + P_H h_T = N \cos \beta h_S + P_H h_T.$$

It is evident that a similar arrangement can be developed for suspension structures in which a suspension cable supports inclined hangers (Figure 10.9). Figure 10.15 depicts two possibilities of the arrangement of the suspension cables anchored at the stiff members protruding above the deck or at the deck supplemented by the prestressing tendons and an additional member.

It is obvious that the above-mentioned basic arrangements can be combined as for the design of the pedestrian bridge that was built inside a museum in Munich, Germany (Section 11.2).

By the arrangement of the cables and prestressing of the deck, it is possible to create an optimum state of stresses in the deck that is primarily stressed by uniform compression stresses. However, depending on the stiffness of individual structural members, plan curvature and span length, the deck can be stressed by significant bending and torsional stresses caused by a live load and wind. The structure then requires not only sufficient moment arm between the deck slab and additional member, but also a torsionally stiff section (Figure 2.24(j)).

10.4. Curved stress ribbon bridge

The idea of balancing the effects of the self-weight was also used in a study of a curved stress ribbon structure. The approach

Figure 10.16 Curved stress ribbon structure: (a) geometry, (b) effects of the bearing tendon, (c) effects of the vertical component of the tendon and (d) effects of the horizontal component of the tendon

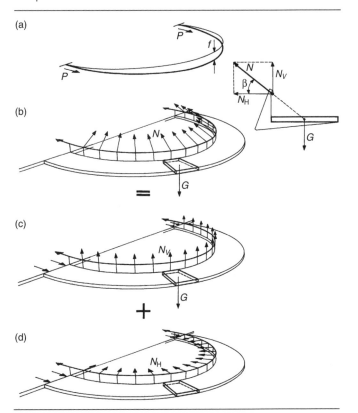

Figure 10.17 Curved stress ribbon structure: cross-section

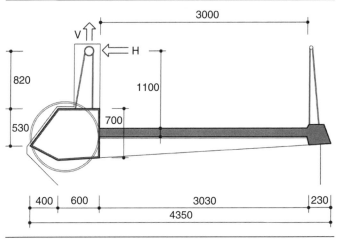

Figure 10.18 Curved stress ribbon structure and flat arch structure

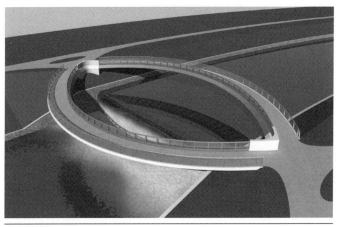

Figure 10.19 Curved stress ribbon structure and flat arch structure

presented in Figures 10.14(a) and 10.15(a) was also used in the design of our stress ribbon structure (Figure 10.16). The structure is formed by a slender deck that is supplemented by steel L-frames supporting the slab. The tops of their vertical portions are connected by steel pipe in which a prestressing cable is lead. Since the pipe is curved, the radial forces load the structure (Figure 10.16(b)). The vertical components of these radial forces balance the dead load (Figure 10.16(c)) and the horizontal components balance the dead load's torsional moment and load the structure by horizontal radial forces (Figure 10.16(d)). Since the stress ribbon is fixed to the abutments, these forces create uniform compression in the deck.

Our design has been developed from the described approach. However, to resist the effects of the live load, additional cables in a complex arrangement were required. These cables were therefore substituted by a torsionally stiff member of the pentagon cross-section. The resulting arrangement of the structure is evident from Figures 10.17–10.19. The pedestrian bridge of span 45 m is in a plan curvature with a radius of pathway axis of 32.212 m. The maximum longitudinal slope at the abutments is 7%. Both the concrete slab and the steel girder are fixed to the anchor blocks. The external cables are situated in the handrail

Figure 10.20 Curved stress ribbon structure and flat arch structure, elevation: (a) stress ribbon, (b) flat arch and (c) stress ribbon and flat arch

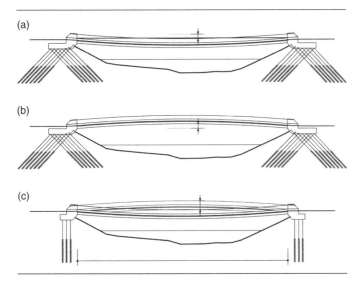

Figure 10.21 Curved stress ribbon structure and flat arch structure, static model: (a) cross-section, (b) plan and (c) elevation

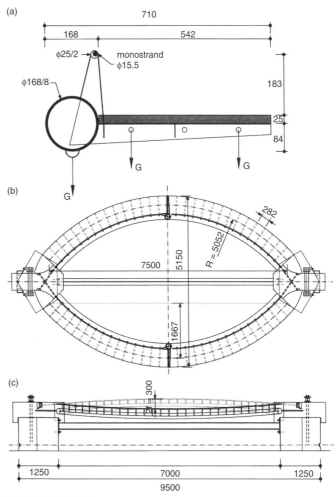

pipe and are anchored at end concrete walls that are fixed to the anchor blocks. Horizontal forces are resisted by battered micropiles (Figure 10.20(a)).

The detailed static and dynamic analyses have proven that the structure is able to resist all design loads. The first bending frequency is $f_1 = 1.386\,\mathrm{Hz}$. The forced vibration has caused a maximum acceleration $a_{max} = 0.578\,\mathrm{m/s^2}$ that is close to the allowable acceleration $a_{lim} = 0.589\,\mathrm{m/s^2}$. It is therefore proposed to use two dampers to ensure that pedestrians do not experience an unpleasant feeling when standing or walking on the bridge.

The function of the studied structure was verified on a static model built to a scale of 1:6 (Figures 10.21–10.23). To reduce the longitudinal horizontal force, the curved stress ribbon was tested together with a structure formed by a curved flat arch of a similar arrangement.

It is evident that the actual structure can be built similarly. The horizontal force from the stress ribbon can be balanced by the horizontal force originating at the flat arch (Figure 10.20(b)). In this way, a very economic structural system can be created. The analyses have proven that the dynamic behaviour of the curved flat arch is similar to the curved stress ribbon structure; the first bending frequency is $f_1 = 1437\,\mathrm{Hz}$.

To simplify construction of the model, the steel box of the stress ribbon and flat arch was substituted by a steel pipe that was fixed to the end anchor blocks. These anchor blocks were supported by concrete pedestals that were mutually connected by steel members and post-tensioned bars. The steel pipes were produced with steel L members. Their horizontal part supported the

Figure 10.22 Curved stress ribbon structure and flat arch structure, static model: load situated on one-half of the stress ribbon length

Figure 10.23 Curved stress ribbon structure and flat arch structure, static model: ultimate load

concrete slab; their vertical parts supported steel pipes in which the monostrands were placed. The slab deck and monostrands were anchored in a common anchor block.

To guarantee a model similarity, the curved steel pipes and transverse L members were loaded by concrete block and steel rods representing the self-weight of the structure. This load was suspended on the steel structure. The live load was represented by additional concrete block and cylinders placed on the stress ribbon deck.

Table 10.1 Static model of curved stress ribbon and flat-arch structure: deformation of the stress ribbon at mid-span

	Load 1: mm	Load 2: mm	Load 3: mm
Measurement	6.5	6.7	13.1
Calculation	6.8	6.8	12.9

The construction of the model was carefully monitored. Both structures were tested for three positions of the live load and for the ultimate load. The live load was situated on the left side (Figure 10.22) and then the right side of the entire length of the deck. The measured deformations and strains were in good agreement with the results of the static analysis (Table 10.1). Finally, the model was loaded by an ultimate torsional load that was situated along the whole length of both structures (Figure 10.23). Both structures demonstrated a sufficient margin of safety.

REFERENCES

Lin TY and Burns NH (1981) *Design of Prestressed Concrete Structures*. John Wiley & Sons, New York.

Schlaich J and Seidel J (1998) Die Fußgängerbrücke in Kelheim. *Bauingeneur* 63.

Troyano LF and Mantreola JA (1995) *Spatial cable-stayed bridges. Spatial structures: heritage, present and future.* Proceedings of IASS International Symposium, Milan, Italy.

Stress Ribbon and Cable-supported Pedestrian Bridges
ISBN 978-0-7277-4146-2

ICE Publishing: All rights reserved
doi: 10.1680/srcspb.41462.179

Chapter 11
Examples

The possibilities of stress ribbon and cable-supported structures are demonstrated in 50 examples of pedestrian bridges. The aim in this chapter is not to present every type of built structure or those of maximum spans. Instead, the aim is to illustrate their development and highlight those examples which are interesting from an architectural, structural solution or construction process viewpoint.

11.1. Stress ribbon structures
11.1.1 Bircherweid Bridge, Switzerland

The Bircherweid Bridge was the first stress ribbon bridge built for the public. It was built in 1965 across the N3 motorway near Pfäfikon, Switzerland (Figure 11.1) (Walther, 1969).

The bridge is formed from a cast-in-place stress ribbon of span 48.00 m (Figure 11.2). The prestressed band is formed by a slender deck of thickness 0.18 m that is frame connected to the abutments. Close to the abutments, the slab is gradually thickened to 0.36 m and slab edges are supported by saddles.

Due to a large longitudinal slope (≈15%) the sag was chosen to be as small as possible. Its average value is 0.40 m. The deck is

Figure 11.1 Bircherweid Bridge, Switzerland (courtesy of Walther Mory Maier Bauingenieure AG)

post-tensioned by six VSL tendons of permissible force 7.02 MN. The horizontal force is transferred to the soil by rock anchors.

The first calculated vertical frequencies are 1.59 Hz and 2.90 Hz and the first torsional frequencies are 5.98 Hz and 10.1 Hz. However, measurements made of unfinished structures proved that the first bending frequency is higher than the calculated frequency; it is close to the walking frequency of 2 Hz. Nevertheless, after the erection of the railings the dynamic response was significantly improved and the oscillation is hardly noticeable during normal pedestrian traffic.

The bridge was designed by Walther Mory Maier Bauingenieure AG, Basel, Switzerland.

11.1.2 Bridge Lignon–Löex, Geneva, Switzerland

The Lignon–Löex Bridge, which transfers pipeline and pedestrian traffic across the river Rhone, is situated in the suburb of Geneva (Figure 11.3) (Wolfensberger, 1974). It is formed by a stress ribbon of one span of length 136.00 m and the sag

Figure 11.2 Bircherweid Bridge: (a) cross-section, (b) partial elevation at abutment, (c) elevation

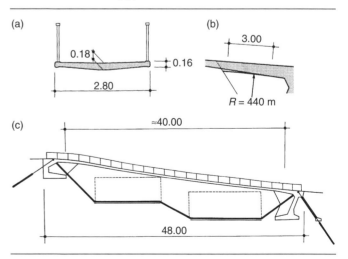

179

Figure 11.3 Lignon–Löex Bridge, Geneva, Switzerland

at mid-span is 5.60 m. The corresponding slope of the structure at the abutments is 17.5% and 15.5% (Figure 11.4). The bridge was built in 1971.

The stress ribbon is assembled from precast segments and short cast-in-place haunches situated at the abutments. The horizontal force is transferred to the soil by rock anchors. The segments are suspended on BBRV prestressing tendons formed by parallel wires encased in polyethylene (PE) tubes that are situated at the segment's troughs. After the erection, the troughs and the joint between segments were cast. When the concrete reached sufficient strength, the BBRV tendons were tensioned to a higher stress. In this way, the structure was prestressed and the required stiffness was achieved.

The dynamic behaviour of the structure was verified in the wind tunnel. The results proved that it is necessary to have at least a free space of depth 1.5 times the pipe diameter between the prestressed band and pipe. In this arrangement, the structure is stable for wind speeds of up to 280 km/hour.

Static and dynamic behaviour was also verified by tests. The first vertical frequency is 0.88 Hz and the first swing frequency is 1.70 Hz.

The bridge was designed by Büro Weisz, Geneva and Büro Wenaweser & Wolfensberger, Zürich.

11.1.3 Freiburg Bridge, Germany

The bridge, which was built in 1970, diagonally crosses a busy junction and connects the city parks to the city centre (Figures 3.5 and 11.5) (Batsch and Nehse, 1972). The bridge is formed by a stress ribbon of three spans of lengths of 33.0, 39.50 and 42.00 m; the sag of the central span varies from 0.26 to 48 m (Figure 11.6).

The prestressed band is formed by a cast-in-place slab of thickness 0.25 m that is fixed at heavy abutments. The horizontal force from the stress ribbon is resisted by passive pressure of the soil. Above the intermediate supports, the stress ribbon is supported by concrete saddles (Figure 11.6(b)). At the bottom of the piers concrete hinges, which allow rotation in the

Figure 11.4 Lignon–Löex Bridge, Geneva, Switzerland: (a) cross-section, (b) partial elevation at abutment, (c) elevation

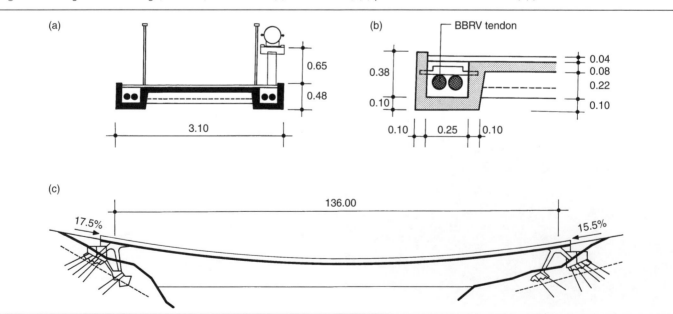

Figure 11.5 Freiburg Bridge, Freiburg, Germany

longitudinal direction of the bridge, were designed. The deck is post-tensioned by Dywidag bars.

The function of the bridge was verified by dynamic tests. Although the first vertical frequencies are in the range 1.2–3.3 Hz, the oscillation of the bridge is hardly noticeable during normal pedestrian traffic.

The bridge was designed and built by Dyckerhoff & Widmann, München.

Figure 11.6 Freiburg Bridge, Germany: (a) cross-section at the abutment, (b) partial elevation at pier, (c) cross section at pier, (d) elevation

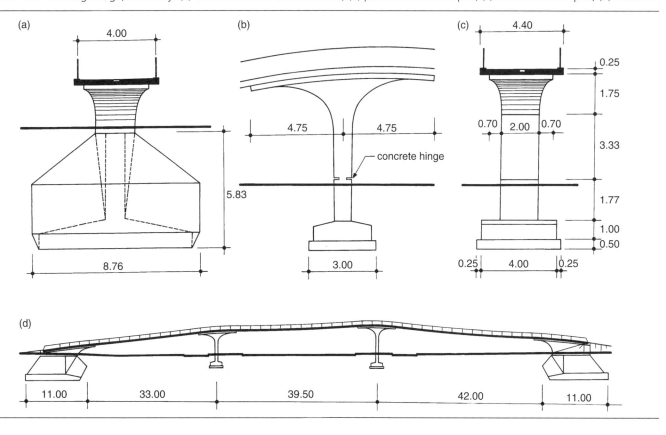

11.1.4 DS-L Bridges, Czech Republic

During the period 1978–1985 the author, as chief designer of the firm Dopravni stavby Olomouc, designed seven stress ribbon bridges of similar arrangement (Strasky and Pirner, 1986; Strasky, 1987a). The firm refer to these structures as 'DS-L Bridges'. The bridges have one, two or three spans; see Table 11.1. All these bridges were assembled from the same precast segments and have a similar structural arrangement (Figure 7.10) demonstrated by the two bridges built in Brno-Komin and Prague-Troja.

The decks of all bridges are assembled of two types of segments: waffle segments form the prevailing part of the deck and solid segments are designed at the abutments. The precast segments are 3.00 m long, 3.80 m wide and have a depth 0.30 m (Figures 7.10, 7.11 and 11.7). The section of the waffle segments is formed by edge girders and a deck slab. At the joint between the segments, the section is stiffened by low diaphragms.

During the erection the segments were suspended on bearing tendons situated at troughs. After the casting of the joints between the segments, the deck was post-tensioned by pre-stressing tendons situated in the deck slab (Figure 11.8). Bearing and prestressing tendons were formed of six 0.6 inch strands. The number of tendons depended on the span length and the sag. The wearing surface of the segments is formed of a 10 mm thick layer of epoxy concrete.

11.1.4.1 Bridge across the Svratka River in Brno-Komin

The bridge connects the residential area of the suburb Brno-Komin with the recreation area situated on the right bank of the Svratka River (Figure 11.9). The deck of the bridge, which is formed by a stress ribbon of length 78.00 m and sag 1.35 m, is assembled of 26 segments (Figure 11.10). The end solid segments were placed on neoprene pads situated on saddles that were formed by the front parts of the abutments (Figure 11.11). The horizontal force from the stress ribbon is resisted by rock anchors.

Table 11.1 DS-L Bridges, Czech Republic

	Bridge	No. spans	Span: m	Sag: m	Year constructed
1	Brno-Bystrc	1	63.00	1.20	1979
2	Kromeriz	1	63.00	1.20	1983
3	Radonice	1	63.00	1.20	1984
4	Brno-Komin	1	78.00	1.35	1985
5	Prerov	2	67.50, 28.50	1.43, 025	1983
6	Prague-Troja	3	85.50, 96.00, 67.50	1.34, 1.69, 0.84	1984
7	Nymburk	3	46.50, 102.00, 70.50	0.41, 1.98, 0.95	1985

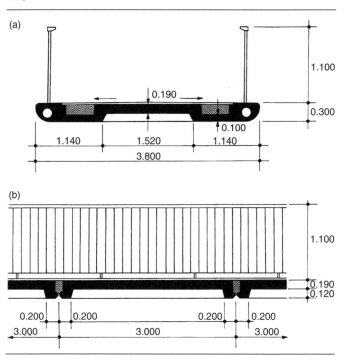

Figure 11.7 DS-L Bridges, Czech Republic – deck: (a) cross-section, (b) partial elevation

After the casting of the end abutment and post-tensioning of the first half of the rock anchors, the solid segments were placed on the neoprene pads situated on the front portion of the abutments. The bearing tendons were pulled across the river and then tensioned to the design stress. The segments were then erected by a mobile crane. The segments were suspended on bearing tendons and shifted along them into the design position (Figure 7.52). After all segments were erected, prestressing tendons were pulled through the deck. The reinforcing steel of the troughs was then placed and the joints and troughs were cast.

To eliminate possible damage to concrete caused by temperature changes and by unauthorised crossings by pedestrians, the post-tensioning was performed in two steps. When the concrete reached 30% of its strength, the tendons were post-tensioned to 30% of the designed stress. When the concrete reached 80% of its strength, the strands were post-tensioned to the designed stress. The second half of the rock anchors was then tensioned.

The static assumptions and quality of the workmanship were checked by a static and dynamic loading test (Section 7.10).

11.1.4.2 Bridge across the Vltava River in Prague-Troja

The bridge of total length 261.20 m crosses the Vltava River in the north suburb Prague-Troja. It connects the Prague Zoo and

Figure 11.8 DS-L Bridges, Czech Republic – bearing and prestressing tendons: (1) bridges 1–5 of Table 11.1, (2) bridges 6 and 7 of Table 11.1, (a) partial elevation, (b) partial cross-section

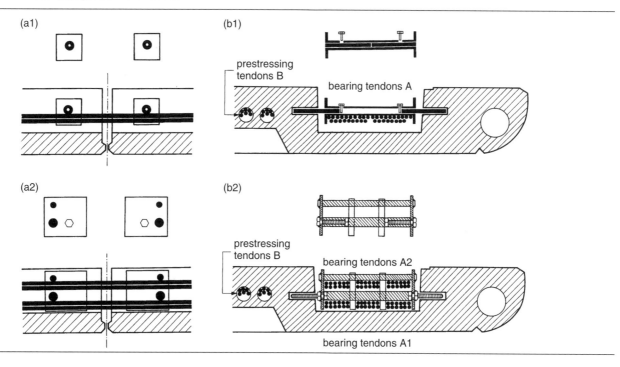

Troja Chateau with sports facilities situated on the Emperor Island with the Stromovka Park (Figures 11.12 and 11.13) (Strasky, 1987b).

The bridge has three spans of lengths of 85.50, 96.00 and 67.50 m; the sags at mid-spans are 1.34, 1.69 and 0.84 m. The stressed ribbon is formed by precast segments and by cast-in-place saddle (pier tables) frames connected to intermediate piers (Figures 7.31 and 7.32). Concrete hinges at the bottom of the piers, which allow rotation in the longitudinal direction of the bridge, were designed (Figure 11.14). The horizontal force from the stress ribbon is resisted by wall diaphragms and micropiles.

After the casting of the end abutment, the solid segments were placed on the neoprene pads situated on the front portion of the abutments (Figure 11.15). The first half of the bearing tendons were then pulled across the river and tensioned to the design stress. The tendons were supported by steel saddles situated on the piers (Figures 7.33 and 11.16).

Figure 11.9 Brno-Komin Bridge, Czech Republic

Figure 11.10 Brno-Komin Bridge, Czech Republic: (a) partial elevation and cross-section, (b) elevation

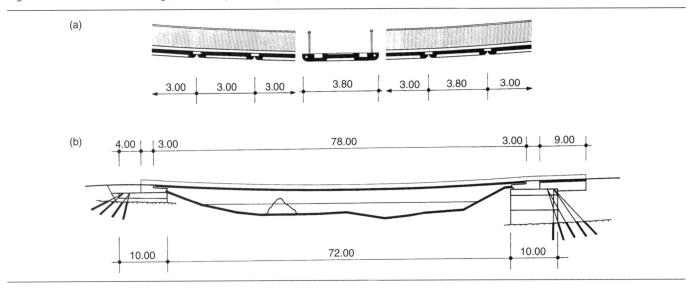

Figure 11.11 Brno-Komin Bridge, Czech Republic, abutment: (a) partial elevation, (b) cross-section

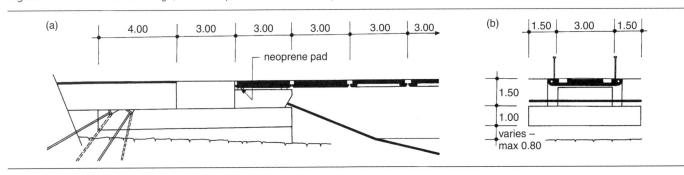

Figure 11.12 Prague-Troja Bridge, Czech Republic

Figure 11.13 Prague-Troja Bridge, Czech Republic: (a) elevation, (b) plan

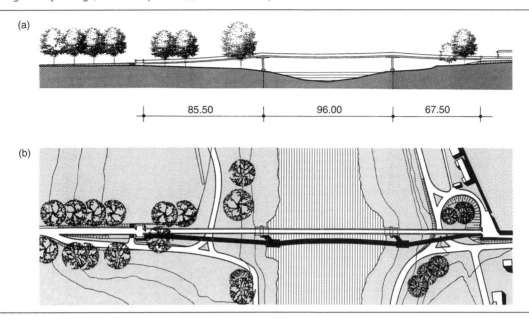

The segments were then erected by a mobile crane. The segments were suspended on bearing tendons and shifted along them into the design position. The segments of the side spans were erected first, followed by the segments of the main span (Figures 7.50 and 7.51).

After all segments were erected, the second half of the bearing tendons was pulled and tensioned to the design stress. In this way the structure reached the design shape. The steel tubes that form the ducts in the joints between the segments were then placed and prestressing tendons were pulled through the deck.

Figure 11.14 Prague-Troja Bridge, Czech Republic – pier table: (a) elevation, (b) cross-section, (c) plan

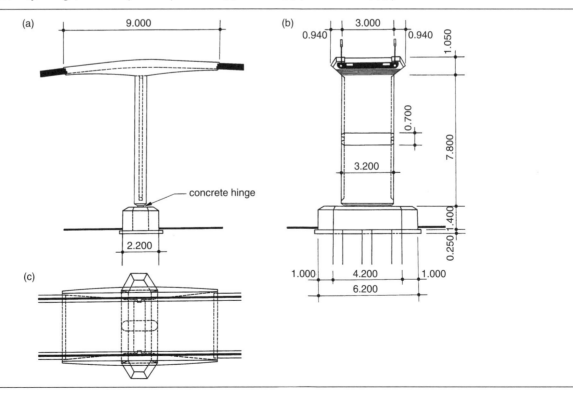

Figure 11.15 Prague-Troja Bridge, Czech Republic – abutment: (a) plan, (b) partial elevation, (c) longitudinal section, (d) cross-section

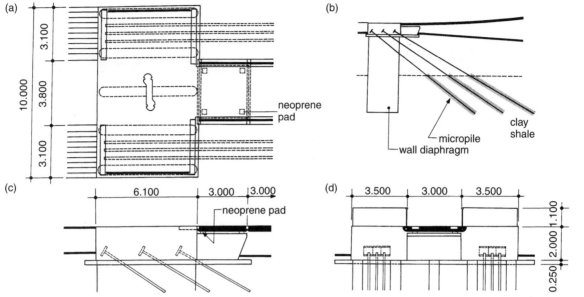

The reinforcing steel of the troughs and saddles was then placed and the joints, troughs and saddles were cast. The side spans were cast first, followed by the central span and saddles. The saddles were cast in formworks that were suspended on the already erected segments and were supported by the piers (Figure 7.34).

The static assumptions and quality of the workmanship were also checked by static and dynamic loading tests (see Section 7.10). When exceptional flooding occurred in Prague in 2001, the pedestrian bridge was totally flooded. Careful examination of the bridge after the flood confirmed that the structure was undamaged.

The DS-L bridges met with public approval; no problems with their static or dynamic performance have so far been reported. Dynamic tests have confirmed that it is not possible to damage the bridges by excessive vibration caused by people (a case of

Figure 11.16 Prague-Troja Bridge, Czech Republic – steel saddle: (a) elevation, (b) cross-section

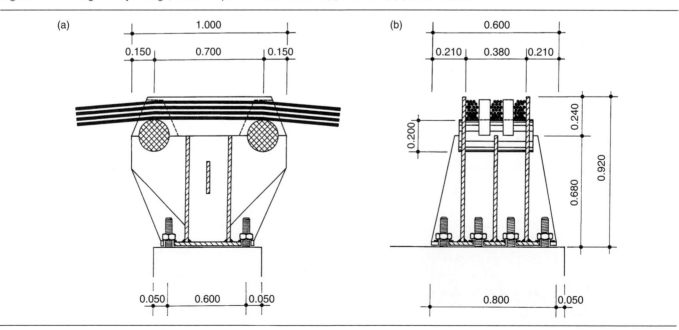

Figure 11.17 Prague-Troja Bridge, Czech Republic

Figure 11.18 Redding Bridge, California, USA: completed structure

vandalism) and that the speed of motion caused by pedestrians is within acceptable limits.

11.1.5 Sacramento River Trail Bridge, Redding, California, USA

The Sacramento River Trail and connecting bridge form part of the City of Redding's park system. The riverbanks have extensive rock outcropping that dramatically increase the beauty of the basin. To preserve this natural terrain and to mitigate adverse hydraulic conditions, it was important to avoid founding any piers in the river basin (Figures 1.10, 1.11 and 11.18) (Redfield and Strasky, 1991, 1992a and 1992b).

The bridge is formed by a stress ribbon of span 127.40 m (Figure 11.19). During bridge service, the sag at mid-span varies from 3.35 m (time 0 with maximum temperature and full live load) to 2.71 m (time infinity with minimum temperature). Apart from at a distance of 4.20 m at each end abutment where the deck is haunched to 0.914 m, the deck has a constant depth of 0.381 m. The stress ribbon is assembled of precast segments suspended on bearing tendons (Figures 1.12 and 7.12) and it

Figure 11.19 Redding Bridge, California, USA: (a) cross-section, (b) bearing and prestressing tendons, (c) elevation, (d) plan

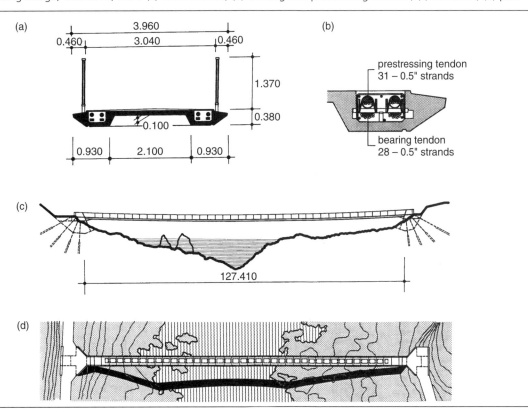

Figure 11.20 Redding Bridge, California, USA – abutment: (a) section A–A, (b) section B–B, (c) section C–C, (d) longitudinal section, (e) plan

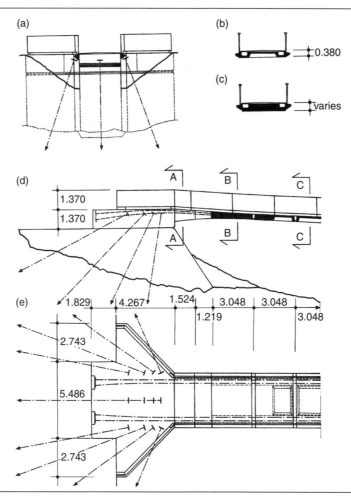

is post-tensioned by prestressing tendons. Both bearing and prestressing tendons are placed in troughs situated at the edges of segments. Horizontal force from the stress ribbon is resisted by rock anchors (Figures 7.40 and 11.20).

The bridge was designed as a geometrically non-linear structure (Figures 7.69 and 7.70). The haunches were designed as partially prestressed members in which tension forces at bottom fibres are resisted by reinforcing steel. Bridge vibration studies were carefully considered in the design for a wide range of pace frequencies, including jogging and the remote possibility of vandals attempting to physically excite the bridge. Because the bridge is an extremely shallow band with a long span, an aeroelastic study was deemed necessary to check the stability under dynamic wind loads.

The construction of the bridge was commenced by casting of the abutments and installation of the rock anchors. The bearing tendons were then pulled across the river and post-tensioned to the design stress. The segments were subsequently suspended on bearing tendons and shifted along them into the design position. All segments were erected within two days. The prestressing tendons were then placed directly above the bearing tendons and the joints and troughs were cast. By post-tensioning of prestressed tendons, the structure received the required stiffness.

Due to the first use of the stress ribbon in the USA, it was considered prudent to load test the bridge and verify the structural behaviour with the design assumptions. A successful test was conducted on the completed bridge with 24 vehicles spaced over the whole length of the structure (Figures 7.117 and 7.118).

The bridge was designed by Charles Redfield and Jiri Strasky (Consulting Engineers) and the contractor was Shasta Constructors Inc.

11.1.6 Umenoki-Todoro Park Bridge, Japan

One of the most beautiful stress ribbon bridges was built in 1989 across a ravine at the entrance to the famous Umenoki-Todoro

Figure 11.21 Umenoki-Todoro Bridge, Japan: completed structure (courtesy of Sumitomo Mitsui Construction Co., Ltd)

Figure 11.23 Yumetsuri Bridge, Japan: completed structure (courtesy of Sumitomo Mitsui Construction Co., Ltd)

Waterfall in Kumamoto Prefecture in Japan (Figure 11.21) (Arai and Yamomoto, 1994). The stress ribbon of length 105.00 m and sag 3.10 m is very narrow. The width between the railings is only 1.30 m. The stress ribbon is assembled of precast slab segments of width 2.00 m and depth of only 0.19 m (Figure 11.22). At the abutments, cast-in-place haunches were created. The horizontal force from the stress ribbon is resisted by rock anchors.

During construction, the segments were suspended on bearing tendons situated in the troughs designed close to the edges of the section. After the casting of the troughs, joints and haunches, the stress ribbon was post-tensioned by tendons situated in the deck slab. The bridge was built within four months without any effect on the environment.

Although the dynamic analysis determined the first bending frequencies in the range 1.5–2.3 Hz, nobody has reported any problems with the performance of the bridge.

The bridge was designed by Maeda Engineering Corporation. The contractor was Sumitomo Mitsui Construction Co., Ltd.

Figure 11.22 Umenoki-Todoro Bridge, Japan: (a) cross-section, (b) bearing and prestressing tendons, (c) elevation

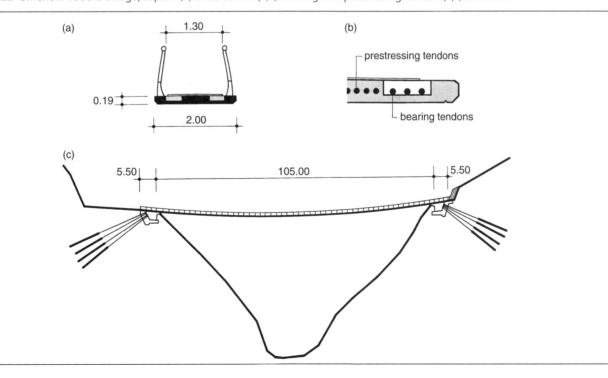

Figure 11.24 Yumetsuri Bridge, Japan: (a) cross-section, (b) bearing and prestressing tendons, (c) elevation

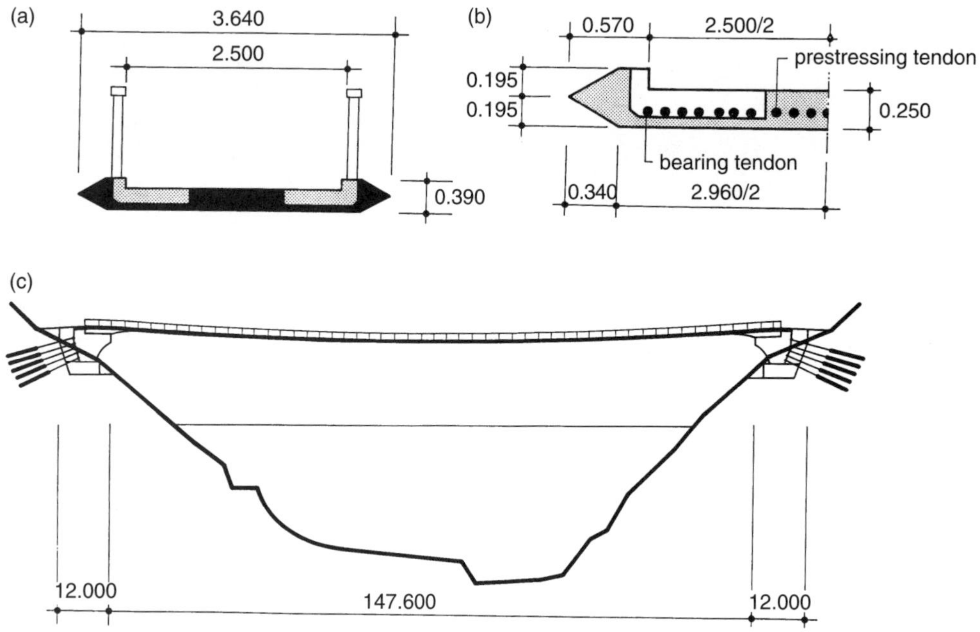

11.1.7 Yumetsuri Bridge, Japan

Yumetsuri Bridge was built in 1996 across the Hattabara Dam Lake in Hiroshima Prefecture in Japan (Figure 11.23) (Hata, 1998).

The stress ribbon of length 147.60 m and sag 3.50 m is assembled from precast segments and cast-in-place haunches situated at the abutments (Figure 11.24). The segments with triangular edges have a width of 3.64 m and depth of 0.39 m. The segments were suspended on bearing un-bonded tendons situated in the troughs designed close to the edges of the section. The stress ribbon was post-tensioned by tendons situated in the deck slab. The horizontal force from the stress ribbon is resisted by rock anchors.

The arrangement of the structure and structural details were developed from the adopted prestressing system and the process of construction (Figure 11.25). The abutments and bottom portion of the special abutment segments that are supported by sliding bearings were first cast. Bearing tendons were then erected (Figure 11.25(a)).

The precast segments were then gradually suspended on an erection frame and moved along the bearing tendons to the design position, where the segments were suspended on the bearing tendons.

After the erection of all segments, the trough, joints, haunches at the abutments and a top portion of the abutment segment were cast (Figure 11.25(b)). When the concrete reached sufficient strength, the structure was post-tensioned by prestressing tendons anchored at the abutment segments (Figure 11.25(c)).

The prestressing tendons were then coupled with the tendons pulled through abutments and the gaps between the abutment

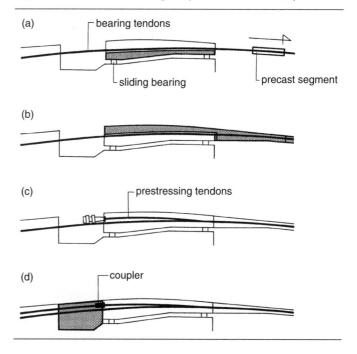

Figure 11.25 Yumetsuri Bridge, Japan: construction sequences

Figure 11.26 Tonbo No Hashi Bridge, Japan: completed structure (courtesy of Dai Nippon Construction)

segments and the abutments were reinforced and cast (Figure 11.25(d)). When the concrete of the gaps reached sufficient strength, the bearing tendons were post-tensioned and anchored at the abutments.

The bridge was designed by Shin Nippon Giken Co., Ltd. The contractor was Sumitomo Mitsui Construction Co., Ltd.

11.1.8 Tonbo No Hashi Bridge, Japan

Tonbo No Hashi Bridge was built in 1996 over Takanabe marshland in a suburb of Takanbe, Miyazaki Prefecture in Japan (Figure 11.26) (Katuyama *et al.*, 1998). The bridge of three spans of lengths 30.00, 80.00 and 30.00 m and maximum sag of 1.70 m is assembled of precast segments of width 2.00 m and depth 0.18 m (Figure 11.27). At the abutments and intermediate piers, cast-in-place haunches were created. The horizontal force from the stress ribbon is resisted by rock anchors.

The segments were suspended on bearing tendons situated in the troughs designed close to the edges of the section. The stress ribbon was post-tensioned by tendons situated in the deck slab. The piers have frame connections to piers and footings.

The bridge was designed by Nissetsu Consultant Co., Ltd. The contractor was Dai Nippon Construction.

11.1.9 Blue Valley Ranch Bridge, Colorado, USA

The Blue Valley Ranch Bridge was built in 2001 on private property across the Blue River in the high country of Colorado, USA. The bridge serves as access for the personnel and for general maintenance on the ranch (Figure 11.28) (Redfield and Strasky, 2002).

The bridge is formed by a stress ribbon of span 76.80 m and sag 1.02 m (Figure 11.29). The stress ribbon of depth 0.381 m is assembled of precast elements with a coffered soffit and by a composite deck slab. The elements are 2.44 m in length and their width is 5.38 m. At the abutments, cast-in-place haunches resist higher bending moments.

Figure 11.27 Tonbo No Hashi Bridge, Japan: (a) cross-section, (b) cross-section at the pier table, (c) elevation at the pier table, (d) elevation

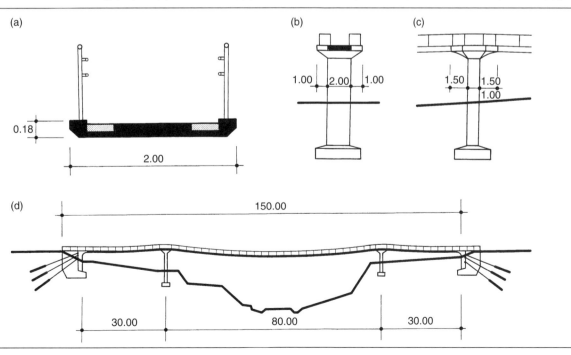

Figure 11.28 Blue Valley Ranch Bridge, Colorado, USA

The bridge was designed not only for pedestrian loading but also for loading by trucks. To check the stresses, the HS20 was considered. Detailed dynamic analyses were also carried out. The natural frequencies and modes were first determined. Since the first two bending frequencies were below 1 Hz, the structure was checked for excited vibration caused by pedestrians. The influence of the dynamic load caused by vehicles was determined by using the impact factor $I = 2$.

Construction of the bridge was commenced by casting the abutments and installation of the rock anchors. The first half of the rock anchors was stressed. The bearing tendons were then pulled across the river and post-tensioned to the design stress. The segments were subsequently suspended on bearing tendons and shifted along them into the design position. The prestressing tendons were then placed, and the composite deck slab and haunches were reinforced. The second half of the rock anchors was stressed.

The *in situ* casting of the deck then took place starting at the span centre and proceeding symmetrically to each abutment. When the concrete reached adequate strength, the final deck tendons were stressed providing the necessary longitudinal prestressing to stiffen the deck.

The bridge was designed by Huitt-Zollars (GNA Inc.) with the collaboration of Charles Redfield, Consulting Engineer and Jiri Strasky, Consulting Engineer; the contractor was Centric/Jones Constructors, Lakewood, Colorado.

11.1.10 Rogue River Bridge, Grants Pass, Oregon, USA

Rogue River Bridge, built in 2000 in Grants Pass, Oregon, connects a major park on one side of the Rogue River to the County Fairgrounds on the other side (Figures 1.9 and 11.30) (Rayor and Strasky, 2001).

Figure 11.29 Blue Valley Ranch Bridge, Colorado, USA: (a) cross-section, (b) bearing and prestressing tendons, (c) elevation

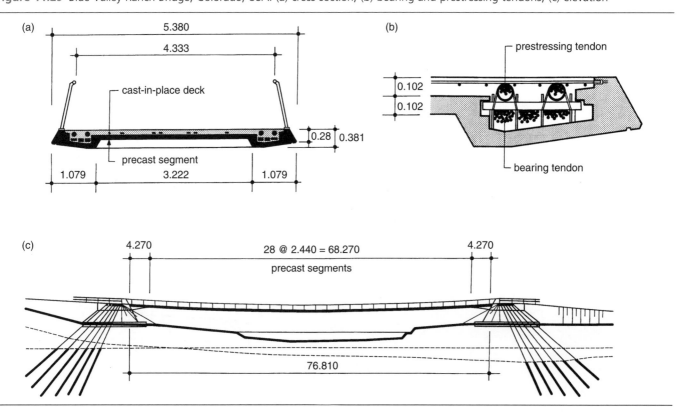

Figure 11.30 Grants Pass Bridge, Colorado, USA: completed structure

The bridge is formed by a stress ribbon of three spans of 73.15, 84.73 and 42.67 m for an overall length of 200.55 m (Figure 11.31). The corresponding sags at mid-spans are 1.10, 1.55 and 0.31 m. The bridge of width of 4.7 m provides a 4.3 m wide multi-use path (Figure 11.32). Observation areas located on widened deck segments at mid-span above the river and wetlands offer the users a location to stop and enjoy the river (Figure 11.33). The bridge also provides access for vital emergency vehicles and carries city water and sanitary sewer pipelines across the Rogue River. The horizontal force from the stress ribbon is resisted by inclined steel piles and tie backs.

At the abutments and above the intermediate supports the stress ribbon is supported by concrete saddles (Figures 7.28–7.30) that allow the pipelines to pass though them.

The bridge deck is formed from precast concrete segments with a coffered soffit that are composite with a cast-in-place deck slab (Figures 7.13, 7.14, 11.32). The segments are suspended on bearing tendons; the composite deck is post-tensioned by

Figure 11.32 Grants Pass Bridge, Colorado, USA: (a) cross-section, (b) deck plan (view from below)

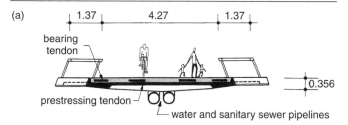

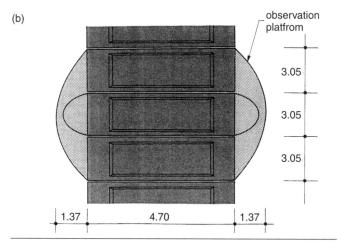

prestressing tendons. Both bearing and prestressing tendons are situated within the deck slab. Above the saddles the segments are 1.00 m long and have a solid cross-section. Before the erection of bearing tendons, these segments were post-tensioned by short internal tendons.

The arrangement of the structure was developed on the basis of very detailed static and dynamic analyses (Figures 7.74–7.76).

Figure 11.31 Grants Pass Bridge, Colorado, USA: (a) elevation, (b) plan

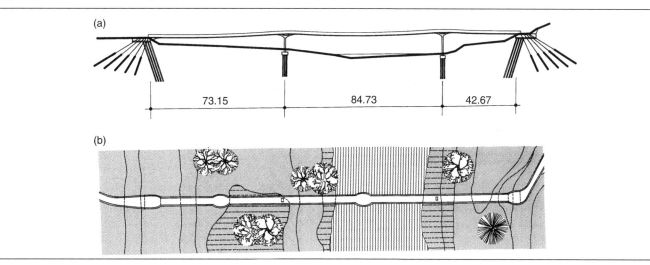

Figure 11.33 Grants Pass Bridge, Colorado, USA: observation platform

Figure 11.34 Lake Hodges Bridge, California, USA

To reduce the dynamic response of the structure to seismic load, the transverse stiffness of piers was reduced as much as possible. Dynamic analysis has confirmed that the speed of motion is within an acceptable limit.

Construction of the bridge was commenced by construction of the end abutments and piers. The solid abutment and pier segments were then erected. After casting of the joints between them, the segments were post-tensioned by short tendons. The bearing tendons were then pulled across the river and post-tensioned to the design stress (Figure 7.48). Subsequently, the segments were suspended on bearing tendons and shifted along them into the design position. The segments of the side spans were first erected, followed by the segments of the main span.

The formwork of the observation platforms was then suspended on the mid-span segments and the prestressing tendons and reinforcement of the deck slab were placed (Figures 7.53–7.55). The observation platforms together with the composite slab were then cast and subsequently post-tensioned.

The bridge was designed by OBEC, Consulting Engineers and Jiri Strasky, Consulting Engineer.

11.1.11 Lake Hodges Bridge, San Diego, California, USA

The world's longest stress ribbon bridge is located in the northern part of San Diego County and is a part of San Dieguito River Valley Regional Open Space Park (Figure 11.34). The bridge is formed by a continuous stress ribbon of three equal spans of length 108.58 m (Figure 11.35). The sag at mid-span is 1.41 m. The stress ribbon of total length 301.75 m is assembled of precast segments and cast-in-place saddles situated at all supports. The stress ribbon is fixed into the end abutments and is frame connected with intermediate piers. The structural solution was developed from the pedestrian bridges built in Prague-Troja and in Redding.

The precast segments of depth 0.407 m are 3.048 m long and 4.266 m wide. Each segment is formed by two edge girders and a deck slab. The segments are strengthened at joints by diaphragms. During the erection the segments were suspended on bearing cables and shifted along them to the design position. After casting of saddles and joints between segments, the stress ribbon was post-tensioned by prestressing tendons. The bearing cables are formed by two sets of three cables, each of 19 0.6-inch strands. The prestressing tendons are formed by two sets of three tendons, each of 27 0.6-inch strands. Both bearing cables and prestressing tendons are placed in the troughs situated at the edge girders.

The saddles have a variable depth and width. The depth varies from 0.407 to 0.910 m and the width from 4.266 to 7.320 m. Viewing platforms with benches were created above the supports. The saddles were cast after the erection of all segments in the formwork were suspended on the already erected segments and supported by piers or abutments (Figures 7.26(b) and 7.27(b)). During the erection the bearing cables were placed on Teflon plates situated on steel saddles – Figure 7.27(e).

A horizontal force as large as 53 MN is directed to the soil at the left abutment by four drilled shafts of diameter 2.70 m at the right abutment by rock anchors.

Very detailed static and dynamic analyses have been performed. The structure was checked not only for service load but also for significant seismic load. The analyses have proven that the bridge will be comfortable to users and will remain elastic under seismic loading. To verify that the bridge will be stable under heavy winds, a special wind analysis was performed by West Wind Labs, CA. Wind tunnel tests on a 1:10 scale model of the bridge section were performed to determine the aerodynamic load characteristics of the bridge deck.

Figure 11.35 Lake Hodges Bridge, California, USA: (a) cross-section of the deck, (b) pier elevation, (c) elevation

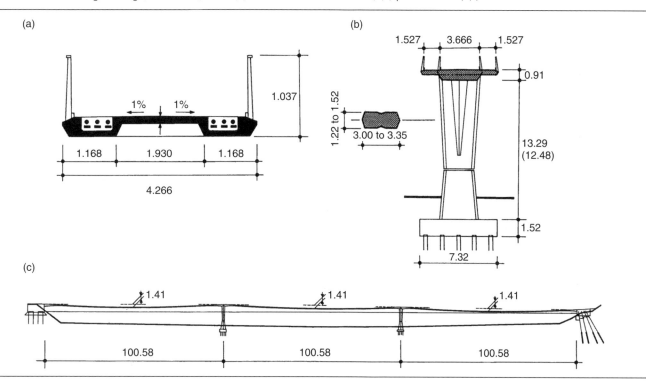

The bridge was completed in May 2009. The bridge was designed by TY Lin International, San Diego with collaborator Jiri Strasky, Consulting Engineer; the contractor was Flatiron, Longmont, Colorado.

11.1.12 Olse River Bridge, Czech Republic

The bridge (Figure 11.36) will be built across the Olse River which forms a border between the Czech Republic and Poland. It connects cities Bohumin and Gorzyce in the spot where a ford was situated for many years. The bridge is formed by a stress ribbon of three spans of length of 38.60, 80.00 and 38.60 m (Figure 11.37(d)). The sag at mid-span of the largest span is 1.38 m. The stress ribbon of total length 157.20 m is assembled of precast segments from high-strength concrete of a characteristic strength of 80 MPa (Figure 11.37(a)). At the abutments and above the Y-shaped piers,

Figure 11.36 Olse River Bridge, Czech Republic (rendering)

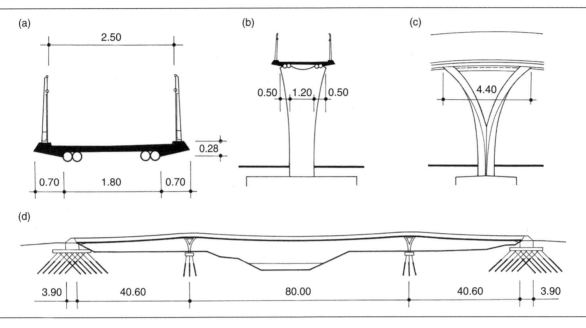

Figure 11.37 Olse River Bridge, Czech Republic: (a) cross-section of the deck, (b) cross-section at the pier, (c) partial elevation at the pier, (d) elevation

where significant positive bending moments originate, the stress ribbon is supported by saddles (Figure 11.37(c)). The piers and end abutments are supported by micropiles. A horizontal force as large as 15.2 MN is transported into the soil by battered micropiles.

The stress ribbon is supported and prestressed by four external cables formed by monostrands grouted in PE pipes. Above the saddles, the pipes are formed of stainless steel and are composite with 1.00 m long strengthened segments. The steel pipes not only resist large radial forces, but also contribute to resistance of bending.

It is assumed that after the construction of piers and end abutments, the external cables will be stretched across the river. The segments will then be placed on the cables by a mobile crane. PE pipes will be provided with stainless sleeves with shear studs that will be inserted at joints between the segments. After the concrete of the joints is cast, the cables will be tensioned up to a design stress. In this way, the stress ribbon will be prestressed.

Very detailed static and dynamic analyses have been performed, demonstrating that the structure has a similar response to dynamic load as previously built structures.

The bridge was designed by the author's design firm Strasky, Husty and Partners, Ltd, Brno, Czech Republic.

11.1.13 Kikko Bridge, Japan

The Kikko Bridge is a three-directional stress-ribbon pedestrian bridge, built in 1991 at the Aoyama-Kohgen golf club in Japan (Figures 11.38–11.40) (Arai and Ota, 1994). It provides a convenient pedestrian link between the clubhouse and the courses, which are arranged around a pond.

The deck is formed by three stress ribbons mutually connected by a central platform formed by a steel frame that is composite with a bottom precast slab and additionally cast top slab

Figure 11.38 Kikko Bridge, Japan (courtesy of Sumitomo Mitsui Construction Co., Ltd)

Figure 11.39 Kikko Bridge, Japan: (a) cross-section, (b) bearing and prestressing tendons, (c) elevation

Figure 11.41 Kikko Bridge, Japan: central link

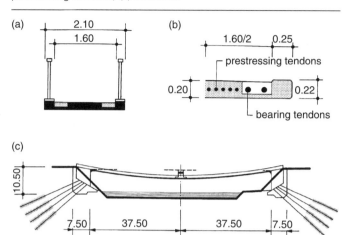

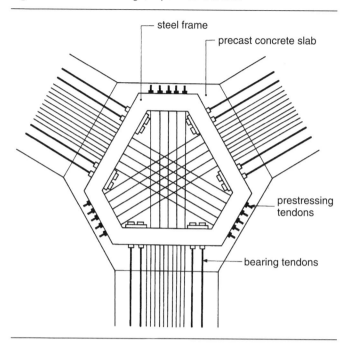

(Figure 11.41). The stress ribbons are assembled of precast segments suspended on bearing tendons that are anchored at the abutments and the central steel frame. The continuity of the structure is given by post-tensioning of prestressing tendons anchored at the abutments and central platform. The

Figure 11.40 Kikko Bridge, Japan: plan

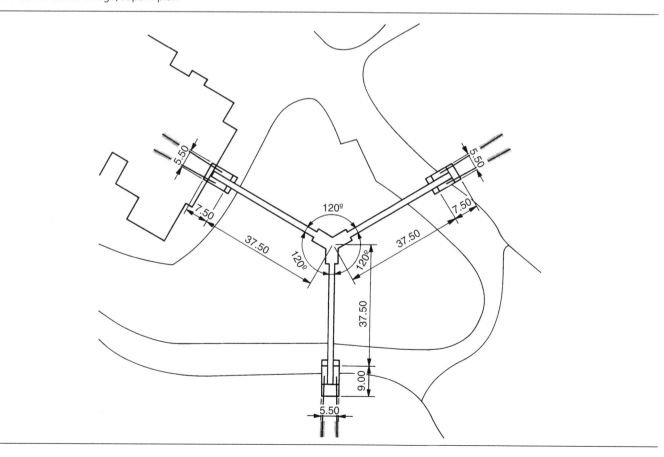

Figure 11.42 Kikko Bridge: construction sequences

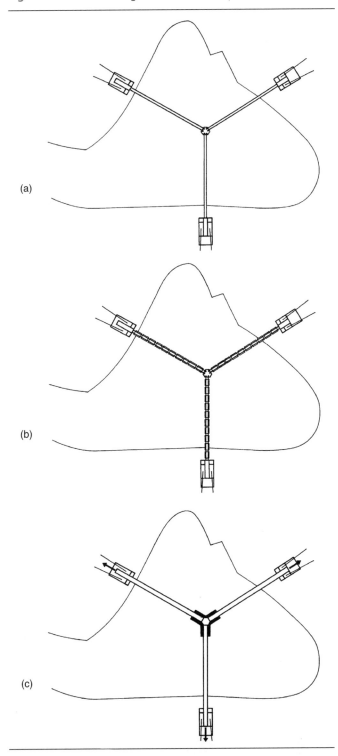

The construction of the bridges began by casting the abutments and post-tensioning the rock anchors. A central link formed by the steel frame and precast concrete slab was then erected and temporarily supported at the central point of the bridge. The bearing tendons were subsequently installed and post-tensioned. By post-tensioning, the central link was lifted from the temporary support to the designed position (Figure 11.42(a)).

The precast segments were then suspended on bearing tendons (Figure 11.42(b)). The troughs, joints between the segments, slab of the central platform and haunches were then cast. By post-tensioning of prestressing tendons, the structure achieved the designed shape and required stiffness (Figure 11.42(c)).

The design of the structure required a complex static and dynamic analysis. The bridge was designed and built by Sumitomo Mitsui Construction Co., Ltd, and received an Fédération Internationale de la Preconstrainte (FIP) outstanding structure award (specially mentioned).

11.1.14 Kent Messenger Millennium Bridge, Maidstone, UK

This pedestrian bridge forms part of a river park project along Medway in Maidstone, UK. The deck of the bridge is formed by a two-span stress ribbon that was – for the first time – designed with a cranked alignment (Figure 11.43) (Bednarski and Strasky, 2003). The length of the bridge is 101.50 m, the span length of the main span bridging the river is 49.5 m and the length of the side span is 37.5 m. The angle plan between the spans is 25 degrees (Figure 11.44).

The stress ribbon is formed by precast segments with a composite deck slab (Figures 7.15–7.16 and 11.45). The segments are suspended on bearing cables and serve as a falsework and formwork for the composite slab which was cast simultaneously with

Figure 11.43 Maidstone Bridge

bearing tendons are placed in the troughs situated close to edges of the section; prestressing tendons are situated in the deck slab. At the abutments, cast-in-place haunches were designed. The horizontal force from stress ribbons is resisted by rock anchors.

Examples

Figure 11.44 Maidstone Bridge: (a) elevation, (b) plan

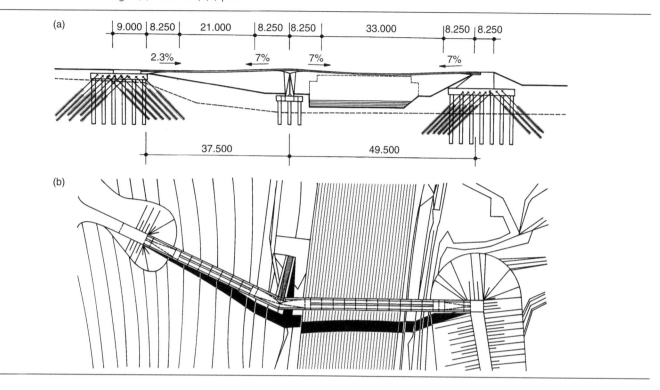

Figure 11.45 Maidstone Bridge – prestressed band: (a) partial elevation, (b) cross-section

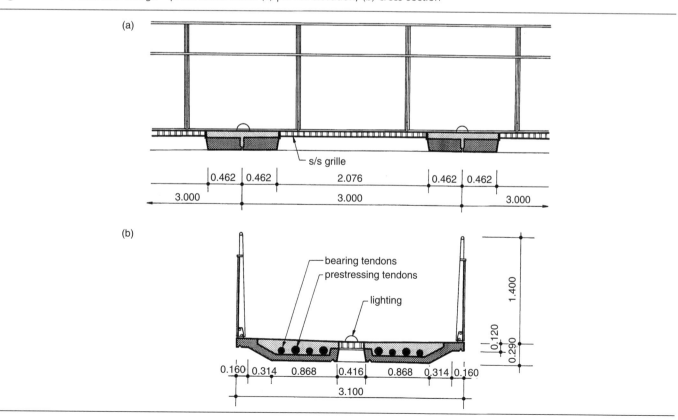

Figure 11.46 Maidstone Bridge – prestressed band at haunch: cross-section

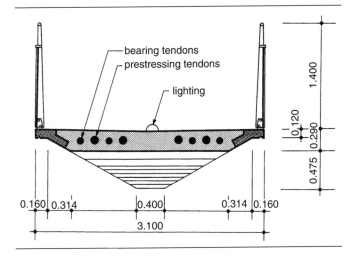

the joints between the segments. The stress ribbon is fixed into the abutments' anchor blocks (Figure 7.38) and is frame connected with the intermediate support. At all support cast-in-place haunches are designed (Figure 11.46).

The precast segments of length 3.00 m are formed by a 80 mm thick slab that gives a form of the soffit and edges of the deck. In the longitudinal axis of each segment, a rectangular opening covered by stainless grid is designed. The opening serves as drainage for the deck and lighting of the ground below. Between the openings, airport runway lights are situated.

Both the precast segments and the composite slab are post-tensioned by prestressing tendons that are situated together with the bearing cables within the cast-in-place slab (Figure 11.45). The bearing cables and prestressing tendons are formed by 7 and 12 monostrands grouted in polyethylene ducts. The arrangement of cables and tendons and their anchoring comply with special UK requirements for post-tensioned structures.

The intermediate support that is situated in the axis of the angle break resists the resultant horizontal force from the adjacent spans to the foundation by a system formed of a compression strut and a tension tie (Figures 7.35–7.37 and 11.47). The compression strut is formed of stairs and the tension tie is formed of a stainless steel tube. Since the value of the horizontal force depends on the position of the live load, the tube also acts as a compression member for several loading cases.

The large tension force from the stress ribbon is transferred to the soil (weald clay) via a combination of vertically drilled shafts and inclined micropiles.

The structure was designed on the basis of very detailed static and dynamic analyses (Figures 7.77–7.79). The structure was modelled by a 3D frame flexibly fixed in the soil. The stress ribbon was modelled by mutually connected parallel members

Figure 11.47 Maidstone Bridge – intermediate support: (a) elevation, (b) section A–A, (c) section B–B, (d) section C–C, (e) section D–D

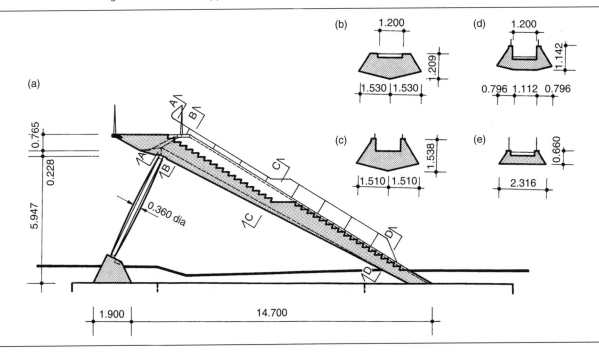

that can express the function of the bearing and prestressing cables, precast deck and cast-in-place concrete. The analysis was performed by the program ANSYS as a geometrically non-linear structure. In the initial stage, bearing-cable forces are in equilibrium with the self-weight of the deck.

The change of direction in the bridge deck plan creates transverse and torsional forces for which the structure was carefully checked. Similarly to all stress ribbon structures, significant bending stresses originate close to the supports. These stresses were reduced by designing cast-in-place haunches that were checked as partially prestressed members.

In a dynamic analysis, natural frequencies and modes were determined first. The speed and acceleration of the deck motion were then checked. The response of the structure to a pulsating point moving along the deck, which represents the pedestrian loading, was checked.

Before the erection of the deck, the abutments and the intermediate support including the haunches were cast (Figure 7.57). Until the tensioning of the bearing cables was complete, the stability of the intermediate support was guaranteed by temporary supports. The bearing cables were erected as stay cables. At first, erection strands were pulled across the river and the left bank. PE ducts were then suspended and moved into the design position. Monostrands were then pushed through the ducts.

After the tensioning of the bearing cables the segments were erected by a mobile crane. The erected segment was first placed on a 'C' frame then transported into the design position under the bearing cables, lifted and suspended on the bearing cables. All segments were erected in one day.

The geometry of the erected segments and horizontal deformation of the abutments were carefully checked. After the erection of the segments, the tension force in the cables was corrected, the prestressing tendons and reinforcing steel of the composite slab were placed and the joints between the segments, closure and slab were cast. The slab was cast simultaneously in both spans symmetrically from the mid-spans to the closures (Figure 7.58). To ensure that the concrete remains plastic until the entire deck is cast, a retarder was used.

After two days the bearing and prestressing tendons were grouted. When the cement mortar reached a sufficient strength, the prestressing tendons were post-tensioned to 15% of the final prestressing force. Partial prestressing prevented cracks arising due to the temperature changes. When the concrete reached sufficient strength, the structure was post-tensioned to the full design level. The function of the bridge was also verified by a dynamic test that demonstrated that the users do not experience any feelings of discomfort when walking or standing on the bridge.

The bridge was designed by Cezary M Bednarski, Studio Bednarski, London, UK and by Strasky Husty and Partners, Consulting Engineers, Ltd, Brno, Czech Republic. UK liaison and checking was completed by Flint & Neill Partnership, London, UK. The bridge was built by Balfour Beatty Construction Ltd, Surrey, UK.

11.1.15 Rosenstein II Bridge, Stuttgart, Germany

The bridge, which is part of a pathway connecting the city parks in Stuttgart, Germany, was built in 1977. It crosses a small cutting containing a double line of tram tracks (Figure 11.48) (Holgate, 1997). The bridge with a span 28.865 m is formed by two sagging cables that support the deck, which is assembled from precast concrete panels of thickness 0.10 m. The cables are stabilised by a cable of the opposite curvature and by the dead load of concrete panels (Figure 11.49).

Although the bridge has the appearance of a stress ribbon structure, it functions as a pre-tensioned space truss. The prestressing is provided by the bottom cable that is connected with the main cable by diagonal ties. The deck is formed from individual precast concrete slabs 0.10 m thick and 1.00 m wide, separated from each other by a 0.01 m gap. The slight friction between the units and their connections to the cables provides some damping of vibrations.

The tensile prestress was carefully calculated to ensure that, with about 60% of the maximum pedestrian load and maximum temperature expansion of the cables, all remain taut. For economical reasons, the designers decided not to prestress the bridge to ensure tautness under the full $5\,kN/m^2$ live load specified in the German code.

Figure 11.48 Rosenstein II Bridge (courtesy of Schlaich, Bergermann and Partners)

Figure 11.49 Rosenstein II Bridge: (a) cross-section, (b) elevation, (c) plan

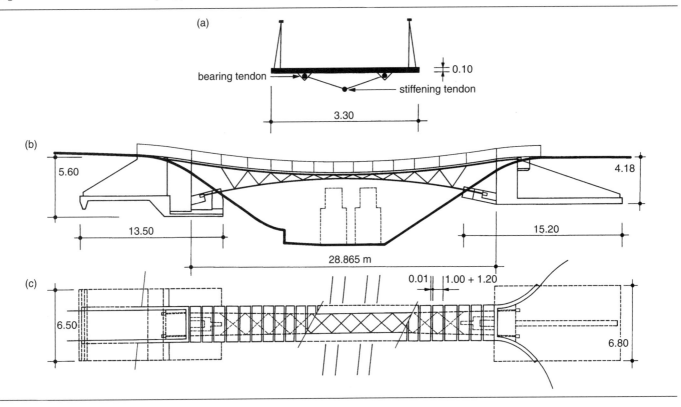

The bridge vibrates more than is customary and in an unusual manner, especially when loaded at the quarter points of the span. However, the citizens of Stuttgart have come to accept this, and tourist information maps of the park include the nickname the 'swinging bridge' for Rosenstein II.

The bridge was designed by Professor Schlaich who was, at that time, a partner of Leonhardt and Andä, Consulting Engineers, Stuttgart.

11.1.16 Phorzheim III Bridge, Germany

The Phorzheim III Bridge was built across the diverted River Enz in 1991 in the city of Phorzheim, Germany (Figure 11.50) (Holgate, 1997). The bridge has one span of length 50 m and sag 0.75 m, and its width is 2.85 m (Figure 11.51).

The deck is not formed by a prestressed band; the structural system consists only of two strips of steel, each with a rectangular cross-section of 0.60 m by 0.04 mm. These are slung between two massive concrete anchor blocks embedded in the banks of the diverted river. Precast planks of lightweight concrete span cross-ways between the steel strips. The planks rest on rubber bearing pads and are slightly separated from each other by rubber spacers in the shape of an inverted 'T' which rest on the steel strips. This ensures that the planks are not stressed by high localised bending of the deck due to a concentration of pedestrians.

The stiffness and damping are provided by the rubber pads and spacers, by the tubular handrail (with specially detailed joints) and even by the wire safety mesh stretched between the deck

Figure 11.50 Phorzheim III Bridge (courtesy of Schlaich, Bergermann and Partners)

Figure 11.51 Phorzheim III Bridge: (a) cross-section, (b) elevation

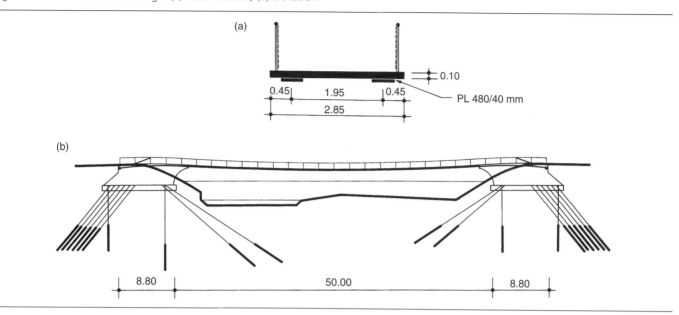

and handrail. These make the bridge quite stiff and no adverse comments have been received from the users.

The bridge was designed by Schlaich, Bergermann and Partners, Consulting Engineers, Stuttgart.

11.1.17 Punt da Suransuns Bridge, Switzerland

The Punt da Suransuns Bridge built in 1999 carries a trail across the Hinterrhein River through the Viamala Gorge in Switzerland (Figure 11.52) (Conzett, 2000). The bridge is formed by a stress ribbon of span 40 m assembled from granite slabs (Figures 11.53 and 11.54). The pathway has a longitudinal slope of 0–20%.

Figure 11.52 Punt da Suransuns Bridge (courtesy of Conzett, Bronzini, Gartmann AG)

The stone chosen was the locally quarried Andeer granite, selected for its outstanding physical properties. As the bridge is within the range of salt-spray mists from the above motorway, V4A stainless steel or duplex steel was chosen for all steel parts.

The granite slab members are supported by two steel strips of cross-section 0.0×0.015 m that are anchored to the abutments. Although the structural arrangement of the bridge looks similar to the arrangement of the Phorzheim III Bridge, its static function is totally different. The joints between the slab members were filled with 3 mm thick aluminium plates that function as mortar and the structure was prestressed by cables wrapped between the granite slabs and steel plates. In this way, a composite section formed by the granite slab and steel was created. The actual flexural stiffness of stone slabs pressed together was measured in a test with five stone slabs. Depending on the accuracy of the joint, the measured flexural stiffness was less than half that of the theoretical value.

The bending stresses at supports were reduced by leaf springs designed at the support area (Figure 11.53(b)). For the stress analyses, the limit cases 'homogenous cross-section' and 'frictionless cross-section' were investigated and compared.

The bridge was constructed using a dry building method, which means that after pouring the abutments there is only stacking, tightening and screwing to be completed. The relatively light tie bars were erected as complete pieces by a helicopter. After placement of the steel ropes, the granite slabs were placed

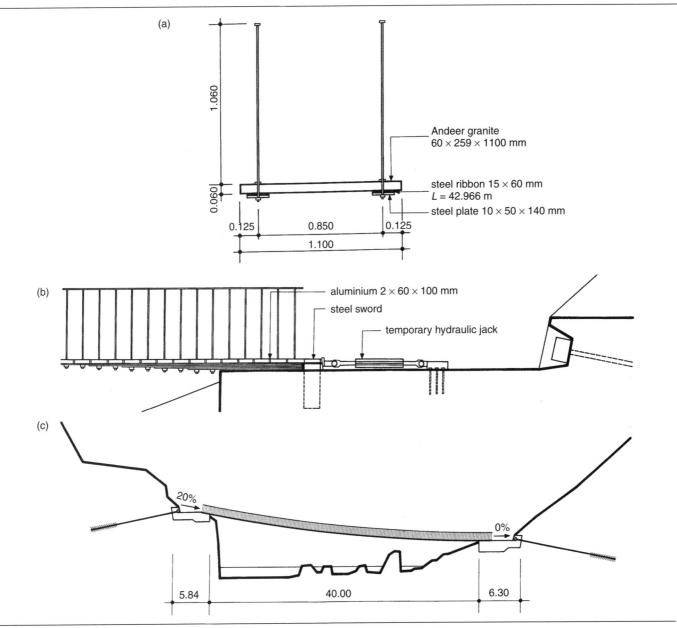

Figure 11.53 Punt da Suransuns Bridge: (a) cross-section, (b) partial elevation at abutment, (c) elevation

piece by piece, starting at the lower abutment. The slabs were attached to the steel ropes via railing posts. After the tensioning of the ropes and wedging of the steel end blocks, the nuts on the railing posts were tightened and the railing was welded to the posts.

During the planning stage, the construction method was tested on a 1:20 model. The deck was modelled with 3 mm thick granite plates. The results of the static calculations could be tested qualitatively, and the torsion stiffness value in the model was found to be very high. Vertical oscillations can be felt when crossing the bridge, but pedestrians have commented that the bridge is not as flexible as it looks.

The bridge was designed by Conzett, Bronzini, Gartmann AG, Chur; the contractor was Romei AG, Granitewerke Andeer, V Luzi.

11.1.18 Olomouc Bridge across the expressway R3508 near Olomouc, Czech Republic

The bridge crosses expressway R3508 near the city of Olomouc (Figure 11.55). The bridge is formed by a stress ribbon of

Figure 11.54 Punt da Suransuns Bridge, Switzerland: structural solution (courtesy of Conzett, Bronzini, Gartmann AG)

Figure 11.55 Bridge Olomouc, Czech Republic

two spans supported by an arch (Figure 11.56). The stress ribbon of length 76.50 m is assembled of precast segments 3.00 m long, supported and prestressed by two external tendons (Figures 11.57 and 11.58).

The precast deck segments and precast end struts consist of high-strength concrete of a characteristic strength of 80 MPa.

The cast-in-place arch consists of high-strength concrete of a characteristic strength of 70 MPa. The external cables are formed by two bundles of 31 0.6-inch diameter monostrands grouted inside stainless steel pipes. They are anchored at the end abutments and are deviated on saddles formed by the arch crown and short spandrel walls (Figures 11.56(b) and 11.56(c)).

Figure 11.56 Bridge Olomouc, Czech Republic: (a) elevation, (b) partial elevation at anchor block, (c) partial elevation at spandrel wall

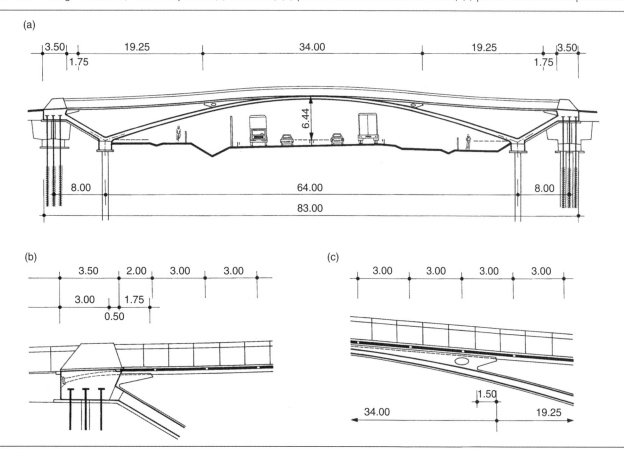

Figure 11.57 Bridge Olomouc, Czech Republic – cross-section: (a) at mid-span, (b) at spandrel wall

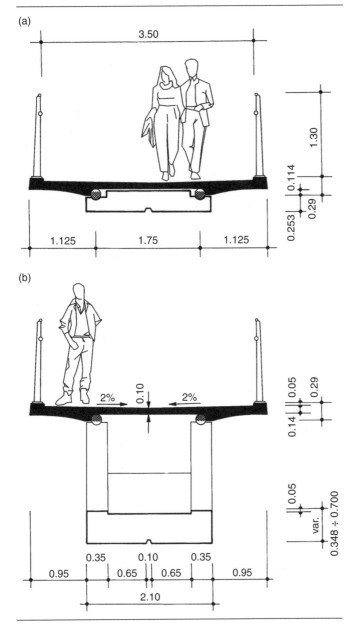

Figure 11.58 Bridge Olomouc, Czech Republic – segment on external cables

end anchor blocks were then cast. The arch was cast in a formwork supported by light scaffolding. When the concrete of the arch had sufficient strength, the external cables were assembled and tensioned. The precast segments were then erected. After the forces in the external cables were adjusted, the joints between the segments were cast and subsequently the external tendons were tensioned up to the design stress.

The structural solution was developed on the basis of the described tests and a very detailed static and dynamic analysis. Great attention was also devoted to the analysis of the buckling of the arch. The stability analysis demonstrated that the structure has a sufficient margin of safety. Although the structure is extremely slender, users do not have an unpleasant feeling when standing or walking on the bridge. The bridge was built in 2007.

The bridge was designed by the author's design firm Strasky, Husty and Partners, Ltd, Brno, and was built by the firm Max Bögl a Josef Krýsl, k.s., Plzeň.

11.1.19 Bridge across the Svratka River Bridge in Brno, Czech Republic

The pedestrian bridge connects a newly developed business area (Spielberk Office Centre) with the old city centre. It is situated in the vicinity of a new international hotel and prestigious office buildings. An old multispan arch bridge with piers in the river is situated close to the bridge. It was evident that a new bridge should also be formed by an arch structure; however, the choice was a bold span without piers in the river bed (Figure 11.59). Due to poor geotechnical conditions, a traditional arch structure which requires a large horizontal force to be resisted would be too expensive. The self-anchored stress ribbon and arch structure has therefore been built (Figure 11.60). Both the stress ribbon and the arch are assembled of precast segments from high-strength concrete

The steel pipes are connected to the deck segments by bolts located in the joints between the segments (Figure 11.58). At the abutments, the tendons are supported by short saddles formed by cantilevers that protrude from the anchor blocks. The stress ribbon and arch are mutually connected at the central point of the bridge. The arch footings are founded on drilled shafts and the anchor blocks on micropiles.

The bridge was erected in several steps. After the piles were placed, the end struts were erected and the arch footings and

Figure 11.59 Svratka River Bridge, Czech Republic

Figure 11.60 Svratka River Bridge, Czech Republic: (a) cross-section, (b) elevation

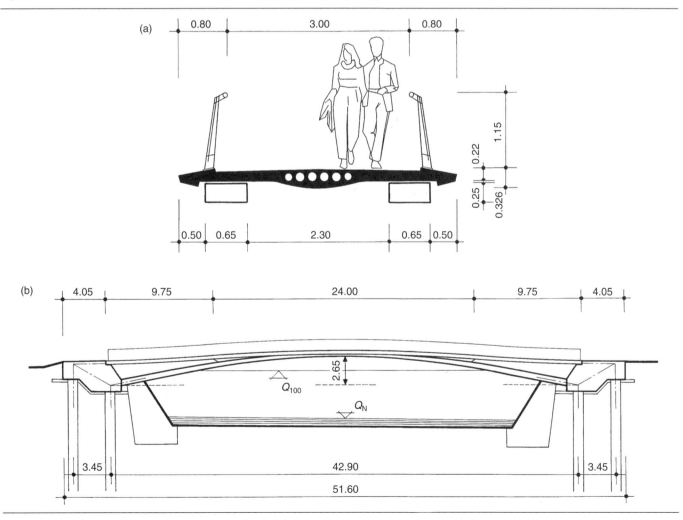

Figure 11.61 Svratka River Bridge, Czech Republic: force diagram

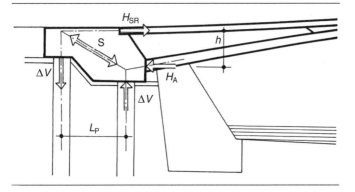

Figure 11.62 Svratka River Bridge, Czech Republic: structural arrangement

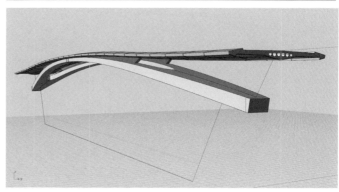

and were erected without any temporary towers. Smooth curves that are characteristic of stress ribbon structures allowed a soft connection between the bridge deck and both banks.

The deck of the bridge is formed by a stress ribbon that is supported by a flat arch. Since both the stress ribbon and the arch are fixed in common end abutments, the structure forms a self-anchored system that stresses its footings by vertical forces only. Because the riverbanks are formed of old stone walls, the end abutments are situated beyond these walls. The abutments are supported by pairs of drilled shafts. The abutments serve as the arch footings, the stress ribbon anchor blocks and the struts (Figure 11.61). The rear shafts are stressed by tension forces and the front shafts are stressed by compression forces. These forces balance a couple of tension and compression forces originating in the stress ribbon and arch. The end abutments serve as compression struts, transferring the tension force from the stress ribbon into the compressed arch.

The arch span is $L = 42.90$ m and its rise $f = 2.65$ m; hence, rise to span ratio is $f/L = 1/16.19$. The arch is formed by two branches that have a variable mutual distance and merge at the arch springs (Figure 11.62). The 43.50 m long stress ribbon is assembled of segments of length 1.5 m. In the middle portion of the bridge, the stress ribbon is supported by low spandrel walls of variable depth (Figure 11.63). At mid-span, the arch and stress ribbon are mutually connected by two sets of three steel dowels that transfer the shear forces from the ribbon to the arch. The stress ribbon is carried and prestressed by four internal tendons of 12 0.6-inch diameter monostrands grouted in PE ducts. The segments have variable depth with a curved soffit. The stress ribbon and the arch were made from high-strength concrete of characteristic strength 80 MPa.

The arch was assembled from two arch segments that were temporarily suspended on erection cables anchored at the end abutments (Figures 11.64(a) and 11.65). The midspan joints were then cast and the erection cables were replaced by external cables that tied the abutments. The spandrel walls were then cast and segments erected. The segments were successively placed on the arch spandrel walls, followed by external cables (Figures 11.63 and 11.64(b)). Subsequently, the internal tendons were pulled through the ducts and tensioned. Finally, the external tendons were removed. In this way, the required geometry of the deck was obtained. After casting the joints between the deck segments, the cables were tensioned up to the design stress and, as a result, the deck was prestressed.

Although the bridge is very slender, the bridge is very stiff and the users do not experience an unpleasant feeling when standing or walking on the bridge. The static function and quality of the workmanship were checked by a loading test. Trucks were situated on the whole- and half-length of the deck.

Figure 11.63 Svratka River Bridge, Czech Republic: segment on the spandrel walls

Figure 11.64 Svratka River Bridge, Czech Republic – erection: (a) arch, (b) stress ribbon

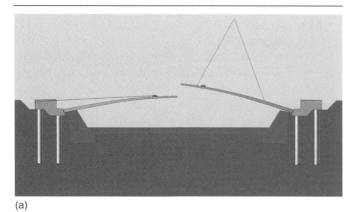

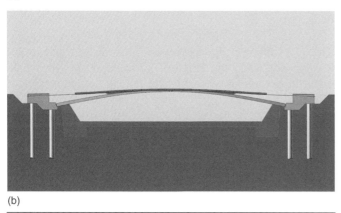

Figure 11.65 Svratka River Bridge, Czech Republic: erection of the arch segment

Construction of the bridge started in February 2007 and was completed in September of that year. The new structure was well received by the public. The bridge was designed by design firm Strasky, Husty and Partners, Brno in collaboration with Acht Architects, Prague, Rotterdam. The bridge was built by SKANSKA DS, Division 77 Mosty.

11.1.20 Tokimeki Bridge, Japan

The Tokimeki Bridge built in 2002 crosses the Takatsuka Pond situated in the centre of Kameyama Sunshine Park which was opened in Mie Prefecture, Japan (Figure 11.66) (Tanaka *et al.*, 2002). The bridge is formed by a two-span stress ribbon that is supported by an arch of span length 50.00 m. Since the stress ribbon's anchor blocks and arch springs are connected by compression struts transferring the horizontal force from the stress ribbon into the arch footings, the bridge forms a self-anchored system that loads the footings by vertical forces only (Figure 11.67).

The stress ribbon of span 27.50 m and sag 0.568 m is assembled of precast slab segments and cast-in-place haunches situated at the end-anchor blocks and above the crown of the arch. The segments are suspended on the bearing tendons placed in the troughs situated close to their edges. The stress ribbon is post-tensioned by prestressing tendons situated within the deck slab. The arch and end struts have solid sections.

Since the end-anchor blocks are not supported, the struts are stressed not only by large compression forces but also by bending moments. They are therefore formed by a stiff solid section and are post-tensioned. The bridge is founded on drilled piles.

The construction of the bridge began with the construction of footings and casting of the struts. The arch was cast in a false-work supported by a temporary fill. The bearing tendons were then installed and post-tensioned. To eliminate the movement of the tendons during the erection of segments, the bearing tendons were temporarily fastened to the arch crown. The precast segments were then suspended from the bearing tendons.

Figure 11.66 Tokimeki Bridge

The prestressing tendons were then installed and reinforcement of the haunches was carried out. The joints between the segments, troughs and haunches were cast and subsequently post-tensioned, and the falsework of the arch was removed. During the construction, horizontal movement of the strut's tops and the stress ribbon sag were carefully monitored.

The bridge was designed by Japan Engineering Consultants Co., Ltd. The contractor was Sumitomo Construction Co., Ltd.

11.1.21 McLoughlin Boulevard Bridge, Portland, Oregon, USA

The McLoughlin Boulevard Pedestrian Bridge (Figure 11.68) is part of a regional mixed-use trail in the Portland, Oregon metropolitan area. The bridge is formed by a stress ribbon deck that is suspended on two inclined arches (Figure 11.69(b)). Since the stress ribbon anchor blocks are connected to the arch footings by struts, the structure forms a self-anchored system that loads the footing by vertical reactions only (Figure 11.70). The deck is suspended on arches via suspenders of a radial arrangement; the steel arches therefore have a funicular/circular shape. The slender arches are formed by 450 mm diameter pipes that are braced by two wall diaphragms.

The stress ribbon deck is assembled from precast segments and a composite deck slab (Figure 11.69(a)). In side spans, the segments are strengthened by edge composite girders (Figure 11.69(c)). The deck tension due to dead load is resisted by bearing tendons. The tension due to live load is resisted by

Figure 11.67 Tokimeki Bridge: (a) cross-section, (b) elevation

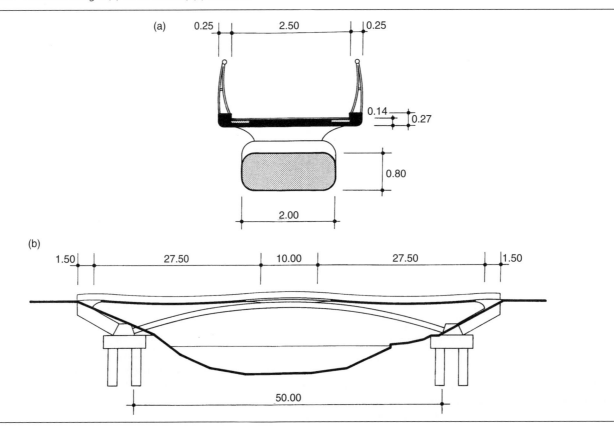

Figure 11.68 McLoughlin Boulevard Bridge, Oregon, USA

the stress ribbon deck being prestressed by prestressing tendons. Both bearing and prestressing tendons are situated in the composite slab. The bearing tendons that were post-tensioned during the erection of the deck are formed by two bundles of 12 × 0.6 inch diameter strands that are protected by the cast-in-place slab; deck prestressing tendons are formed by six bundles of 10 × 0.6 inch diameter tendons that are grouted in ducts.

Edge pipes and rod suspenders make up part of the simple hanger system (Figure 11.71). The suspenders connect to 'flying' floor beams cantilevered from the deck panels to provide the required path clearance. The edge pipes contain a small tension rod that resists the lateral force from the inclined suspenders on the end of the floor beams. Grating is used to span the gap between the edge pipe and the deck panels. Protective fencing is placed in the plane of the suspenders to open up the deck area, the rail is cantilevered in the plane of the suspenders to open up the deck area and the rail is cantilevered from the suspenders.

The structural solution was developed on the basis of very detailed static and dynamic analyses. Great attention was also devoted to analysis of the buckling of the arch and the dynamic analyses. The non-linear stability analysis has proved that the completed structure has a very large margin of safety. However, during construction when the load is resisted by the arches only, the margin of safety was relatively low. The

Figure 11.69 McLoughlin Bridge, Oregon, USA – cross-sections: (a) main span, (b) bridge, (c) side spans

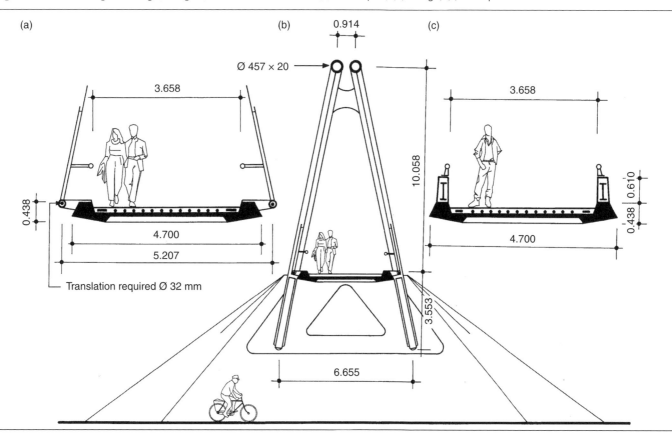

Figure 11.70 McLoughlin Boulevard Bridge: elevation

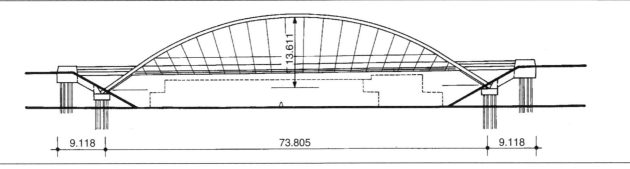

erected structure was therefore stiffened by a mid-span erection tower that loaded the arch by a controllable force.

Although the first bending frequency is below the walking frequency, the pedestrians do not have an unpleasant feeling when standing or walking on the bridge.

Construction of the bridge commenced in March 2005 and was completed in September 2006. The bridge was designed by OBEC, Consulting Engineers and Jiri Strasky, Consulting Engineer. The bridge was built by Mowat Construction Company, Vancouver, Washington.

11.1.22 Morino-Wakuwaku Bridge, Japan

Morino-Wakuwaku Bridge, built in 2001, crosses a park situated in the centre of the city of Iwaki in Prefecture Iwaki, Japan. The bridge is formed by a stress ribbon that is post-tensioned by external cables situated below the deck (Figure 11.72) (Kumagai et al., 2002).

Figure 11.71 McLoughlin Boulevard Bridge: stress ribbon deck

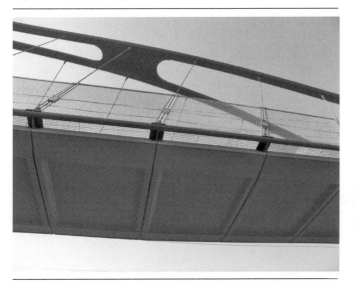

The stress ribbon of length 128.50 m and sag 2.57 m is assembled of precast segments and cast-in-place haunches situated at the abutments (Figure 11.73). The segments with triangular edges have a width of 4.40 m and depth of 0.328 m. The segments were suspended on bearing tendons placed in the troughs situated close to edges of the section. The stress ribbon was post-tensioned by external parabolic prestressing tendons situated under the deck. These tendons have a variable eccentricity with a maximum of 5.425 m at mid-span. They are connected to the stress ribbon by steel struts of a variable length (Figure 11.74) and are anchored at the abutments. The horizontal force from the stress ribbon is resisted by rock anchors.

Since the prestressing tendons have a larger curvature than the stress ribbon, the amount of tendons compared to classical stress ribbon structure is reduced. The torsional stiffness of the structure is also higher. On the other hand, the structure required perfect corrosion protection of steel members and the erection of the structure is more complicated.

Figure 11.72 Morino-Wakuwaku Bridge, Japan (courtesy of Oriental Construction Co., Ltd)

Figure 11.73 Morino-Wakuwaku Bridge: (a) crosssection, (b) bearing tendons, (c) elevation

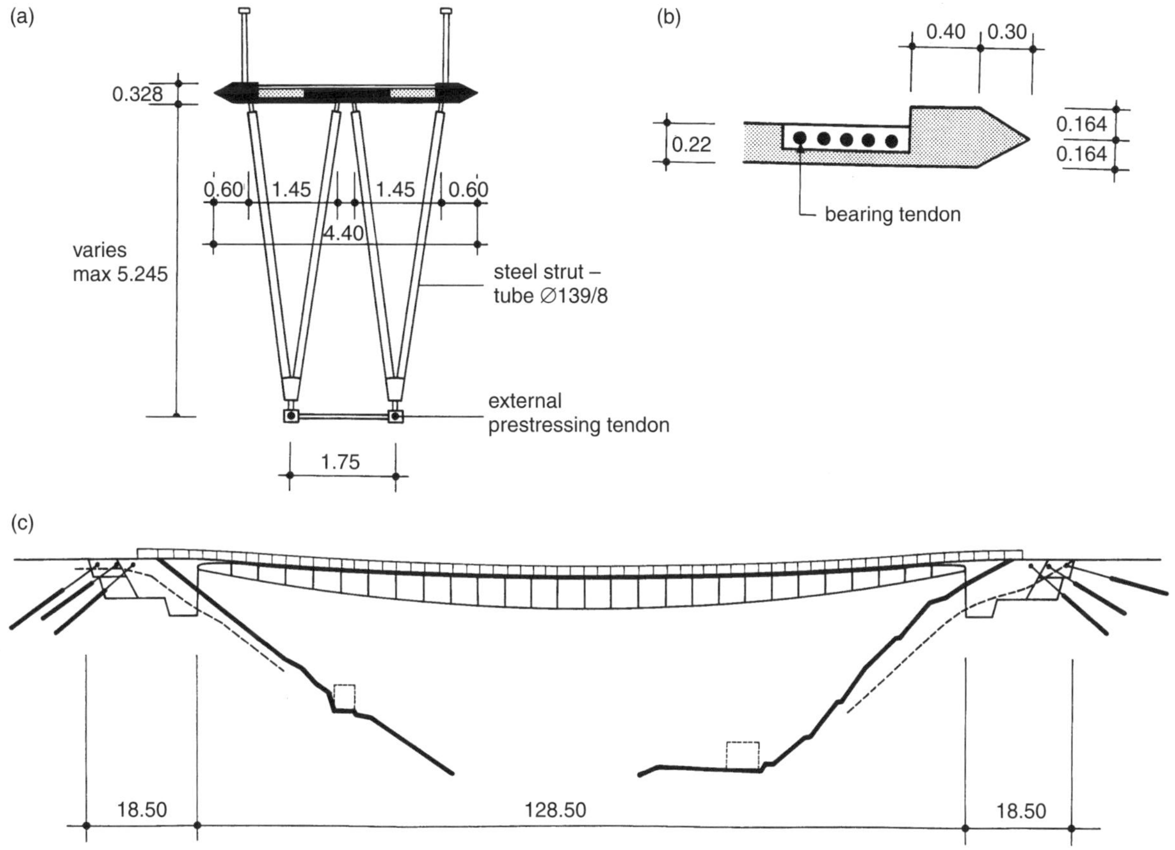

The stresses in the external cables due to the dead load are $0.44 f_u$, the live load $2.00\,kN/m^2$ creates additional stresses of $40\,N/mm^2$. The steel struts are pin connected to both the deck and external tendons.

Figure 11.74 Morino-Wakuwaku Bridge: deck, struts and external prestressing tendons (courtesy of Oriental Construction Co., Ltd)

Construction of the bridge started with the casting of the abutments and post-tensioning of the rock anchors. The bearing tendons were then pulled across the valley and were post-tensioned (Figure 11.75(a)).

The precast segments with steel struts were then suspended on bearing tendons and shifted along them into the designed position (Figure 11.75(b)). The hanging scaffolding was then erected. Subsequently, the troughs, joints between the segments and haunches were cast (Figure 11.75(c)). By post-tensioning of external prestressing tendons, the structure achieved its designed shape and required stiffness (Figure 11.75(d)).

The function of the structure was confirmed by static and dynamic loading tests. The bridge was designed by Kyoryo Consultants Co., Ltd and the contractor was Oriental Construction Co., Ltd.

11.2. Suspension structures
11.2.1 Bridges over the Segre River in Alt Urgell, Lérida, Spain

Four suspension bridges were built in 1984 across the Segre River in Alt Urgell, Lérida, Spain to replace structures which

Figure 11.75 Morino-Wakuwaku Bridge: construction sequences

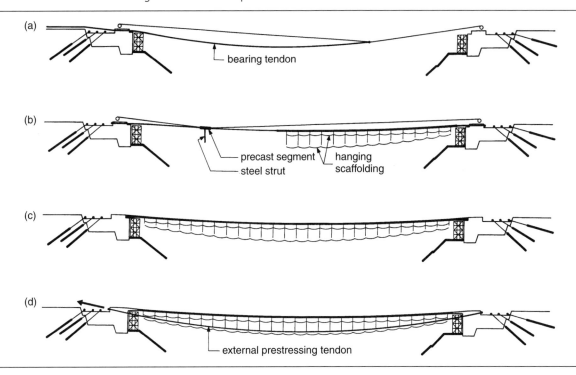

had been damaged by flooding in 1982. Three of these bridges (Reula, Basella and Peramola) were constructed for pedestrians. The Reula Bridge has a span of 70 m, the Basella Bridge has a span of 90 m and the Peramola Bridge has a span of 102 m (Figures 11.76 and 11.77) (Troyano *et al.*, 1986).

The three bridges have a similar arrangement. Their deck is assembled of precast ribbed segments 3.66 m wide and 0.40 m deep. The 3.60 m long segments are suspended on two

Figure 11.76 Peramola Bridge across Segre River (courtesy of Carlos Fernandez Casado, SL, Madrid)

suspension cables formed by locked coil strands that are deviated on steel saddles supported by precast towers. The distance between the hangers is 2.00 m.

The towers are trapezoidal with the maximum width at the head. Due to the size of the saddles that required a wide radius, the top width was 2–3 m. The anchor blocks of the main cables were also prefabricated. They were fixed into the foundation formed by slurry walls by active ground anchors.

The main cables are stiffened by negative stays (that extend from the foot of the tower to the main cables) in order to reduce the bending moments in the deck and the vibrations produced by traffic.

The prefabrication of all precast members was carried out in a central plant from which they were then transported to construction sites. The towers were erected first, followed by the main cables. After that, the prefabricated members were hung on the main cables (Figure 11.78), the negative stays were connected and finally the 0.40 m wide joints between the segments were concreted and the deck was longitudinally prestressed.

The bridge was designed by Carlos Fernandez Casado, SL, Madrid, Spain.

Figure 11.77 Segre River Bridges – Peramola Bridge: (a) cross-section, (b) elevation

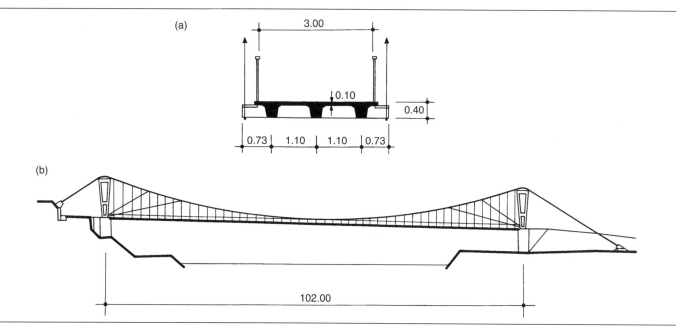

11.2.2 Max-Eyth-See Bridge, Stuttgart, Germany

The bridge across the Neckar River built in 1987 provides a direct link from the suburbs above the vineyards situated on one side of the river to the recreation area situated on the other side (Figure 11.79) (Schlaich and Schober, 1994).

The bridge deck is suspended on two inclined suspension cables deviated at saddles of two single masts (Figure 11.80). The masts have different heights: 24.50 and 21.50 m. The slender deck of thickness 0.42 m is suspended on a criss-crossed arrangement of diagonal hanger cables at very close spacing (Figure 2.27).

The main portion of the deck above the river is straight. On the vineyard side, the end of the deck is curved away from the centreline just before it reaches the mast so that it can pass over the tow-path and join the path through the vineyards on a convenient promontory of earth. On the park side, the deck is bifurcated just before reaching the mast and passes on either side of it.

The main portion of the deck above the river is assembled of precast segments that are composite with a cast-in-place deck slab. The curved portions are cast-in-place. The deck has variable depth and it is fixed to the abutments. While the

Figure 11.78 Peramola Bridge across Segre River: deck during the erection (courtesy of Carlos Fernandez Casado, SL, Madrid)

Figure 11.79 Max-Eyth-See Bridge, Stuttgart, Germany (courtesy of Schlaich, Bergermann and Partners)

Figure 11.80 Max-Eyth-See Bridge: (a) cross-section of the deck, 1 – at ramps, 2 – at main span, (b) tower elevation, (c) elevation, (d) plan

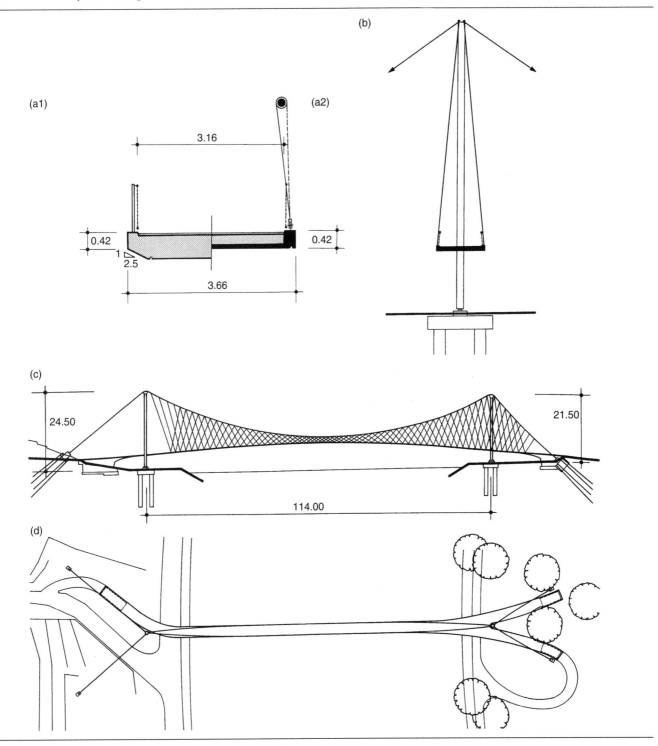

abutments of the bifurcated rams are fixed into the soil, the abutment on the vineyard side is movable. The abutment is supported by a foundation slab that restricts its rotation.

To ensure that the visual impact of the handrail was not greater than that of the deck it was conceived as a wire mesh clamped to two longitudinal tensioned cables, one strung at handrail height and the other at deck level (Figure 2.27).

Figure 11.81 Phorzheim I Bridge, Germany (courtesy of Schlaich, Bergermann and Partners)

The construction began by casting the abutments and rams (Figure 8.19). The masts and suspension cables with hanger were then erected and anchored at anchor blocks. Subsequently, the segments were progressively suspended on the main cables (Figure 8.20) and a cast-in-place composite slab and closures were cast.

This beautiful structure is ideal for those who cross it, view it and live with it. The deck splits and curves, which is welcoming for the pedestrians. Although the bridge is light and transparent, it behaves very well.

The bridge was designed by Schlaich, Bergermann and Partners, Stuttgart, Germany.

11.2.3 Phorzheim I Bridge, Germany

The Phorzheim I Bridge was built across the river Enz in 1991 in the city of Phorzheim, Germany (Figure 11.81) (Shlaich and Bergermann, 1992). The bridge is formed by a self-anchored suspension structure of two spans of 51.80 and 21.80 m (Figure 11.82).

The bridge forms a simple and clear structural system. The slender deck of a depth of only 0.30 m is suspended on two inclined suspension cables deviated at a steel saddle situated at the top of a single mast. Since the mast is situated in the bridge axis, the deck has a variable width of 3.40–6.29 m. At the shorter span, the deck is fixed to the abutments while at the longer span it is supported by pendulum steel columns. The columns are directly connected to steel plates in which the suspension cables are anchored. The force from the plate is transferred into the deck via steel 'teeth' welded to the plate.

The bridge deck was cast on the falsework and the masts and suspension cables were then installed.

The bridge was designed by Schlaich, Bergermann and Partners, Stuttgart, Germany.

11.2.4 Vranov Lake Bridge, Czech Republic

The suspension bridge, which was built in 1993, is located in a beautiful, wooded recreation area where Lake Vranov was created by a dam in the 1930s (Figures 1.20, 8.1 and 11.83–11.85) (Strasky, 1994, 1995, 1998a and 1998b). The structure replaced a ferry service carrying people between a public beach on one side of the lake and accommodation, restaurants and shops located on the other side. The structure was also designed to carry water and gas lines.

Figure 11.82 Phorzheim I Bridge: (a) cross-section of the deck, (b) tower elevation, (c) elevation, (d) plan

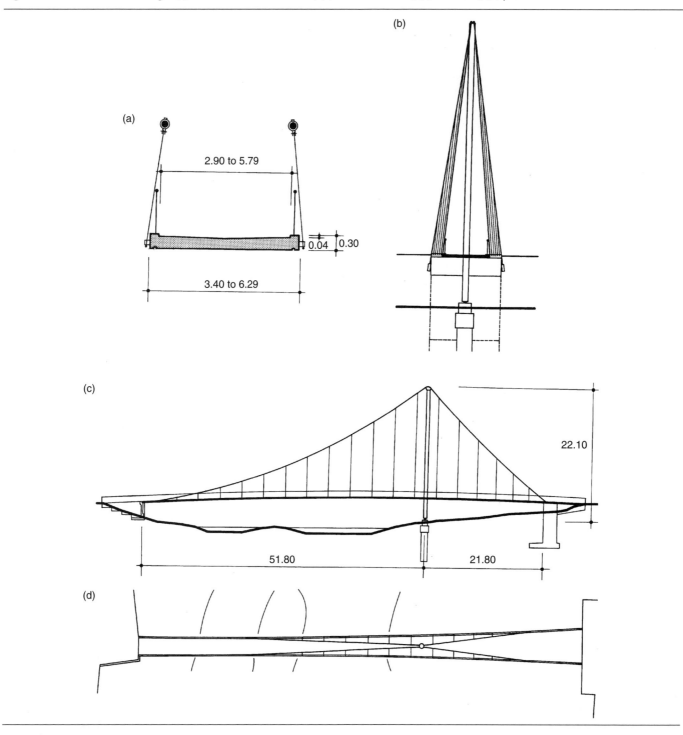

A very slender deck of depth only 0.40 m (Figures 1.21 and 11.86) is suspended on two inclined suspension cables of spans 30, 252 and 30 m. The cables are deviated in steel saddles situated at the diaphragms of the concrete pylons and anchored in anchor blocks (Figure 11.87). The pull from the cables is transferred to the ground by rock anchors. The anchor blocks and the abutments are mutually connected by prestressed concrete ties.

Figure 11.83 Vranov Lake Bridge, Czech Republic

To stiffen the structure for the effects of the wind load, the deck is widened from mid-span toward the pylons. The deck is suspended at its outer edges on hangers that are perpendicular to the longitudinal axis. It was assembled from precast segments of a double-T cross-section stiffened by diaphragms at the joints. The 3.00 m long segments have a variable width corresponding to the variable width of the bridge deck. The two end segments are solid. Steel pipe conduits for gas and water lines were placed on the outer, but not mutually connected, overhangs. The deck was post-tensioned by four internal tendons that are led through the whole deck and anchored at the end segments. The vertical and horizontal curvatures allow the structure to be stabilised by stiffening the external cables situated within the edges of the deck; the cables pass across the expansion joints and are anchored at the end abutments.

The deck is supported at both ends by two multi-directional pot bearings situated on the diaphragms of the pylons (Figures 11.87(c) and 11.88). The horizontal force due to wind is transferred by steel shear keys.

The main cables are formed by 2×108 15.5 mm diameter strands grouted in steel tubes. To eliminate tension stresses in the cement mortar of the suspension cables, the deck was temporarily loaded before the cables were grouted. The load was created by radial forces caused by the tension of the external and internal cables temporarily anchored at the abutments. The suspension cables are fix connected to the deck at mid-span. The hangers are formed from solid steel rods of 30 mm diameter which are pin connected to the deck and main suspension cables.

The inclined pylons have an A-shape with curved legs connected by top and bottom diaphragms. The legs of the pylon were post-tensioned by draped cables to balance the bending stresses due

Figure 11.84 Vranov Lake Bridge, Czech Republic: (a) elevation, (b) plan

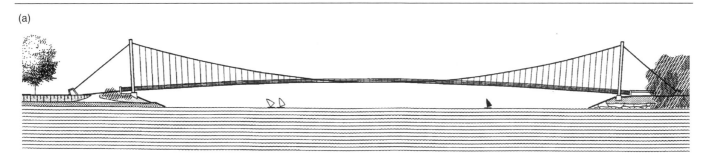

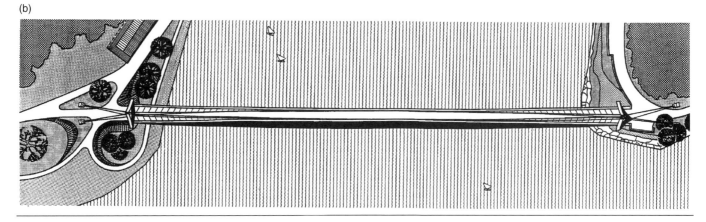

Figure 11.85 Vranov Lake Bridge: elevation

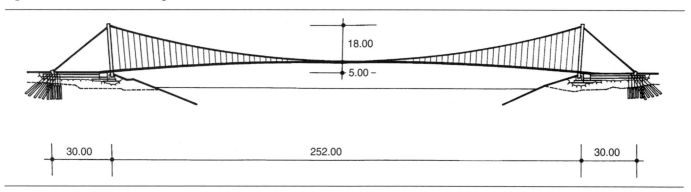

to the curvature of the legs. During the erection of the structure (Figures 8.25–8.28), the pylons were supported by pins. After erection, the pylons were cast in the footings. The anchor blocks protruding above the grade were post-tensioned to the anchor foundation slabs, where rock anchors are anchored, by prestressing rods.

The bridge forms a partly self-anchored system in which the arched deck is suspended on the cables and is flexibly connected to the abutments that in turn are mutually connected to the anchor blocks by prestressed concrete tie rods (Figure 8.17).

The construction of the bridge began in spring 1991 and was finished in spring 1993 (Figures 8.38–8.40). Due to the recreation season from June to mid-September and severe winter conditions, construction work could only be carried out during spring and autumn months.

Although the structure has a very slender deck, the users feel no unpleasant motion of the bridge when walking along it or when standing still to look at the surroundings. The bridge is widely used not only to cross the bay, but also as a local meeting point and for bungee jumping.

Figure 11.86 Vranov Lake Bridge – deck: (a) cross-section, (b) partial elevation

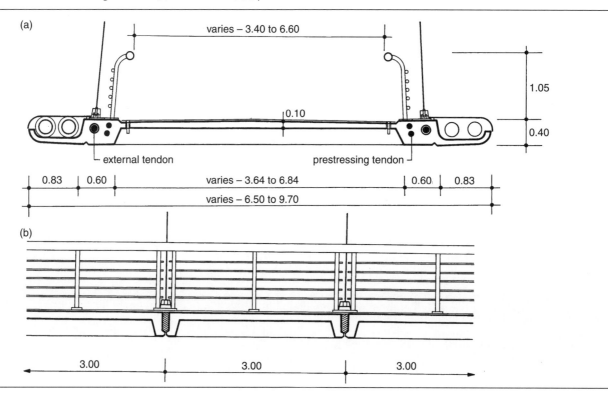

Figure 11.87 Vranov Lake Bridge – tower elevation: (a) section E–E, (b) section A–A, (c) section D–D, (d) section B–B, (e) section C–C

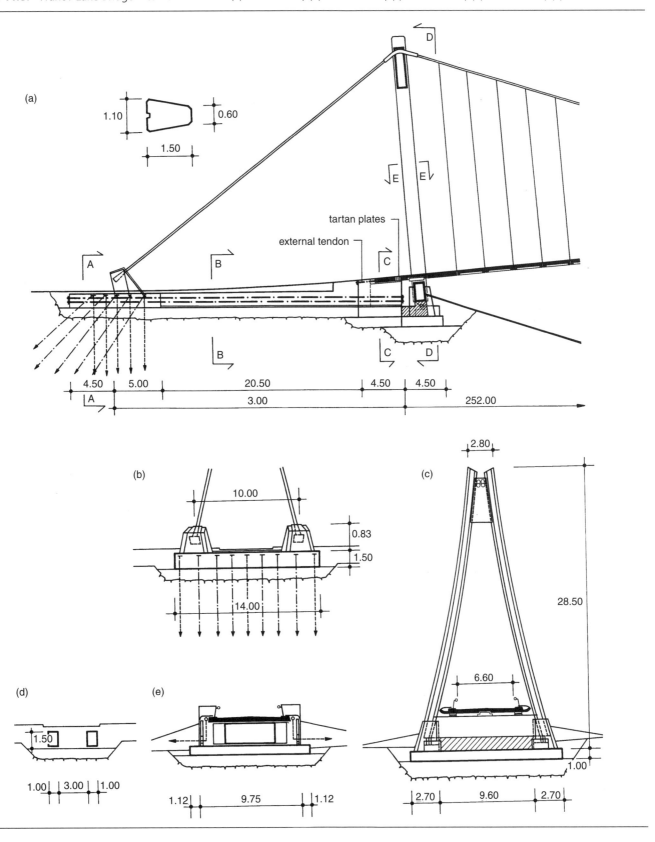

Figure 11.88 Vranov Lake Bridge – deck

Figure 11.89 Willamette River Bridge, Oregon, USA

platform and stairs (Figures 11.89 and 11.90) (Strasky and Rayor, 2000). The suspended spans are 23 m and 103 m for side spans and main span, respectively. The entire bridge consists of five spans including a curved ramp into a park on the east end. The typical deck width is 6.5 m.

The bridge was designed by the author and the contactor was Dopravni Stavby & Mosty, Olomouc.

11.2.5 Willamette River Bridge, Eugene, Oregon, USA

The bridge of overall length 178.8 m consists of two parts: precast suspended spans and cast-in-place approach spans, a

The typical deck section of suspended spans consists of precast concrete segments 3.0 m long, longitudinally post-tensioned together after erection (Figure 11.91). Two edge girders and the deck slab form the cross-section of the segments. Transverse diaphragms stiffen the segments at joints. The design of the bridge included an observation platform situated at the

Figure 11.90 Willamette River Bridge: (a) plan, (b) elevation

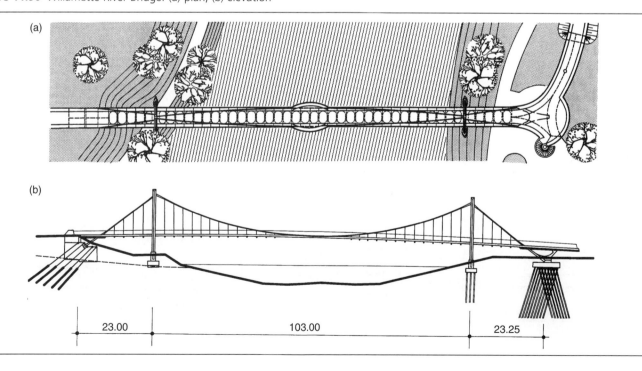

Figure 11.91 Willamette River Bridge – deck: (a) cross-section, (b) partial elevation

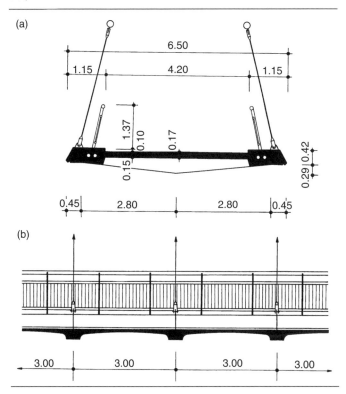

Figure 11.92 Willamette River Bridge – tripod

middle of the main span. The main cables therefore have to be threaded through the central wider deck panels of the main span.

The approach cast-in-place spans are formed by a solid slab of a shape which corresponds to the shape of transverse diaphragms of the segments. The side spans are fixed to the end abutment. Significant parts of these spans are the viewing platform and stairs that form an important structural member contributing to the stiffness of the system. The approach spans are post-tensioned in the direction of the suspension spans and ramp spans with crossing of the tendons above the strut of the tripod.

Suspended side spans are monolithically connected to the west abutment and east approaches. Expansion joints are designed at the towers. The main span is connected to side spans by dampers; these allow slow movement of the deck due to temperature changes and due to concrete creep and shrinkage, but also minimise sudden movement due to live and wind load.

Supports (or bents) 1 and 4 are fixed at structural steel anchors that resist longitudinal main cable forces. The use of the structural steel column/anchors allowed termination of the main cable above high water without extending the concrete anchor block to the level of the deck. This was particularly important

at bent 4 where the steel anchorage consists of a tripod-shaped support that anchors the main cables and supports a saucer-shaped viewing deck above (Figure 11.92). The tripod is formed by two tension ties connected by a top diaphragm and by a compression strut founded in one common footing (Figures 8.13 and 8.14). The tension from the main suspension cables is transferred via ties into the ranked steel piles.

The pylons are A-shaped (Figures 2.26 and 8.24). The slightly curved legs are connected by a top diaphragm and by a cross-beam situated in the deck level. During construction (Figure 8.24) the pylons were pin-connected. The pylons are fixed into the foundation after the erection of the deck.

The main suspension cables are formed by mono-strands that are grouted in steel tubes. They overlap at the pylons and are anchored in anchors that are welded to the plates of the leg (Figure 8.11). Rods with threaded ends form the suspenders.

The bridge was designed by OBEC Consulting Engineers with the collaboration of the author. The contractor was Mowat Construction Company, Vancouver, Washington.

11.2.6 McKenzie River Bridge, Eugene, Oregon, USA

The bridge was built by Wildish Company, Eugene who excavate gravel on both sides of the McKenzie River (Figure 11.93). The bridge carries a conveyor belt and an adjacent truck lane. It is assumed that after excavation is completed, the whole area will be converted into a park and the bridge will serve as a pedestrian bridge. The bridge was therefore designed for both a present and future load, and was completed in 2003.

The bridge consists of three spans of 36.576, 131.064 and 36.576 m for an overall bridge length of 204.216 m

Figure 11.93 McKenzie River Bridge, Oregon, USA

(Figure 11.94). The deck is assembled from precast concrete deck segments formed by two edge girders and a deck slab stiffened by integral floor beams located at the suspenders (Figures 8.35 and 11.95).

The tower consists of A-shaped concrete (Figures 8.23 and 11.96). The deck is directly suspended on the towers and vertical neoprene pads are placed between the segments and the legs of the towers. Since the tower legs are transversally post-tensioned by a PT bar running between the diaphragms of the segments, the legs and the deck form a transverse frame resisting transverse horizontal forces.

The main suspension cables are formed by PE-covered and greased mono-strands grouted inside the steel tubes. Solid steel rods form the suspender hangers with standard clevises at each end (Figures 2.36, 8.36 and 8.37). The steel pipe (Figure 8.30) used in the main cables and the steel rods in the suspension hangers are readily available galvanised sections and provide a long-term corrosion-resistant solution to maintenance problems experienced by traditional suspension bridge construction. The abutments at which the main suspension cables are anchored as well as the bents of the towers are founded on steel piles.

The bridge forms a partly self-anchored system in which the arched deck is suspended on the cables and is flexibly connected to the abutments (Figure 8.18).

The bridge was designed by OBEC Consulting Engineers with the collaboration of the author.

11.2.7 Halgavor Bridge, UK

The Halgavor Bridge is a 47 m span suspension bridge carrying a footpath, cycle track and bridleway over the busy A30 dual carriageway south of Bodmin, Cornwall (Figure 11.97) (Firth and Cooper, 2002). It is the first publicly funded bridge in the UK to use glass-fibre-reinforced polymer composites (GRP) as the principal structural material.

A glass-reinforced vinyl ester resin composite deck is suspended from a conventional primary support system comprising steel masts, steel spiral strand main cables and stainless steel hangers. The steel suspension cables, with the radial pattern of stainless steel hangers and parapet posts, are attached to inclined steel masts and tied back to concrete anchorages by pairs of steel backstays (Figure 11.98). The 3.5 m wide deck comprises two C-shaped edge beams, with a 37 mm thick sandwich construction deck plate supported on transverse plates and on a longitudinal central spine.

The parapet is formed by stressed strands and a special stressed stainless steel wire mesh to maximise transparency. Red cedar panels provide the necessary low-level visual solidity and reflect the wooded nature of the surrounding environment. They also deliberately introduce damping to the structural system to control vibration amplitudes. The finished bridge surface is formed of interlocking rubber blocks made from recycled car tyres bonded to the GRP deck. These also add damping and give a maintenance-free and attractive finish suitable for horses, pedestrians and cyclists.

Figure 11.94 McKenzie River Bridge: (a) plan, (b) elevation

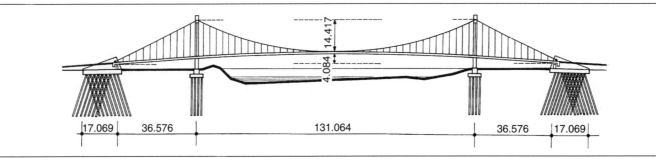

Examples

Figure 11.95 McKenzie River Bridge deck: (a) cross-section, (b) partial elevation

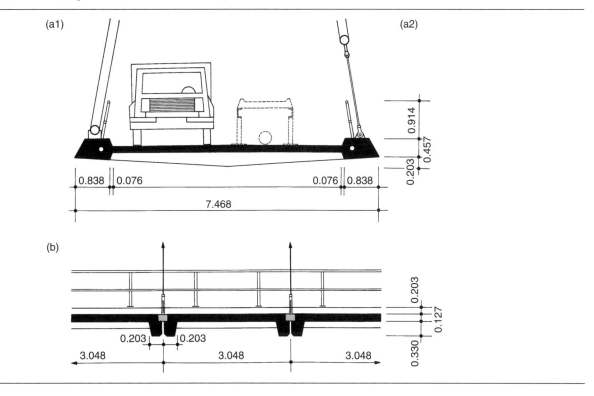

Figure 11.96 McKenzie River Bridge tower: (a) cross-section, (b) partial elevation

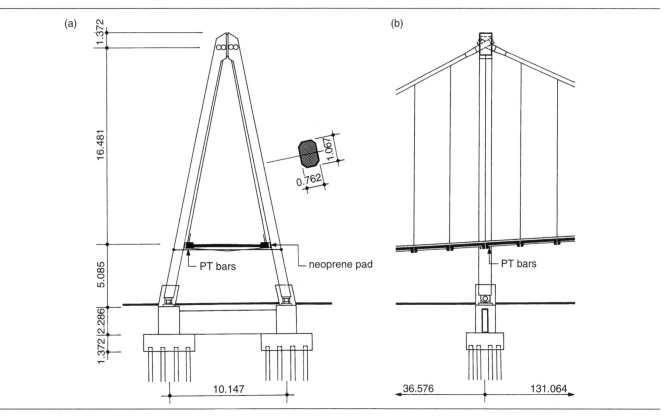

Figure 11.97 Halgavor Bridge, UK

The principal GRP structural elements are as follows. The C-shaped edge beams, which are up to 12 mm thick, were formed by conventional hand lay-up methods in a purpose-built timber mould. A variety of flat plates for the internal stiffeners and soffit panels were fabricated using vacuum infusion techniques. The sandwich deck panel, comprising two 3.5 mm GRP skins with a 30 mm plastic foam core, was manufactured by vacuum infusion.

The deck was assembled upside-down on a curved set of supports to match the vertical profile of the bridge. The 31 m central section was transported to a lay-by close to the site by road where it was fitted with hangers, parapet posts and the main cables. The whole assembly was lifted into place by crane in a single overnight operation. With the assembly hanging on the crane above its final position, the ends of the main cables were picked up by small mobile cranes and slotted into the masthead anchorages.

The assembly was then lowered until it was hanging on the main cables. The cable and hanger lengths were predetermined so that, at this point, the deck assumed the correct geometry. Pinned connections were then made between the ends of the deck and the steel mast arms, enabling the crane to be released. The *in situ* splice at each end of the centre section was made by band later.

The Halgavor Bridge complies with current UK requirements for vertical response due to wind or pedestrian excitation as

Figure 11.98 Halgavor Bridge: (a) cross-section, (b) elevation

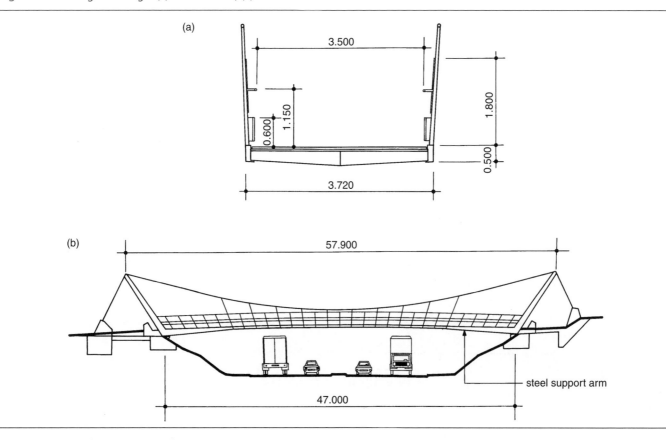

Figure 11.99 Bridge Kelheim (courtesy of Schlaich, Bergermann and Partners)

defined in BD37/88, and horse-riders, cyclists and pedestrians have used it without complaint. The bridge exhibits no perceptible lateral vibration under pedestrian footfall.

The bridge, which opened in July 2001, was designed by Flint & Neill Partnership, London, UK and the main contractor was Balfour Beatty.

11.2.8 Bridge Kelheim, Germany

The bridge, which was completed in 1987, was built in a small city of Kelheim, Germany. It crosses a new 47 m wide Rhine-Main-Danube Canal in smooth curves, naturally connecting the pedestrian traffic between the banks (Figures 10.8, 11.99 and 11.100) (Schlaich and Seidel, 1998).

The deck, which is curved on the plan with radius 18.89–37.79 m, is suspended on one suspension cable situated inside the plan curvature (Figures 10.9 and 11.101). Two inclined masts situated on each bank support the suspension cables with hangers. The deck, a hollowed out cross-section of depth 1.00 m, combines high bending and torsional resistance. The curved deck that is fixed to the abutments is internally prestressed in the circumferential direction by an eccentric tendon.

The geometry and initial stresses in the cables were designed in such a way that the vertical components of the hanger forces balance the dead load (Figure 11.102). The horizontal components of the hanger force, together with the radial forces from prestressing cables situated close to the top fibre of the cross-section, create a moment that balances a torsional moment caused by vertical forces.

The bridge deck was cast on the falsework and then the masts and suspension cables were installed. By full understanding of

Figure 11.100 Bridge Kelheim: (a) plan, (b) elevation

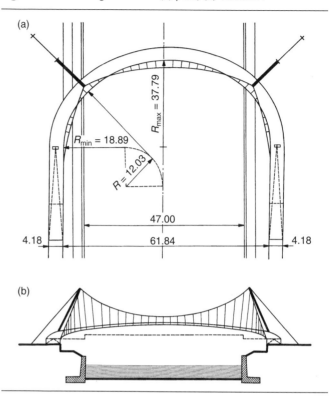

prestressing and clever arrangement of the suspension cables, a true structure was developed.

The bridge was designed by Schlaich, Bergermann and Partners, Stuttgart, Germany.

11.2.9 Bridge in Deutsches Museum in Munich, Germany

Inside the Deutsches Museum (the German science and technology museum in Munich), a circular pedestrian bridge was

Figure 11.101 Bridge Kelheim cross-section of the deck

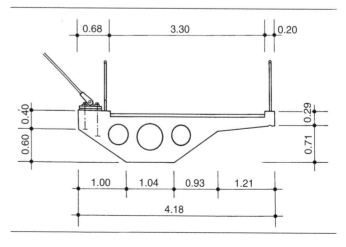

Figure 11.102 Bridge Kelheim – balancing of moments

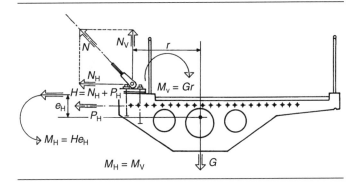

Figure 11.103 Museum Bridge structural arrangement (courtesy of Schlaich, Bergermann and Partners)

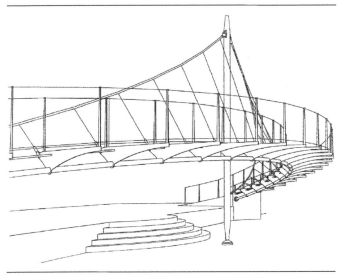

needed to provide access to some exhibits while simultaneously highlighting the art of bridge engineering (Schlaich, 2000). The bridge follows up an earlier structure built at Kelheim. The structure demonstrates that a circular bridge deck can be suspended only on one side and is able to counteract the torsion by an inner pair of forces.

In Munich, these forces are visible in the form of circular cables above a circular arch beneath a glass deck (Figures 11.103–11.106). The bridge is formed of the circular arch which comprises a round bar suspended on one side of the suspension cable. The arch is fixed to the abutments. The arch supports steel diaphragms in which the hangers are anchored and horizontal strands are deviated. The radial forces from the strands create a torsional moment that balances the moment created by the glass deck.

Figure 11.104 Museum Bridge, Munich, Germany: (a) elevation, (b) cross-section, (c) plan

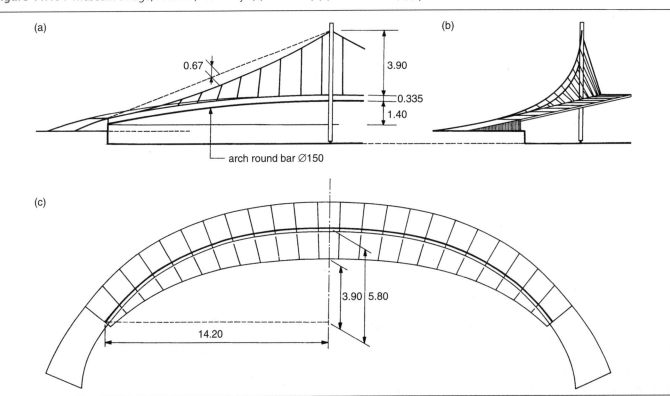

Figure 11.105 Museum Bridge cross-section of the deck

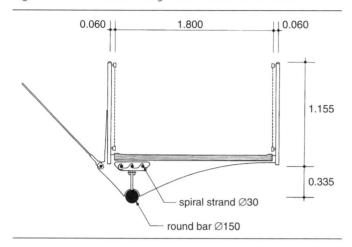

The static function of the bridge is evident from Figure 11.107. The structural solution was developed from the structures discussed in Figure 10.14(b).

The single mast situated at the centre of the bridge is supported by a pin. The interested visitor of the museum may study the stabilisation of the mast. Although it is guided by only two in-plane suspension cables, it is stable due to the fact that their anchorages are placed above and not on the same level or even below the base hinges of the mast.

The bridge, which was built in 1998, was designed by Schlaich, Bergermann and Partners, Stuttgart, Germany.

11.2.10 Harbor Drive Bridge, San Diego, California, USA

The bridge that crosses Harbor Drive and several railroad tracks connects a new downtown ballpark with San Diego

Figure 11.106 Museum Bridge – deck (courtesy of Schlaich, Bergermann and Partners)

Figure 11.107 Museum Bridge – calculation model

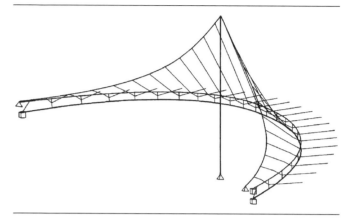

Convention Center and a parking garage (Figure 11.108). The City Development Corporation (the San Diego Redevelopment Agency) needed a pedestrian structure that would also serve as a landmark for the New Downtown, and was prepared to invest in aesthetic considerations. A curved suspension was therefore accepted.

The bridge, which is curved in plan with a radius of 176.80 m, forms a self-anchored suspension structure suspended by the hangers on the inside of the curve (Figure 11.109). The suspended span of length 107.60 m is monolithically connected to stairs at both ends. The stairs of length 13.54 and 21.97 m form part of the structural system that transfers the stresses to the abutments supported on piles. The ramp to the parking structure on the south side, as well as the elevator on the north side, is structurally independent from the deck.

The 39.80 m tall pylon which supports the main cable is founded on the convex side of the deck, leans over the deck and supports the main cable on the inside of the curve. It is stabilised with two

Figure 11.108 Harbor Drive Bridge, California, USA

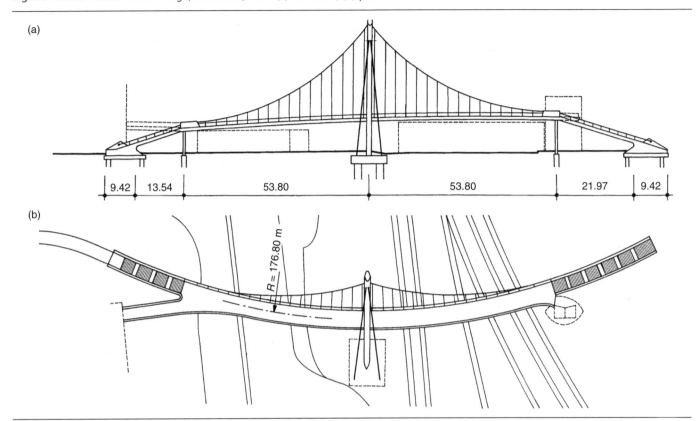

Figure 11.109 Harbor Drive Bridge, California, USA: (a) elevation, (b) plan

backstays and internal post-tensioning. To take the overturning moment, the footing consists of 12 tie-downs on the back and 1.40 m diameter piles in front. The pylon itself was constructed using four cast-in-place segments and was prestressed.

The main cable stretches from the abutment to a deviator at the top of the stairs to the anchorage at the top of the pylon. It is made by prestressing strands encased in stainless steel pipes. The hangers are attached to the steel pipe of the main cable and to the handrail on the bottom. The top of the handrail also carries a large post-tensioning cable which is anchored at the deviators at the top of the stairs. This cable is overlapped by the internal cables that prestressed the stairs.

The suspended deck is formed by a non-symmetrical box girder with overhangs on one side supported by ribs (Figure 11.110). The girder is prestressed not only by internal tendons situated at the top slab, but also by horizontal components of the hanger forces and external cables. The inner railing, in which the hangers are anchored, is therefore a part of the structural system. The geometry of the deck, position of the anchoring of the hanger and position of the external cable and internal tendons were determined in such a way that the horizontal forces balance the moment created by eccentricity of the suspension.

The main suspension cables and hangers were erected after casting of the deck and pylon. The required shape was obtained by tensioning of the main cable and adjusting the length of the hangers. Since the deck also moved in a horizontal direction due to prestressing, the deck was cambered not only in the vertical but also in the horizontal direction. The formwork of the deck was placed on Teflon plates.

The structural solution was developed on the basis of very detailed static and dynamic analyses. The multi-mode analysis completed for a given response spectrum has demonstrated that the structure has a satisfactory response to seismic load. The aerodynamic stability of the bridge was tested in the wind tunnel on a geometrically and aerodynamically similar model of the structure built to a scale of 1:70. The tests were carried out by M. Pirner (ITAM). Detailed analyses of the response of the structure to people and wind were also performed.

Construction of the bridge was completed in March 2011. The bridge was designed by TY Lin International, San Diego, California. The conceptual design is the work of Strasky and Anatech (Jiri Strasky and Tomas Kompfner) who also checked the final design. The bridge was built by Reyes Construction Inc., National City, California.

Figure 11.110 Harbor Drive Bridge, California, USA: cross-section

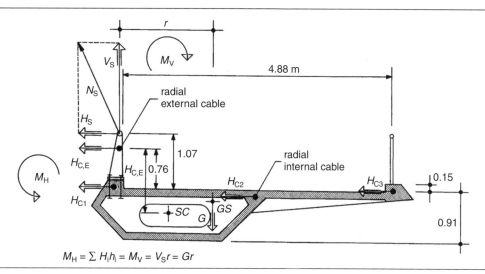

$M_H = \Sigma H_i h_i = M_V = V_S r = Gr$

11.2.11 Ishikawa Zoo Bridge, Japan

The Ishikawa Zoo Bridge built in 1999 in Tatsu-No-Kuchi Town in Ishikawa, Japan is formed by a slender deck slab that is supported by a suspension chord. The chord is formed by a one-span stress ribbon (Figures 11.111 and 11.112) (Sinohara, 1999).

The stress ribbon is assembled of precast segments and cast-in-place haunches designed at supports. The segments are suspended on bearing tendons situated in the troughs and are post-tensioned by prestressing tendons situated within the segments. The segments were erected with steel struts that support the deck (Figure 8.41). The deck segments were shifted to the design position along the temporary steel

Figure 11.111 Ishikawa Zoo Bridge, Japan: completed structure (courtesy of Sumitomo Mitsui Construction Co., Ltd)

girders that were supported by steel struts (Figure 8.42). The deck segments are also post-tensioned by internal tendons situated within the segments.

After all precast members were erected, the joints between the stress ribbon and deck segments, troughs and haunches were cast and additionally post-tensioned.

The bridge was designed by Natural Consultant Co., Ltd and the contractor was Sumitomo Mitsui Construction Co., Ltd.

11.2.12 Shiosai Bridge, Japan

The Shiosai Bridge built in 1995 across the Kikugawa River in Shizuoka Prefecture, Japan, is formed of a prestressed concrete deck of four spans that is supported by a continuous suspension chord (Figure 2.29). The chord is formed by a stress ribbon of four spans (Figures 11.113 and 11.114) (Horiuchi *et al.*, 1998).

To minimise the horizontal force in the stress ribbon, the sag to span ratio was set at 1/10. For the same reason, a lightweight concrete of specified concrete strength 40 MPa and density of 18.5 kN/m³ was used. The tendons are formed by epoxy-coated strands.

The stress ribbon is assembled of precast segments of variable depth from 0.250 m at mid-span to 0.480 m at piers. During the erection, the segments were suspended on bearing tendons situated in the troughs. After the erection of the segments and casting of the joints between them, the stress ribbon was post-tensioned by prestressing tendons situated in the segments. These tendons were anchored at the abutments and at the top of the piers where they overlapped.

Figure 11.112 Ishikawa Zoo Bridge: (a) cross-section, (b) elevation

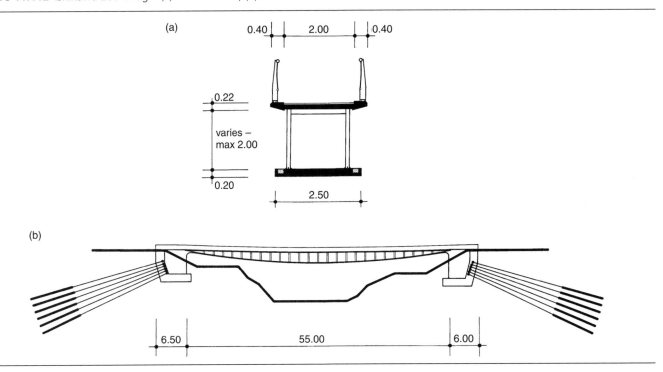

The support columns were precast, divided into three sections and connected by prestressing 32 mm diameter epoxy-coated prestressing bars. The slab decks consisted of 0.40 m deep hollow prestressed concrete girders. These girders were rigidly connected to the tops of the piers and columns. Two columns situated close to the abutments are pin-connected both to the stress ribbon and the decks.

Between the deck and abutment, horizontal neoprene bearings and prestressing tendons were installed. In this way, a partially self-anchored structure of increased stiffness was created.

Figure 11.113 The Shiosai Bridge, Japan: completed structure (courtesy of Sumitomo Mitsui Construction Co., Ltd)

The construction of the bridge began with the casting of the abutments and piers. The bearing tendons were then installed and precast segments of the stress ribbon were erected. After casting the joints between the segments, the stress ribbon was post-tensioned. The columns and precast girders were then erected.

The function of the bridge was verified by static and dynamic loading tests. The structure was loaded by a truck and excited either by a mechanical rotation exciter, human force or by a moving truck. For example, the logarithmic decrement of damping varied from 0.04 to 0.06 for the longitudinal bending modes.

The bridge was designed by Shizuoka Construction Technology Center and the contractor was Sumitomo Mitsui Construction Co. Ltd, PS Corporation.

11.2.13 Ganmon Bridge, Japan

The Ganmon Bridge was built in 2001 in a recreation area of Ishikawa Prefecture, Japan. It forms a self-anchored suspension structure in which the tension cord is formed by a stressed ribbon. The technology of the erection was also developed from the stress ribbon technique (Figure 11.115) (Komatsubara et al., 2002).

The bridge consists of a deck slab, steel struts and curved bottom slab (stress ribbon) (Figure 11.116). The deck slab

Figure 11.114 The Shiosai Bridge: (a) cross-section, (b) elevation

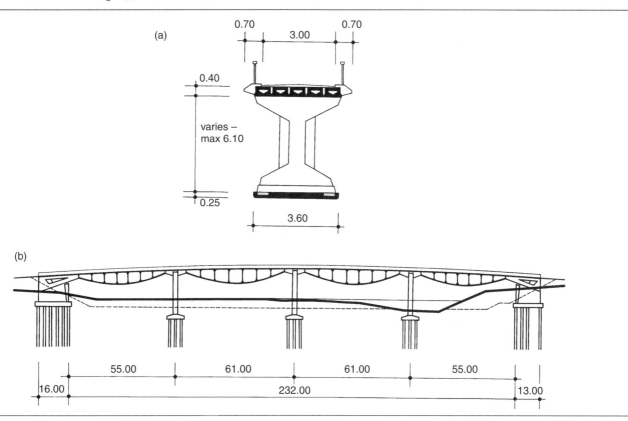

and bottom slab are assembled of precast segments. The span length is 37.0 m and the sag of the bottom slab is 2.80 m. The structural arrangement was developed from the technology of the erection.

The abutments were first cast and rock anchors were installed. Special anchor segments were then placed on elastomeric bearings on the abutments. Their position was secured by horizontal and vertical prestressing rods anchored in the abutments.

The bearing tendons with couplers were then pulled across the valley and tensioned to the design stress. The precast segments with steel struts were then erected. The deck slab segments were subsequently erected. They were shifted to the design position

Figure 11.115 Ganmon Bridge, Japan: completed structure (courtesy of Sumitomo Mitsui Construction Co., Ltd)

Figure 11.116 Ganmon Bridge: (a) abutment during erection, (b) abutment during service, (c) cross-section, (d) elevation

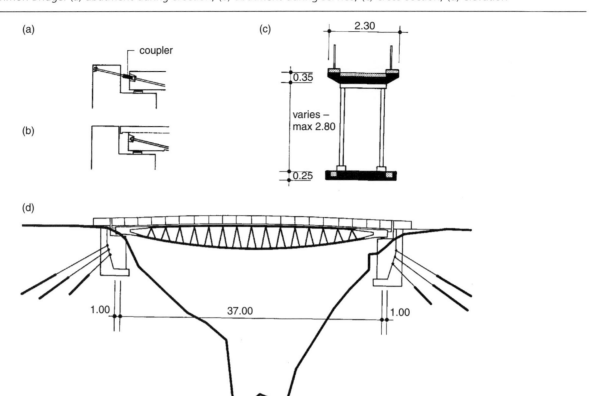

along the rails that were supported by the steel struts. When all precast members were assembled, the joints between the precast members and trough in which the bearing tendons were placed were cast.

The prestressing tendons situated in the deck slab and in the stress ribbon were post-tensioned. Ring nuts of the couplers of the bearing tendons were tightened and the tension in bearing tendons was released by hydraulic jacks at their anchors. Compression stresses were therefore generated both in the deck slab and in the stress ribbon. In this way, the structure changed its static system from an externally suspended structure into a self-anchored structure.

The design assumptions and function of the bridge were studied using a model of the structure. The completed bridge was also checked by dynamic tests which verified the calculated natural modes and frequencies. The logarithmic decrement of damping was determined for the vertical modes and found to vary from 0.03 to 0.04.

The bridge was designed by Nihonkai Consultant Co., Ltd. and the contractor was Sumitomo Construction Co., Ltd.

11.2.14 Ayumi Bridge, Japan

The Ayumi Bridge built in 1999 crosses the Kanogawa River in Numazu City in Japan (Figure 11.117) (Uchimura et al., 2002). The bridge is formed by a continuous girder of four spans of lengths 16.63, 79.50, 41.00 and 41.00 m; the spans cross a levee suspended on a pylon. In the spans above the river, the girder is post-tensioned by external cables situated below the deck (Figure 11.118). In this way, the structure forms a multi-span self-anchored suspension structure.

The deck is formed by a box girder that was assembled of precast segments. The stay cables that are situated in the bridge axis are anchored in anchor blocks situated under the deck. They overlap with external suspension cables that support the deck via steel struts.

The geometry and forces in the stay and suspension cables were determined in such a way that they balance the effects of the dead load. The structure was analysed as a geometrically non-linear structure. The aerodynamic stability of the structure was checked on the sectional model in the wind tunnel. The function of the bridge was also checked by static and dynamic loading tests.

Figure 11.117 Ayumi Bridge, Japan: completed structure (courtesy of Sumitomo Mitsui Construction Co., Ltd)

The bridge was designed by Architects & Planners League and CTI Engineering; the contractor was Sumitomo Construction Co., Ltd.

11.2.15 Tobu Bridge, Japan

The Tobu Bridge built in 1998 in Tobu Recreation Resort in the Nagano Prefecture, Japan forms a self-anchored suspended structure (Figure 11.119) (Tsunomoto and Ohnuma, 2002). It is formed by a slender deck slab that is supported via steel struts by external cables (Figure 11.120). The cables are anchored in the deck.

The geometry of the external cables, force in the cables and position of the struts were determined in such a way that the deck slab acts as a continuous girder of equal spans for the dead load. However, for all other effects, the structure acts as

Figure 11.118 Ayumi Bridge: (a) cross-section, (b) elevation

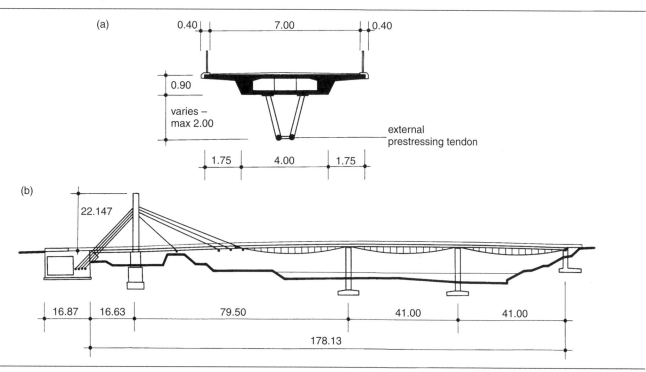

Figure 11.119 Tobu Bridge, Japan: completed structure (courtesy of Oriental Construction Co., Ltd)

self-anchored suspension bridge that was analysed as a geometrically non-linear structure. Although the external cables create large compressive stresses in the deck slab, the slab was also post-tensioned by internal tendons. The effect of creep and shrinkage of concrete were also carefully analysed.

Since the full live load causes tensile stresses of 136 MPa in the cables, the allowable stresses in the cables were set to $0.4f_u$ and their anchors were designed for fatigue. To verify the function of the bridge, a model of the actual structure at a scale of 1:5 was also tested until failure.

The deck slab was cast *in situ* on the scaffolding with an opening for the erection of the steel struts and external cables (Figure 8.44). The geometry of the deck slab was carefully checked during the post-tensioning of the cables. Dynamic tests of the complete structure also confirmed very good function of the structure.

The bridge was designed by Toyo Ito & Associates, Architects and the contractors were Kajima Co., Ltd and Oriental Construction Co., Ltd.

11.2.16 Inachus Bridge, Japan

The Inachus Bridge built in 1994 in the city of Beppu in southern Japan provides access across the river to a park (Figure 11.121) (Kawaguchi, 1996). This overpass combines the tensile strength of modern suspension structures with the fine beauty of the traditional stone arch bridges.

Figure 11.120 Tobu Bridge: (a) cross-section, (b) elevation

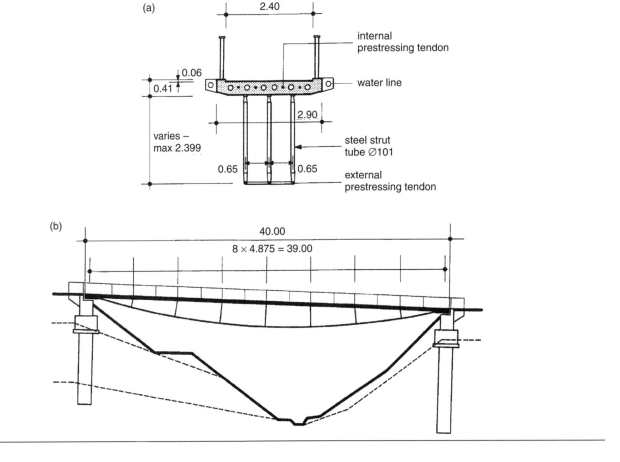

Figure 11.121 Inachus Bridge, Japan: completed structure (courtesy of Kawaguchi)

The bridge was designed to have a lenticular shape with an arched upper chord and a suspended lower chord. The span length of the bridge is 34.00 m and the distance between the top and bottom chord is 2.20 m (Figure 11.122).

The upper chord is a circular arc with a maximum slope of 12% at the ends. For aesthetic reasons, it is narrower in plan at the centre and widens toward the ends. It consists of 78 granite blocks 40 cm wide and 25 cm deep of graduating lengths 2.6–3.6 m. The granite chord serves not only as a principal structural member, but it also forms the deck for pedestrian traffic.

The upper chord is post-tensioned by prestressing tendons running through holes drilled in the centre of the granite blocks. The prestressing guarantees that there is no tension in the joints caused by live load. A transverse reinforcing bar was also placed in each joint between the blocks to resist transverse bending stresses (Figure 11.123).

The lower chord has the longitudinal shape of a funicular polygon, almost symmetrical to the upper chord. It consists of steel plates arranged in a chain. To provide the bridge with torsional resistance, the lower chord is forked at both ends and anchored to end blocks of reinforced concrete (in which the prestressing tendons of the upper chord are also anchored). The upper and lower chords are mutually connected by struts formed by steel tubes arranged to form inverted pyramids (Figure 11.124).

The visual effect of the bridge is one of simplicity and strength, and the natural granite material fits well with the site. The only ornamentation on the bridge is in the form of ceramic covers on the hinge bolts of the lower chord.

The bridge was designed by M. Kawaguchi and the main contractor was Maeda Corporation.

11.2.17 Johnson Creek Bridge, Oregon

The proposed bridge across the Johnson Creek is situated on the Springwater Trail in Milwaukee, Oregon. The bridge is formed by a partially self-anchored suspension structure of span 60.80 m (Figures 11.125 and 11.126). The deck has a variable

Figure 11.122 Inachus Bridge: (a) elevation, (b) plan

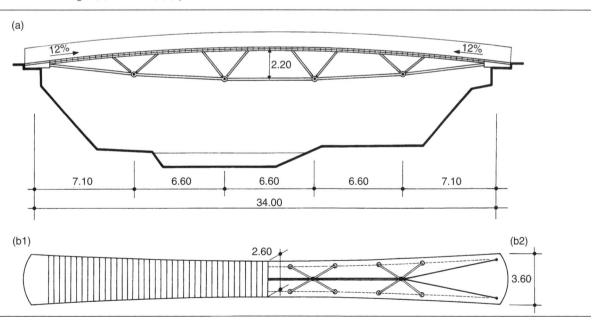

Figure 11.123 Inachus Bridge: structural arrangement

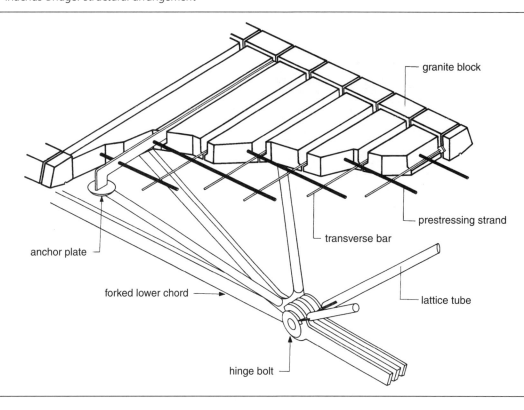

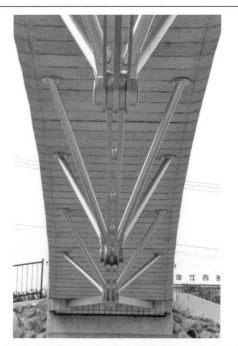

Figure 11.124 Inachus Bridge: view from below (courtesy of Kawaguchi)

longitudinal slope with maximum slope at abutments of 5%. The proposal tries to solve the problem of the erection of the self-anchored structures whose decks are usually cast or assembled on falsework before the suspension cables are installed and tensioned.

The deck is formed by precast segments and a composite deck slab. The segments are identical to the segments that were

Figure 11.125 Johnson Creek Bridge, Oregon, USA: rendering

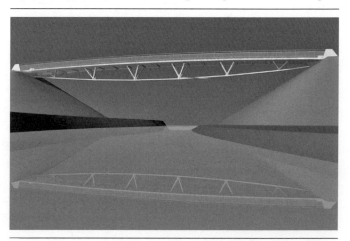

Figure 11.126 Johnson Creek Bridge, Oregon, USA: (a) cross-section, (b) elevation

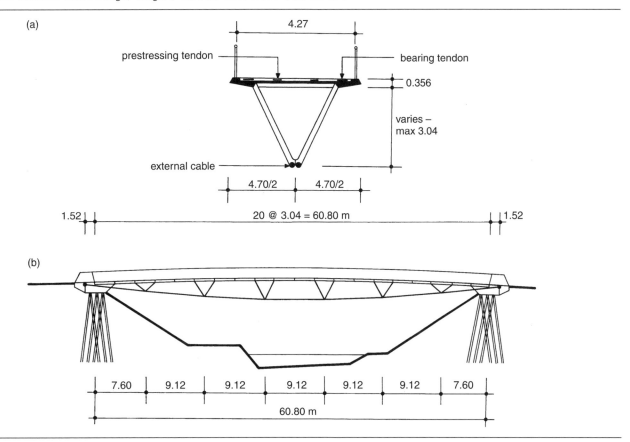

used in the construction of the Grants Pass Pedestrian Bridge across the Rough River in Oregon (see Section 11.1.10). Each third segment is connected to triangular steel struts that transfer the radial forces from the external cables into the deck (Figure 11.126(a)). The deck is fixed to end abutments that are founded on battered steel piles. The external cables are also anchored to the end abutments. The piles are economically designed to resist the erection horizontal force only.

The erection procedure (Figures 8.45 and 8.46) was developed from the erection of stress ribbon structures. The erection-bearing cables that are situated within a composite portion of the deck are erected and tensioned first. The cables are then anchored to the end abutments that resist the horizontal force corresponding to the weight of segments and relatively large erection sag.

11.2.18 Seishun Bridge, Tsumagoi Village, Gunma Prefecture, Japan

The bridge across the Ohorigawa River connects a junior high school and adjacent athletic field. Since the permanent rock anchors were restricted, it was not possible to build a classical stress ribbon structure. A self-anchored suspension structure with two rows of suspension cables (a double suspension structure) was therefore built (Figure 11.127).

The bridge of single span 57.50 m is straight and in a longitudinal slope of 3.5%. The deck is assembled of precast U-shaped segments that are suspended on two suspension cables situated on the outer sides of the webs (Figure 11.128). These primary cables were used for the erection of the segment. The secondary cables are situated under the deck; their position is determined by steel struts connecting the segments with the cables.

The erection of the deck began by placing the end segments and stretching and tensioning both cables. The cables were provided with couplers that allowed connection of their anchors to short cables anchored at the end abutments, which were temporarily anchored into the soil. The span segments were then placed on the primary cables and shifted along them into the design position. When all segments were in place, the struts were erected and connected to the deck and secondary cables.

The geometry of the deck was adjusted by the tensioning of the secondary cables. The joints between the segments were then

Figure 11.127 Seishun Bridge, Japan

cast. By releasing the short connection cables, the earth-anchored structure was transformed into a self-anchored suspension structure in which the tension forces of the cables are resisted by the compression capacity of the concrete deck.

Construction of the bridge, which was designed and built by Sumitomo Mitsui Construction Co. Ltd, was completed in 2006.

11.3. Cable-stayed structures
11.3.1 Neckar River Bridge, Mannheim, Germany

The bridge built in 1975 across the river Neckar connects the historic centre with the commercial residential zone Collini (Figure 9.1) (Völkel *et al.*, 1977).

The bridge has three spans with a main span of 139.50 m (Figure 11.129). It crosses the river at an angle of 70°. Two lateral planes of stay cables that have a semi-harp pattern are suspended on two single steel pylons. The deck that is hinge-connected to the piers has a central movable hinge to allow a longitudinal displacement.

The deck consists of a reinforced concrete slab of trapezoidal cross-section of maximum thickness 0.60 m. The deck is widened at pylons and strengthened to 1.20 m. Widening of the deck allows the pedestrians to pass on either side of the pylons.

Figure 11.128 Seishun Bridge, Japan: (a) cross-section, (b) elevation

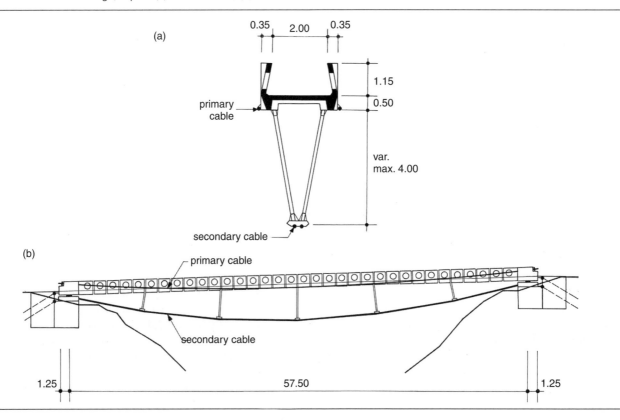

Figure 11.129 Neckar River Bridge, Mannheim, Germany: (a) cross-section of the deck, (b) elevation, (c) plan

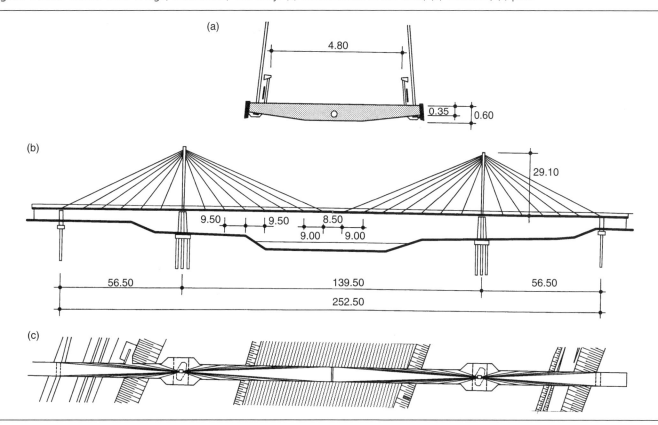

The pylons are formed by square section masts. The stay cables are anchored in the steel section by means of metal plates. The stay cables are formed by parallel wires grouted in PE ducts.

The side spans were cast on falsework while the main span was progressively cast in and suspended on the pylons in two cantilevers starting from the pier tables.

The bridge was designed by Leonhard and Adrä, Stuttgart and was built by Bilfinger and Berger, Mannheim.

11.3.2 Scripps Crossing at UCSD, La Jolla, CA, USA

The Scripps Crossing forms a key link in the University of California, San Diego (UCSD) master plan to provide pedestrian and wheelchair access from the research pier to the upper campus (Figure 11.130) (Seible and Burgueno, 1994).

The adopted solution consists of a cable-stayed bridge with a single eccentric pylon on the uphill side (Figure 11.131). The cross-section of the deck consists of two edge girders and recessed 200 mm deck providing a closed smooth soffit. The cable-stay concept consists of a total of 15 stay cables. There are five pairs of two stays on the front side straddling the tapered walkway from the single-tapered pylon with octagonal cross-section, and a single plane of cable stays penetrates the back span of the slotted deck with anchorage beam between the abutment and pylon shaft.

The flexible downhill support is formed by a steel frame also supporting the elevator and stairs. The deck and pylon form a

Figure 11.130 Scripps Crossing at UCSD, La Jolla, California, USA (courtesy of Frieder Seible)

Figure 11.131 Scripps Crossing at UCSD, La Jolla, California, USA: (a) cross-section of the deck, (b) elevation, (c) cross-section of the pylon, (d) plan

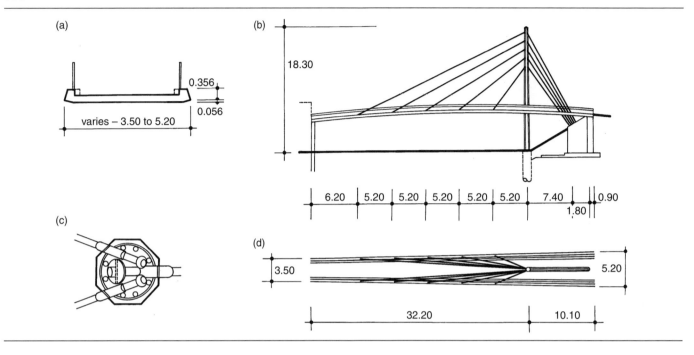

closed-force system which itself can be free-standing on the single pile shaft which is independent of soil contribution.

The stay cables consist of high-strength bars encased in stainless steel tubes and grouted by cement mortar. The bridge was cast on the falsework and the stay cables were tensioned and grouted when the concrete had sufficient strength.

The bridge was designed by F. Seible of UCSD.

11.3.3 Hungerford Bridge, London, UK

The twin bridge built on each side of the railway bridge enables easy crossing of the River Thames between Charring Cross station and the South Bank Centre (Figures 2.28 and 11.132)

(Parker et al., 2003). The bridge, which was opened in 2002, is the result of a bridge competition.

The bridge is formed by a multi-span cable-stayed structure supported by Macalloy bar stays (Figures 2.34 and 2.39). The pylons are formed of tapered steel cylinders held at an angle to the vertical by additional Macalloy bars and supported on concrete foundations (Figures 11.133 and 11.134). The deck is formed by two edge girders and a deck slab and is stiffened by transverse diaphragms at anchors.

Due to the possibility that some unrecorded, unexploded World War II bombs could be present, the foundation near the North Bank was moved out of the river. The original asymmetric

Figure 11.132 Hungerford Bridge, London, UK: elevation

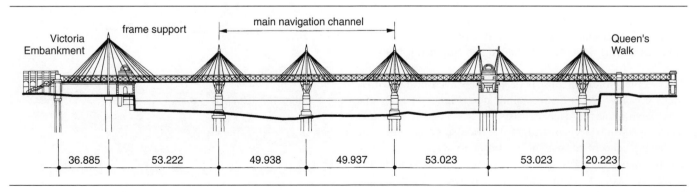

Figure 11.133 Hungerford Bridge, London, UK: tower elevation

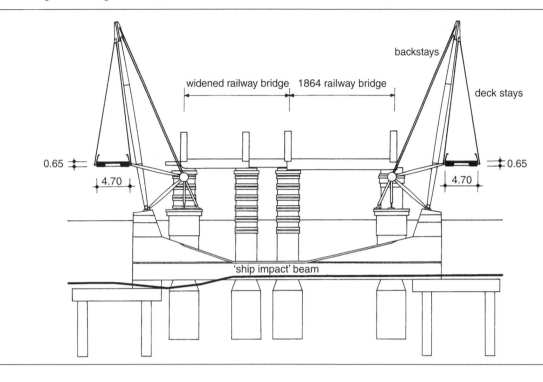

Figure 11.134 Hungerford Bridge, London, UK: anchoring of the stay cables

pylon arrangement was not possible at this location, so an A-frame support was introduced.

The main bridge decks were constructed using the incremental-launch technique. A casting bed was built in a non-navigable span and the section of the deck approximately 50 m long was cast. A temporary stiffening truss was bolted to the top of the deck section and hydraulic jacks were used to pull the deck northwards. The whole process was repeated until the full length of the deck was completed (Figure 11.135(a)). A temporary support was provided at each pier position.

A floating crane was used to lift the pylons into place. Each pylon was erected with its backstays attached and the stays were kept straight during the lifting by attaching them to steel strong-backs.

On completion of the launch, the deck was jacked up. The deck stays were fabricated to the present length, temporarily supported by strong-backs and installed with a preset sag (Figure 11.135(b)).

After the installation of all deck stays, the deck was slowly lowered until it was supported entirely by the stays (Figure 11.135(c)). The temporary truss and tower were then removed (Figure 11.135(d)).

Figure 11.135 Hungerford Bridge, London, UK: construction sequences

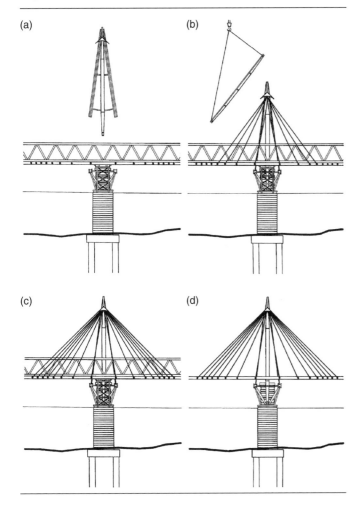

The dynamic behaviour of the bridge was studied in detail using a computer model, wind tunnel test and measurements at the site. The tests demonstrated the good function of the bridge.

The bridge was designed by WSP and Lifschutz Davidson, the contractor was Costain Norwest Holst joint venture and the contractor's designer was Gifford and Partners.

11.3.4 Bridge across motorway D47, Czech Republic

The bridge crosses the motorway D47 near the city of Bohumin in the Czech Republic (Figure 11.136). The bridge, which is used both by pedestrians and bicycles, is curved in plan with a radius of 220 m. The freeway, which is now under construction, is situated in the northeast part of the country and the bridge will be the first flyover on the route from Poland.

The bridge of two spans of 54.937 and 58.293 m is suspended on a single mast situated in the area between the freeway and local roads (Figure 11.137). The bridge deck is fixed to the end abutments formed by front inclined walls and rear walls forming the anchor blocks.

At the preliminary design stage, the deck of effective width 6.00 m was suspended on two inclined planes of stay cables (Figure 9.11). The deck was formed by a slender deck slab stiffened by transverse diaphragms and edge girders protruding above a sidewalk. The stay cables were anchored at anchor block situated outside the edge girders.

Due to heavy bicycle traffic, the city of Bohumin has to separate the pedestrian and bicycle pathways. The deck was therefore modified and is formed by a central spine girder with non-symmetrical cantilevers carrying the pedestrians and bicycles

Figure 11.136 Bohumin Bridge, Czech Republic, before completion

Figure 11.137 Bohumin Bridge, Czech Republic: elevation

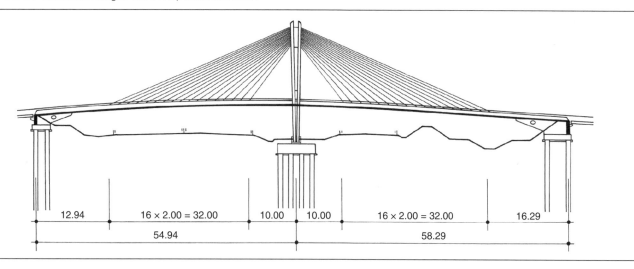

(Figures 9.12 and 11.138). To balance the load, the shorter cantilever is solid while the longer is formed by a slender slab stiffened by transverse ribs.

The mast is formed by two inclined columns of two cell box sections that are tied by top and bottom steel plates connecting the central webs of the boxes. The boxes are filled with concrete that was pressed from the footing of the columns to their tops. The stays are anchored on the central webs.

To reduce the torsional stresses due to the dead load, the deck was cast in two steps. The central portion of the deck supported

Figure 11.138 Bohumin Bridge, Czech Republic – tower elevation: (a) suspension on two inclined planes of stay cables, (b) suspension on one central plane of stay cables

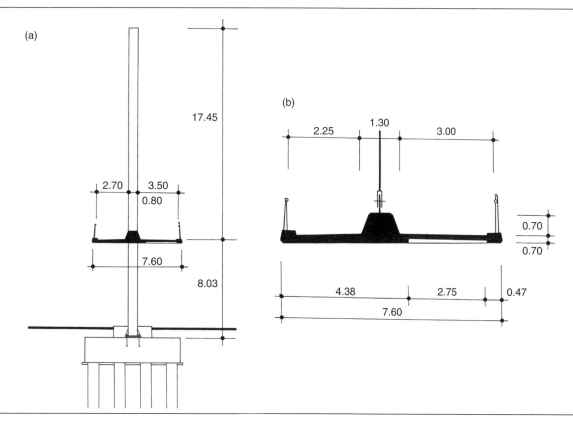

Figure 11.139 I-5 Gateway Bridge, Oregon, USA

by stay cables was first cast and after suspending it on the stay cables, the end sections were cast. The structure was then prestressed by continuous cables situated in the central web. The forces in the stay cables, together with forces and a layout of the prestressing cables, balance the effects of the dead load.

Although the bridge is very slender, it is very stiff and users do not have an unpleasant feeling when standing or walking on the bridge. The static function and quality of the workmanship were checked by a loading test when trucks were parked on the whole and half length of the deck.

The construction of the bridge was completed in autumn 2010. The bridge was designed by Strasky, Husty and Partners, Brno and was built by **SKANSKA DS**, Division 77 Mosty.

11.3.5 I-5 Gateway Bridge, Eugene, Oregon, USA

The bridge connects the Eugene neighbourhoods west of the motorway to a regional shopping centre located east of the motorway in Springfield (Figure 11.139). In the preliminary study of bridge concepts it was confirmed that, while there is a small cost premium for the cable-stayed bridge type, it was nearly offset by the cost of the additional approach ramp length required to meet the Americans with Disabilities Act (ADA) standards for a much deeper conventional deck-girder-type bridge.

The bridge of total length 161.60 m forms a main two-span cable-stayed structure (Figure 11.140) that is monolithically connected with curved ramps. The main cable-stayed spans consist of segmental precast deck panels with composite topping slab. The approaches consist of a series of 9 m

Figure 11.140 I-5 Gateway Bridge, Oregon, USA: elevation

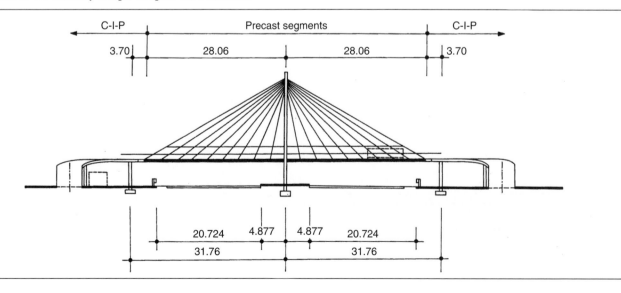

Figure 11.141 I-5 Gateway Bridge, Oregon, USA cross-sections: (a) main span, (b) approaches, (c) at pylon

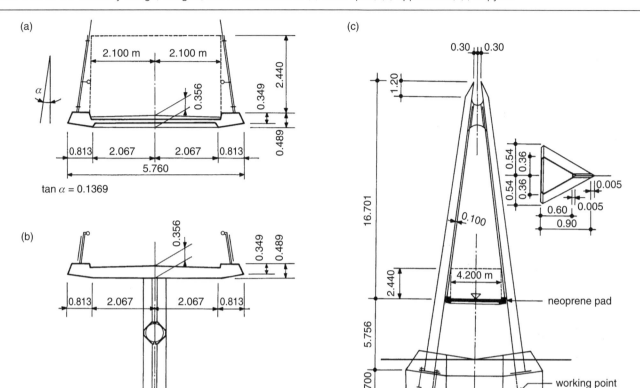

cast-in-place spans matching the deck section dimensions of the main spans (Figure 11.141). The A-shaped tower was cast in the horizontal position at the freeway median and subsequently erected as one precast member. The stays are formed by solid bars that are connected to the segments with standard clevises.

To simplify production, the segments were not match-cast (Section 9.2). To allow cantilever erection, the segments were cast with projecting structural tubes that mated with adjacent segments (Figures 9.23 and 9.24). The precast deck was erected in balanced cantilevers from the central tower. The closures and composite slab was also placed in balanced cantilevers directed away from the tower. The suspended spans together with curved spans were then post-tensioned.

The structural solution was developed on the basis of very detailed static and dynamic analyses. The multi-mode analysis completed for a given response spectrum has demonstrated that the structure has a satisfactory response to seismic load. Although the deck is very slender, pedestrians do not have an unpleasant feeling when standing or walking on the bridge.

Construction of the bridge was completed in April 2009. The bridge was designed by OBEC, Consulting Engineers, Eugene, Oregon and Jiri Strasky, Consulting Engineer. The bridge was built by Mowat Construction Company, Vancouver, Washington.

11.3.6 Delta Ponds Bridge, Eugene, Oregon, USA

The bridge that crosses the Delta Ponds Highway connects the Delta Ponds pedestrian and bicycle trails with a residential area. The bridge of the total length of 231.48 m is assembled from the main cable-stayed structure crossing the highway and approaches (Figures 11.142 and 11.143). Both the main bridge and approaches have a similar deck arrangement as the I-5 Gateway Bridge. The main structure of length 133.35 m is formed by a continuous slab that is suspended on a single pylon in spans 3, 4 and 5. Limited clearance required that the deck be as slender as possible; a cable-stayed structure therefore proved to be the most appropriate solution.

The V-shaped pylon is assembled of two precast columns frame-connected to the deck and stiffened by a bottom diaphragm that was cast after the erection of the deck (Figure 11.144). The stay

Figure 11.142 Delta Ponds Bridge, Oregon, USA

cables have a semi-fan arrangement; they are formed by strands that are pin-connected to the legs of the pylons and the deck.

The main span crossing the highway is assembled of precast segments and a cast-in-place deck slab. To simplify production, the segments were not match-cast (see Section 9.2). To allow cantilever erection, the segments were cast with projecting structural tubes that mated with adjacent segments.

The main span was built after the back spans were cast. The precast segments were erected in a cantilever which started at the pylon (Figure 11.145). The segments were erected during the night when only two lanes of the highway were closed. The composite slab was also placed in the direction from the tower to the closure. The spans of the main bridge were then prestressed.

The structural solution was developed on the basis of very detailed static and dynamic analyses. The multi-mode analysis completed for a given response spectrum has demonstrated that the structure has a satisfactory response to seismic load. Although the deck is very slender, the pedestrians do not have an unpleasant feeling when standing or walking on the bridge.

Figure 11.143 Delta Ponds Bridge, Oregon, USA – elevation: (a) bridge, (b) main bridge

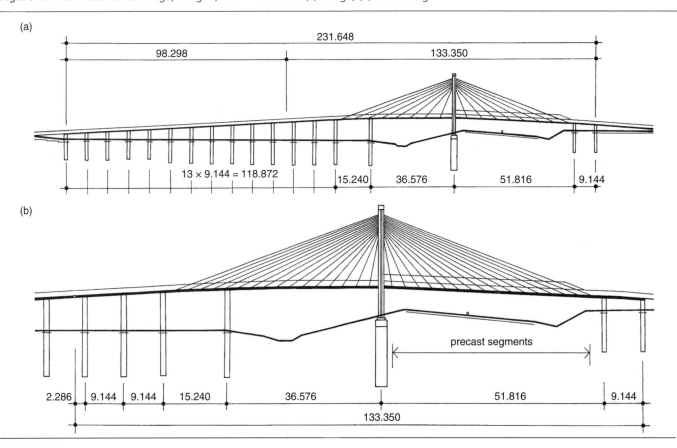

Figure 11.144 Delta Ponds Bridge, Oregon, USA – pylon: (a) cross-section, (b) partial elevation

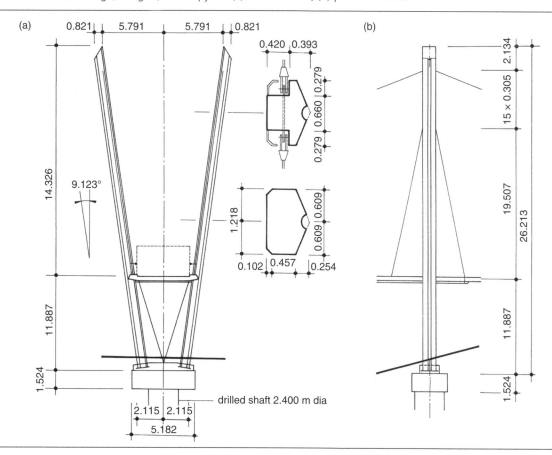

Construction of the bridge was completed in October 2010. The bridge was designed by OBEC, Consulting Engineers, Eugene, Oregon and Jiri Strasky, Consulting Engineer. The bridge was built by Mowat Construction Company, Vancouver, Washington.

Figure 11.145 Delta Ponds Bridge, Oregon, USA: erected cantilever

11.3.7 Lockmeadow Bridge, Maidstone, Kent, UK

The Lockmeadow Bridge which was opened in 1999 crosses a bend of the River Medway in the centre of Maidstone, Kent, UK. The bridge is adjacent to the Archbishop's Palace in a historically and archaeologically important area of the town. It provides pedestrian access from the town to a new leisure development on the west bank, and forms part of the Maidstone Millennium River Park (Figure 11.146) (Firth, 1999).

The two-span bridge is formed by a slender aluminium deck that is suspended on a pair of masts. These are inclined outwards from a sculptured stair support situated on the west bank of the river (Figure 11.147). The splayed twin masts of the cable-stayed system reduced the effective span of the deck to about 16 m between the stay attachments, allowing a structural depth of approximately 300 mm.

The curved deck structure is an assembly of interlocking longitudinal aluminium extrusions stressed together transversely in a laminated arrangement (Figure 11.148). There are no secondary structural elements or added finishes, because the extrusions are designed to fulfil all the functional requirements of the deck including the non-slip top surface.

Figure 11.146 Lockmeadow Bridge, Kent, UK (courtesy of Flint & Neill Partnership)

Figure 11.148 Lockmeadow Bridge, Kent, UK: cross-section of the deck

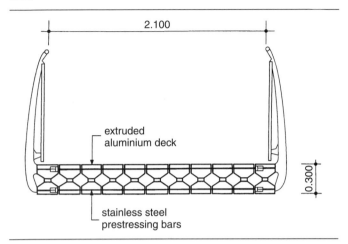

The maximum length of extrusion was about 7 m, and most finished lengths were approximately 6.4 m. Splices between the individual extrusions were made by simple staggering of the joints at a pitch that matched the 1.6 m spacing of the transverse prestressing bars from stainless steel.

The cable stays are 45 mm locked coil ropes with cast steel sockets.

The contractor elected to erect the bridge by incremental launching, assembling the deck in an area at the west abutment and pulling it over a series of rollers. In order to span the river without the assistance of the stay cables, a temporary support was placed at mid-river and a king-post arrangement was used to stiffen the aluminium deck during launching. Erection of the masts and stays was carried out in a single operation. The deck was then lifted off the temporary supports by stressing the stays.

The bridge was designed by Chris Wilkinson Architects Ltd and by Flint & Neill Partnership; the contractor was Christiani & Nielsen Limited.

Figure 11.147 Lockmeadow Bridge, Kent, UK: (a) plan, (b) elevation

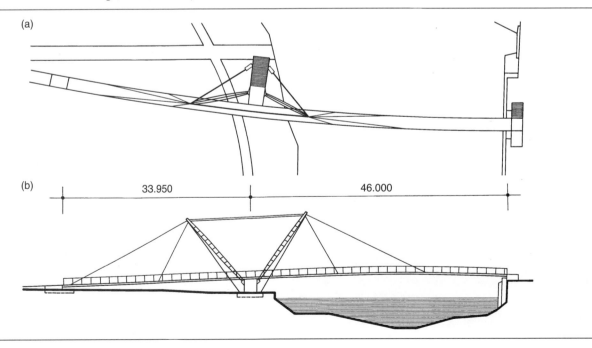

Figure 11.149 Glorias Catalanas Bridge, Barcelona, Spain: (a) elevation, (b) plan, (c) cross-section of the main span, (d) cross-section of the ramp, (e) cross-section of the tower

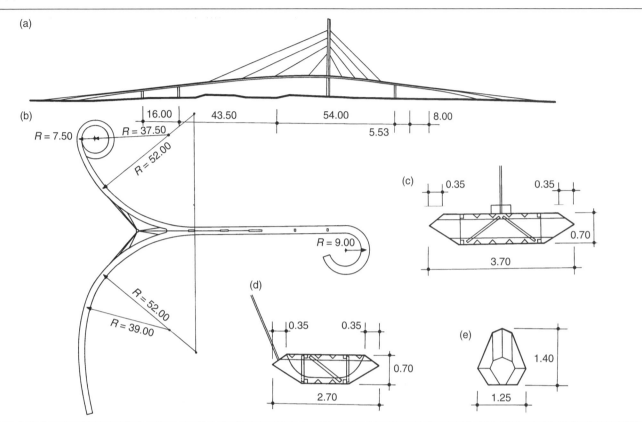

11.3.8 Glorias Catalanas Bridge, Barcelona, Spain

The original bridge was built at the Glorias Catalanas Square in Barcelona, Spain in 1974. Due to the transformation of the square for the 1992 Olympics, the bridge was dismounted and rebuilt on the coastal stretch.

The original bridge was designed to connect three points separated by two traffic lanes. In plan, the structure is Y-shaped which naturally solved the traffic problem (Figure 11.149) (Casado et al., 1976). The deck is suspended on a single tower situated near the intersection of the three branches. Three stay-cable fans extend from the tower, one for each branch. The main branch is suspended in the bridge axis while the others are on their edges (Figures 10.7 and 11.150).

The suspended deck is formed by torsionally stiff box girders of streamline cross-sections that convert into concrete girders at the approaches. Two approaches are formed by snail-shell-shaped cantilevers; the third is formed by a straight cantilever. At the new location, the bridge has almost the same arrangement.

The bridge was designed by Carlos Fernandez Casado, SL, Madrid, Spain.

11.3.9 Rosewood Golf Club Bridge, Japan

The Rosewood Golf Club Bridge was completed in 1993 in Ono city, Hyogo Prefecture in Japan. Its alignment corresponds to the course layout where there is no straight line. The curved bridge deck is suspended on the stay cables anchored at its

Figure 11.150 Glorias Catalanas Bridge (courtesy of Carlos Fernandez Casado, SL, Madrid)

Figure 11.151 Rosewood Golf Club Bridge (courtesy of Shimizu Corporation, Japan)

outer edge and at the inclined tower (Figures 10.6 and 11.151) (Okino *et al.*, 1993).

The bridge has two symmetrical spans of length 37.763 m (Figure 11.152). The deck is formed by a non-symmetrical concrete girder of hollow cross-section of depth 1.20 m. The deck is frame connected to the inclined pylon and is supported by neoprene pads at the abutments.

The suspension of the deck on the outer edge enables the torsional moment to be reduced; however, the transverse moments caused by eccentric anchoring of the stay cables have significant values. The eccentricity of the anchoring was reduced by inclining the pylon.

A detailed static and dynamic analysis demonstrated a good function of the bridge both for the service and ultimate load and good resistance to wind and seismic load. The bridge has become the symbol of the golf course.

The bridge was designed and built by Shimizu Corporation, Japan.

11.3.10 Bridge across the motorway D1, Czech Republic

The proposed bridge should connect two large rest areas known as the 'Nine crosses'. Due to existing utilities, there is no space for piers in the medians or in the space between the freeway and local roads. A cable-stayed structure of span 94.00 m is therefore proposed for the bridging (Figure 11.153).

Figure 11.152 Rosewood Golf Club: (a) elevation, (b) plan, (c) cross-section of the deck, (d) tower elevation, (e) cross-section of the tower – upper part, (f) cross-section of the tower – bottom part

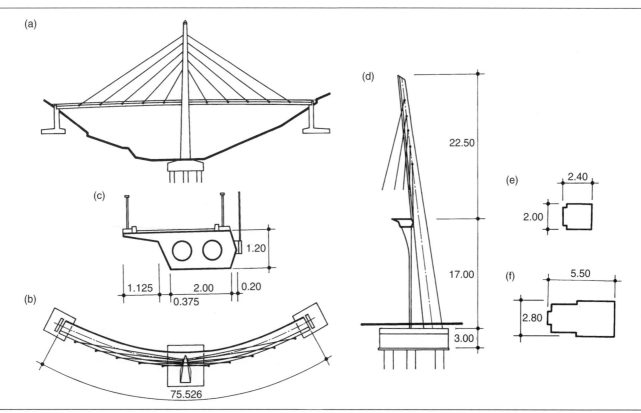

Figure 11.153 Bridge across the motorway D1, Czech Republic: model of the bridge

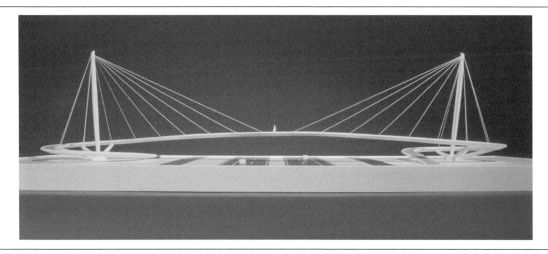

The bridge deck is suspended on two inclined masts situated in the centre of the approach ramps (Figure 11.154). In the main span above the freeway, the deck is formed by two steel edge tubes mutually connected by transverse diaphragms and by a composite deck slab. The edge tubes of the ramps are filled with concrete and the space between the tubes is filled with concrete (Figure 11.155). The deck is suspended at the inner edges of the plan curve. The ramps are also supported by inclined struts anchored into the pedestals of the pylons (Figure 11.156).

Figure 11.154 Bridge across the motorway D1, Czech Republic: (a) elevation, (b) plan

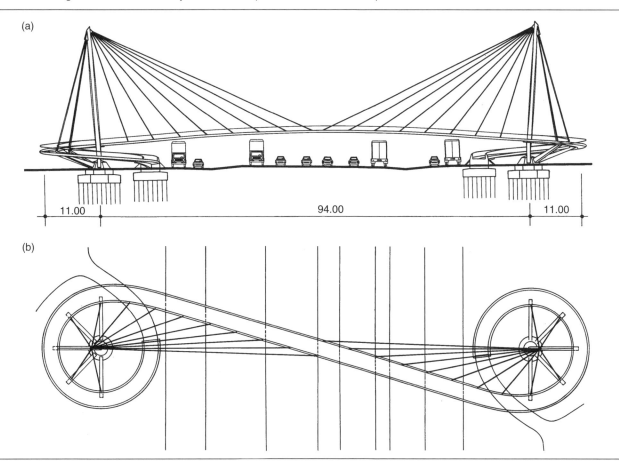

Figure 11.155 Bridge across the motorway D1, Czech Republic: (a) cross-section at the main span, (b) cross section at the ramps, (c) tower elevation

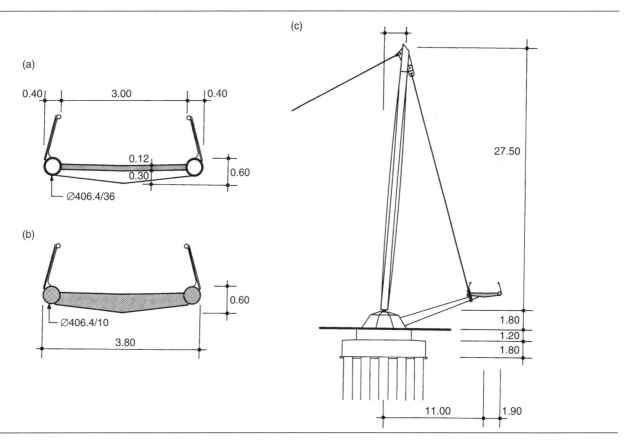

Figure 11.156 Bridge across the motorway D1, Czech Republic: model of the bridge

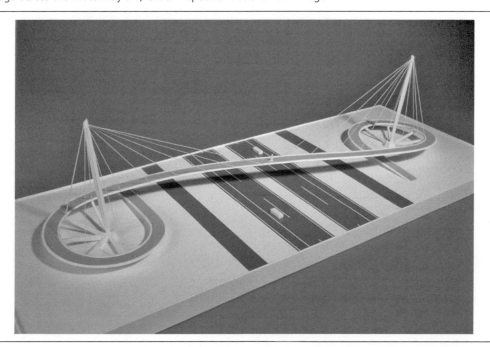

The masts that are supported by pot bearings situated at the top of the pedestals are formed by steel tubes stiffened by external plates of a variable depth. The structural solution was developed on the basis of a very detailed static and dynamic analysis that proved that the structure has a sufficient margin of safety. The analysis also confirmed that the edge steel tubes have sufficient torsional stiffness. The speed of motion and/or acceleration of the deck were within the recommended limits.

The bridge was designed by Strasky, Husty and Partners, Czech Republic.

REFERENCES

Arai H and Ota Y (1994) Prestressed concrete stress ribbon bridge: Kikko Bridge. *Proceedings of Prestressed Concrete in Japan 1994, XII FIP Congress, Washington, DC*. Japan Prestressed Concrete Engineering Association.

Arai H and Yamomoto T (1994) Prestressed concrete stress ribbon bridge: Umenoki Todoro Park Bridge. *Proceedings of Prestressed Concrete in Japan 1994, XII FIP Congress, Washington, DC*. Japan Prestressed Concrete Engineering Association.

Batsch W and Nehse W (1972) Spannbandbrücke als Fussgängersteg in Freiburg im Breisgau. *Beton-und Stahlbetonbau* 3.

Bednarski C and Strasky J (2003) The Millennium Bridže. *L'Industria Italiana del Cemento* N 792, Rome, Italy.

Casado CF, Troyano LF and Mantreola JA (1976) Passerelle haubanee a Barcelona. *Acier-Stahl-Steel*, 2.

Conzett J (2000) Punt da Suransuns Pedestrian Bridge, Switzerland. *Structural Engineering International* **10(2)**: 104–106.

Firth I (1999) Lockmeadow Footbridge, Maidstone, UK. *Structural Engineering International* **9(3)**: 181–183.

Firth I and Cooper D (2002) New materials for new bridges. Halgavor Bridge, UK. *Structural Engineering International* **12(2)**: 80–83.

Hata K (1998) Single-span prestressed concrete stress ribbon bridge: Yumetsuri Bridge. *Proceedings of Prestressed Concrete in Japan 1998, XIII FIP Congress Amsterdam*. Japan Prestressed Concrete Engineering Association.

Holgate A (1997) *The Art of Structural Engineering. The Work of Jöger Schlaich and his Team*. Edition Axel Menges, Stuttgart/London.

Horiuchi S, Watanabe K and Kondoh S (1998) Four-span stress ribbon bridge with roadway slab decks: Shiosai Bridge. *Proceedings of Prestressed Concrete in Japan 1998, XIII FIP Congress, Amsterdam*. Japan Prestressed Concrete Engineering Association.

Katuyama T, Kitsuta T and Ito T (1998) Three-span continuous prestressed concrete stress ribbon bridge: Tonbo No Hashi. *Proceedings of Prestressed Concrete in Japan 1998, XIII FIP Congress, Amsterdam*. Japan Prestressed Concrete Engineering Association.

Kawaguchi M (1996) Granite pedestrian bridge, Beppu, Japan. *Structural Engineering International* **3(96)**: 148–149.

Komatsubara T, Kondoh S and Itoh K (2002) Prestressed concrete curved cord truss bridge: Ganmon Bridge. *Proceedings of Recent Works of Prestressed Concrete Structures, First FIP Congress, Osaka*. Japan Prestressed Concrete Engineering Association.

Kumagai T, Tsunomoto M and Machi T (2002) Stress-ribbon bridge with external tendons: Morino-Wakuwaku Bridge. *Proceedings of Recent Works of Prestressed Concrete Structures, First FIP Congress, Osaka*. Japan Prestressed Concrete Engineering Association.

Okino K, Inazumi T and Watanabe Y (1993) Design and construction of one-sided cable stayed bridge. *Proceedings of Modern Prestressing Techniques and their Applications, FIP Symposium*, Kyoto, Japan.

Parker JS, Hardwick G, Carroll M, Nicholls NP and Sandercock D (2003) Hungerford Bridge millennium project: London. *Civil Engineering* **156**: 70–77.

Rayor G and Strasky J (2001) Design and Construction of Rogue River (Grants Pass) Pedestrian Bridge. *Proceedings of Western Bridge Engineers' Seminar, Sacramento*, California.

Redfield C and Strasky J (1991) Sacramento River Pedestrian Bridge, USA. *Structural Engineering International* **1(4)**.

Redfield C and Strasky J (1992a) Sacramento ribbon. *Concrete Quarterly*, British Cement Association, Autumn.

Redfield C and Strasky J (1992b) *Stressed Ribbon Pedestrian Bridge Across the Sacramento River in Redding, CA, US*. *L'Industria Italiana del Cemento* **62(2)**: 82–99.

Redfield C and Strasky J (2002) Bleu River Ranch Bridge. *Proceedings of IABSE Symposium*, Vancouver.

Schlaich J (2000) Urban footbridges. *Proceedings of 16th Congress of IABSE*, Lucerne.

Schlaich J and Bergermann R (1992) *Fußgängerbrücken. Ausstellung und Katalog*. ETH, Zürich.

Schlaich J and Schober H (1994) A suspended pedestrian bridge crossing the Neckar river near Stuttgart. *Proceedings of Cable-stayed and Suspension Bridges*, Deauville, France.

Schlaich J and Seidel J (1998) Die Fußgängerbrücke in Kelheim. *Bauingeneur* 63.

Seible F and Burgueno R (1994) Cable-stayed bridges at UCSD. *Proceedings of International Symposium on Cable-Stayed Bridges*, Shanghai.

Sinohara O (1999) *Landscape and Civil Design Report. Works of Engineer – Architects in Japan*. Institute of Landscape & Civil Design. Tokyo.

Strasky J (1987a) Precast stress ribbon pedestrian bridges in Czechoslovakia. *PCI Journal* **32(3)**: 52–73.

Strasky J (1987b) The stress ribbon footbridge across the river Vltava in Prague. *L'Industria Italiana del Cemento* N 615.

Strasky J (1994) Suspension pedestrian bridge across the Swiss bay of Vranov lake. *Space & Society* N 67. Milano, Italy.

Strasky J (1995) Pedestrian Bridge at Lake Vranov, Czech Republic. *Civil Engineering* **108**: 111–122.

Strasky J (1998a) Pedestrian Bridge Suspended over Lake Vranov, in the Czech Republic. *L'Industria Italiana del Cemento*, N 736.

Strasky J (1998b) Design-construction of Vranov Lake Pedestrian Bridge, Czech Republic. *PCI Journal* 42(6): 60–75.

Strasky J and Pirner M (1986) *DS-L Stress ribbon footbridges*. Dopravni stavby, Olomouc, Czechoslovakia.

Strasky J and Rayor G (2000) Technical innovations of the Willamette River pedestrian bridge, Oregon. *Proceedings of International Bridge Conference*, Pittsburgh.

Tanaka T, Kawakami M, Teramoto Y, Kuribayashi T and Shimizu K (2002) Tokimeki Bridge (provisional name) flat arch bridge with suspended deck (self-anchored structure). *Proceedings of the 1st fib Congress: Concrete Structures in the 21st Century*, Osaka.

Troyano LF, Mantreola J and Astiz MA (1986) Puentes ligeros, en el Alt Urgell, sobre el rio Serge. Articulo publicado en el no. 158 de la Revista Hormigon y Acero. Madrid 1986.

Tsunomoto M and Ohnuma K (2002) Self-anchored suspended deck bridge: pedestrian bridge of the Tobu recreation resort. *Proceedings of Recent Works of Prestressed Concrete Structures, 1st fib Congress, Osaka, Japan*. Japan Prestressed Concrete Engineering Association.

Uchimura T, Miyzaki M, Kondoh S and Okumura K (2002) Prestressed concrete deck bridge supported from below by cables: Ayumi Bridge. *Proceedings of Recent Works of Prestressed Concrete Structures, 1st fib Congress, Osaka, Japan*. Japan Prestressed Concrete Engineering Association.

Völkel E, Zellner W and Dornecker A (1977) Die Schrägkabelbrücke für Fußgänger über den Neckar in Mannheim. *Beton und Stahlbetonbau* 72, S. 29–35, 59–64.

Walther R (1969) Spannbandbrücken. *Schweizerische Bauzeitung* **87(8)**.

Wolfensberger R (1974) SAPPRO Fussgängersteg Lignon: Löex, Genf. Spannbeton in der Schweiz. *Proceedings of VII FIP Congress*, New York.

Stress Ribbon and Cable-supported Pedestrian Bridges
ISBN 978-0-7277-4146-2

ICE Publishing: All rights reserved
doi: 10.1680/srcspb.41462.257

Index

ASHTO, 31, 34, 159
Abutments *see* piers and abutments
Acht Architects, 209
aerodynamic behaviour, 39–40, 99, 112, 113, 122–126, 148, 151–152, 234
   aerodynamic model in wind tunnel, photograph, 125
   aerodynamic model of stress ribbon structure, diagram, 125
   flutter, 39
   vertical response of model, diagram, 125
Almond River Bridge, Scotland, 10
Americans with Disabilities Act (ADA), 246
Ammann & Whitney Engineers, 3
anchoring of cables, 26, 27, 28, 46–47, 57, 59, 70
   anchors for cable-stayed structures, 155, 157, 166
   anchors for stress ribbons, 81, 84–86, 87, 91, 92, 107, 112
   anchors for suspension structures, 127, 128, 129, 132, 127, 133, 134–141, 142–146, 153, 217, 220
Angrand, L., 2
ANSYS program, 20, 21, 22, 35, 66, 73, 97, 98, 108, 109, 147, 150, 201
Arai and Ota, 196
arch, natural, Utah, USA, 14
arch, stress ribbon supported by, 106–113
   aerodynamic model of the structure, cross section and elevation, diagram, 112
   analysed bridge (Radbuza pedestrian bridge), cross section and elevation, diagram, 109
   buckling of the arch, photograph, 112
   calculation model, diagram, 109
   load, normal stresses, deflection of arch, deflection of prestressed band, diagram, 111
   load, service load, ultimate load, diagram, 110
   loading, ultimate: the structure before failure, photograph, 111
   model of bridge, diagram, 110
   model of bridge, photograph, 110
   natural modes, diagram, 113
      natural modes and frequencies, table, 113
   stress ribbon supported by arch, diagrams, 106, 107
   stress ribbon suspended on arch, diagram, 108
   vertical and horizontal response of the model, diagram, 113
arch structure, flat, and curved stress ribbon structure, 176, 177, 178
arch structures, diagram of cable and, 13

Architects & Planners League, 235
auto-excited oscillations, 39
Ayumi Bridge, Japan, 234–235

Bachman, H., 34, 35
Bachman *et al.*, 34
Badhomburg Bridge, Germany, 156
Balfour Beatty Construction Ltd., 201, 227
bamboo ropes, bridge with, 1, 2
Basella Bridge, Spain, 214
Bednarski, C., 104, 105, 201
Bednarski and Strasky, 198
Berlin Congress Hall suspended roof, Germany, 3, 4, 5
bi-stayed bridge, 17
bicycles, 31, 33, 122, 157, 224, 227, 244
Bilfinger and Berger, 241
Bircherweid bridge over motorway N3, near Pfäfikon, Switzerland, 4, 179
Blue Valley Ranch Bridge, Colorado, USA, 99, 191–192
Bohumin, Czech Republic, bridge over D47 freeway, 38, 157, 158, 244–246
Bosporus bridge, Turkey, 1, 2
Brno-Komin Bridge, Czech Republic, 6, 73, 89, 119, 120, 121, 182, 183, 184
Brooklyn Bridge, New York, USA, 7, 8, 17
Brown, D.J., 1, 3, 127

cable analysis, 41–53
   bending of the cable, 47–52; bending moments, 49, 51; comparison of the static effects, table, 52; deformation and bending at beam and stress ribbon, 51; deformation and bending moments of the tested cable, 51; finite difference method, 50; geometry and internal forces at the cable, 47; geometry and internal forces at flexibly supported cable, 49; loading test of the stay cable, 50, 51
   natural modes and frequencies, 52–53
   single cable, 41–47; basic characteristics, 42; determination of D for a uniform load and an arbitrary load, 45; determination of the horizontal force H_i, 46; elastic deformations of the cable at the anchor blocks, 47; elongation of the cable, 45–46; influence of deformation of supports and elongation of the cable at the anchor blocks, 46–47; initial and final stage of the cable, 44; non-tension length of the cable, 43; uniformly loaded cable, 42

257

cable-stayed structures, 155–169, 172, 174, 175
  erection of the structures, 159–164; erection of the deck in a cantilever, diagram, 160; erection of the deck on a falsework, diagram, 160, erection of the Diepoldsau Bridge, diagram, 160; Gateway Bridge, 162–163; Židlochovice Bridge, connection of the longitudinal and transverse members, elevation and longitudinal normal stresses in the deck, diagram, 161; Židlochovice Bridge, progressive erection of the deck, diagram, 162
  examples, 240–255; Bohumin, bridge over D47 freeway, 244–246; D1 expressway, bridge across, 252–255; Delta Ponds Bridge, 247–249; Gateway Bridge (1–5), 246–247; Glorias Catalanas Bridge, 251; Hungerford Bridge, 242–244; Lockmeadow Bridge, 249–250; Neckar River Bridge, 240–241; Rosewood Golf Club Bridge, 251–252; Scripps Crossing Bridge, 241–242
  static and dynamic analysis, 164–169; redistribution of stresses due to creep and shrinkage, 166–169; stay cables, 164–166
  structural arrangement, 155–159; arrangement of the cable-stayed structure, diagram, 156; arrangement of the stay cables, 156; cable-stayed structure, classical and extradosed, diagram, 157; cable-stayed structure supported by stay cables, one stay and multiple stays, diagram, 158; length of the side spans, diagram, 156; pin connection of the stay cable, diagram, 159; typical arrangement of the stay cable, diagram, 159
cables, anchoring of see anchoring of cables
cables, stress ribbon structures
  analysis of structure as a cable, 94–96; erection stage, 94–96; service stage, 96
  stress ribbon suspended on stay cable, 104
cables, structural systems and members see structural members; structural systems
cables, suspension structures
  erection of suspension cables, 134, 135, 137–138; McKenzie River Bridge, erection cables, 137; McKenzie River Bridge, erection of the main cables, 137; Willamette River Bridge, erection of the island segments suspended on the erection cables, 138; Willamette River Bridge, island segments suspended on the erection cables, 138; Willamette River Bridge, island segments suspended on the main suspension cables, 138
  geometry of the cable, 128–130, 143
  grouting of cables, 134, 138–139
  initial stage of a structure suspended on cables situated in vertical planes, 142–143; diagrams, 143, 144
  initial stage of a structure suspended on inclined cables, 143–145; diagram, 144
  structures supported by cables, 139–140
  structures suspended on cables, 130–134
Canada, Calgary Winter Olympic Games, Saddledome structure, 3, 4
cantilever erection, 17, 160, 162, 163, 164, 245, 248, 249, 251
Casado, C.F., 158, 172, 173, 214, 215, 251
CEB-FIP, 69, 99, 167
Centric/Jones Constructors, 192
chain bridge, James Finley, 17
Chris Wilkinson Architects Ltd., 250
Christiani & Nielsen Ltd., 250
Collins and Mitchell, 10
Conzett, J., 203
Conzett, Bronzini, Gartmann, 203, 204, 205

Costain Norwest Holst, 244
cranked alignment of the stress ribbon, 104
  diagram, 105
creep and shrinkage of concrete, 63–70, 132, 133
  creep functions, diagram, 63
  final value of creep coefficient, table, 64
  final value of shrinkage strain, table, 64
  redistribution of stresses, static and dynamic analysis of cable-stayed structures, 166–169
  redistribution of stresses between members of different age, 66–69; area of progressively cast concrete member m^2, 69; progressively cast concrete member, 66; redistribution of forces of pre-cast segment and concrete slab, 67; redistribution of forces of pre-cast segment, concrete slab and prestressing tendon, 68; redistribution of forces of pre-cast segment, concrete slab, prestressing and bearing tendons, 68
  redistribution of stresses in structures with changing static systems, 69–70; dead-load balancing, 70; dead-load balancing at a two-span beam, 69; redistribution of bending moments in a two-span beam, 69
  shrinkage strain and creep coefficient, diagrams, 64
  time dependent analysis, 63–66; time-dependent modelling of cable-supported structures, 65
CTI Engineering, 235
curved structures, 171–178, 244, 249, 251, 253–254
  curved stress ribbon bridge, 175–178; curved stress ribbon structure, cross-section, diagram, 176; curved stress ribbon structure, effects of the bearing tendon, effects of the horizontal component of the tendon, effects of the vertical component of the tendon and geometry, diagram, 176; curved stress ribbon structure and flat arch structure, photograph 176; curved stress ribbon structure and flat arch structure, static model, cross-section, elevation and plan, diagram, 177; curved stress ribbon structure and flat arch structure, static model, load situated on one-half of the stress ribbon length, diagram, 177; curved stress ribbon structure and flat arch structure, static model, ultimate load , diagram, 178; static model of curved stress ribbon and flat arch structure, table, 178
  suspension of the deck on both edges, 171–172; model of the curved cable-stayed bridge, 172; suspension of the curved deck, outer edge and inner edge, diagram, 172
  suspension of the deck on an inner edge 173–175; curved cable-stayed structure, cable-stayed structure, continuous beam, effects of the horizontal components of the stay and effects of the vertical components of the stay, diagram, 174; curved cable-stayed structure, suspension above deck, suspension at the deck, the effects of the post-tensioning and the effects of the effects of the stays, diagram, 175; curved suspension structure, suspension above deck and suspension at the deck, diagram, 175; internal forces in the curved deck, diagram, 174
  suspension of the deck on an outer edge, 172
Czech Republic
  Bohumin, bridge over D47 freeway, 38, 157, 158, 244–246
  Brno-Komin Bridge, 6, 73, 89, 119, 120, 121, 182, 183, 184
  D1 expressway, bridge across, 252–255
  DS-L stress ribbon structures, 75, 122, 123–124, 182–187
  Elbe Bridge, 16, 50, 51, 165, 166

Nymburk Bridge, 74
Olomouc, bridge over R35 expressway, 36, 76, 204–206
Olse River Bridge, Bohumin, 36, 195–196
Prague-Troja Bridge, 14, 36, 83, 88, 89, 119, 120, 121, 122, 182–183, 184, 185–186, 187
Radbuza River Bridge, Plzen, 107–113
Svratka River Bridge, Brno, 36, 206–209
Vltava River Bridge, Ceske Budejovice, 36, 38
Vranov Lake Bridge, 8, 9, 10, 37, 127, 132, 133, 136, 139, 140, 149, 150, 151, 152, 217–222
Židlochovice Bridge, 157, 158, 161, 162

D1 expressway, bridge across, Czech Republic, 252–255
Dai Nippon Construction, 191
decks, 23, 24, 25, 39, 129, 130, 132, 141, 156, 157, 158, 159, 210–211, 218, 224, 227, 228, 229, 230, 232–234, 239–240, 241, 242, 247, 248, 251, 252
   curved structures, 171–178, 244, 249, 251, 253–255
   D1 expressway, bridge across, 253–255
   diagram of bending moment in the deck, suspension structures, 146
   diagram of deck supported by cables, static function, earth-anchored suspension structure, 146
   diagram of deck suspended on cables, static function, earth-anchored suspension structure, 144
   diagram of deck width and railings height, 31
   diagram of modelling the cable-stayed structure, deck modelled by 3D bars and deck modelled by shell elements, 164
   diagram of modelling the suspension structure, deck modelled by 3D bars and deck modelled by shell elements, 142
   DS-L bridges, deck, cross-section and partial elevation, diagram, 182
   erection of the deck, stress ribbon structure, 86–90
   erection of the deck in a cantilever, diagram, 160
   erection of the deck on a falsework, diagram, 160
   erection of segments, suspension structures, 134, 135, 138, 139, 140
   Gateway Bridge (1–5), Oregon, USA, construction sequences, diagram, 163; segment connection, diagram, 163; segment erection, photograph, 162
   geometry of the cable, suspension structures, deck supported by cables, diagram, 129
   geometry of the cable, suspension structures, deck suspended on cables, diagram, 130
   glass deck, Deutsches Museum bridge, 228, 229
   Grants Pass Bridge, cross-section and deck plain, diagram, 193
   Ishikawa Zoo Bridge, Japan, erection of a deck segment, photograph, 140
   Johnson Creek Bridge, construction sequences, diagram, 141
   Johnson Creek Bridge, model of the erection, 141
   Lake Hodges Bridge, cross-section of the deck, diagram, 195
   Lockmeadow Bridge, 249–250; diagrams of cross-section of the deck and plan and elevation, 250
   Max-Eyth-See Bridge, cross-section, diagram, 216; erection of the deck, photograph, 134
   McKenzie River Bridge, erection of island segments, island segments suspended on erection and main cables, 138; lifting of a typical segment, 139; suspension of a typical segment, 139
   McKenzie River Bridge, cross-section and partial elevation, 225; partial fixing of the deck, diagram, 134; diagram of pin connection, diagram, 139
   model test of a cable-stayed structure with slender concrete deck, 10
   Morino-Wakuwaku Bridge, deck, struts and external prestressing tendons, 213
   Nozomi Bridge, Japan, erection of a deck segment, photograph, 140
   Olse River Bridge, cross-section of the deck, diagram, 196
   post-tensioning of the deck, 134, 138–139
   Vranov Lake Bridge, 219; cross-section and partial elevation, diagram, 220; geometry of the structure during erection, 139; partial fixing of the deck, diagram, 133; photograph of the deck, 222; structure during erection of segments, photographs, 1, 40
   Willamette River Bridge, 222; cross-section and partial elevation, diagram, 223; erection of a typical segment, 138
   Židlochovice Bridge, connection of the longitudinal and transverse members, elevation and longitudinal normal stresses in the deck, diagram, 161
   Židlochovice Bridge, progressive erection of the deck, diagram, 162
   see also grades; lighting; prestressed band; railing; surfacing; vibrations
Delta Pond Bridge, Oregon, USA, 37, 163, 247–249
Department of Transport, UK, 31, 34, 35
design criteria, 31–40
   dynamics, 34–40
   geometric conditions 31–33
   loads, 33–34
Deutsches Museum, Munich, Germany, pedestrian bridge built inside, 175, 227–229
Diepoldsau Bridge, Switzerland, 160–161, 169
Dishinger, F., 8
Dopravni Stavby & Mosty, 222
DS-L stress ribbon structures, Czech Republic, 75, 121, 122, 123–124, 182–187
   bearing and prestressing tendons, diagram, 183
   deck, cross-section and partial elevation, diagram, 182
   table, 182
Dulles International Airport terminal, Washington DC, USA, 1, 2
   aerodynamic behaviour, 39–40
   geometry of the deck, diagram, 71
   physiological effect of vibrations, 34–39
   seismic design, 40
Dyckerhoff & Widmann, 181
dynamic analysis *see* static and dynamic analysis; static and dynamic analysis, stress ribbon structures; static and dynamic analysis, suspension structures
dynamics, 34–40, 112–113, 119
dynamic loading tests, loading tests, 119, 120–126
Dywidag bars, 181

earth-anchored structures, suspension structures 127, 134–141, 142–146
   erection and service, diagram, 128
   structures supported by cables, 139–140
   structures suspended on cables, 134–139
   typical arrangement of the earth-anchored suspension structure, diagram, 129

earthquake, 99–100, 150
Elbe Bridge, Czech Republic, 16, 50, 51, 165, 166
Endo et al., 7
Ernst, H.J., 165
ESA Prima Win, 63
Eurocodes, 11, 31, 35
  action on structures, 34

falsework, 91, 141, 142, 160
Favre and Markey, 69
finite difference method, 50
Finley, J., 6, 17
Finsterwalder, U., 1, 2
FIP, 55, 63, 158
  CEB-FIP, 69, 99, 167
Flatiron, 195
Flint and Neill Partnership, 201, 227, 250
flutter, phenomenon of, 39
Freiburg Bridge, Germany, 33, 180–181
Fribourg Bridge, Switzerland, 8

Ganmon Bridge, Japan, 141, 232–234
gate effect, 131
Gateway Bridge (1–5), Oregon, USA, 37, 162–163, 246–247
geometric conditions, 31–33
  deck width, 31
  grades, 31–32
  surfacing, railing and lighting, 32–33
geometrically non-linear structure, analysis of the stress ribbon structure as a, 96–99
geometry, stress ribbon structures, 71–72
  geometry of the deck, diagram, 71
geometry of the cable, suspension structures, 128–130, 143
  deck supported by cables, diagram, 129
  deck suspended on cables, diagram, 130
geometry of the structure during erection, Vranov Lake Bridge, 139
Germany
  Badhomburg Bridge, 156
  Berlin Congress Hall suspended roof, 3, 4, 5
  Deutsches Museum, Munich, pedestrian bridge built inside, 175, 227–229
  Freiburg Bridge, 33, 180–181
  Kelheim Bridge, 173, 227, 228
  Max-Eyth-See Bridge, Stuttgart, 25, 134, 215–217
  Neckar River Bridge, 155, 240–241
  Nordbahnhof Bridge, Stuttgart, 27, 131, 132
  Phorzheim 1 Bridge, 217, 218
  Phorzheim 111 Bridge, 202–203
  Rosenstein 11 Bridge, Stuttgart, 201–202
  Werrekuss Bridge, 15
  Weser River Bridge, 131
Gifford and Partners, 244
Gimsing, N.J., 1, 127
glass deck, Deutsches Museum Bridge, Munich, Germany, 228, 229
Glorias Catalanas Bridge, Barcelona, Spain, 172, 251
Golden Gate Bridge, San Francisco, 7, 8, 15
grades, 31–32
  longitudinal grades, diagram, 32

Grants Pass Bridge, Oregon, USA, 5, 32, 71, 76, 82, 83, 88, 89, 100, 101, 102, 122, 124, 192–194, 239
grouting, 26, 27, 29–30, 134, 138–139, 143, 144
  of stay cables, diagram, 29
  of suspension cables, diagram, 29
Guidelines for the Design of Pedestrian Bridges (Task Group), 31
Gulvanessian et al., 11, 31, 35

Hadrian Bridge, Scotland, 104, 105
Halgavor Bridge, UK, 224, 226–227
Hampe, E., 10
Harbor Drive Bridge, San Diego, California, 37, 229–231
Hata, K., 190
Hendy and Smith, 11
Hokusai Bridge, Japan, 14
Holgate, A., 201
Horiuchi et al., 231
horse-riding, 122, 224, 227
Hungerford Bridge, London, UK, 25, 27, 28, 160, 242–244

Inachus Bridge, Japan, 236–237, 238
inclined struts, 101, 103
Institute of Theoretical and Applied Mechanics, (ITAM) Academy of Sciences, Czech Republic, 112, 120, 124, 152, 230
Institution of Civil Engineers, 39
Ishikawa Zoo Bridge, Japan, 140, 231, 232

Japan
  Ayumi Bridge, 234–235
  Ganmon Bridge, 141, 232–234
  Hokusai Bridge, 14
  Inachus Bridge, 236–237, 238
  Ishikawa Zoo Bridge, 140, 231, 232
  Kikko Bridge, 196–198
  Koushita Bridge, 72
  Morino-Wakuwaku Bridge, 212–213, 214
  Nozomi Bridge, 140
  Rosewood Golf Club Bridge, 172, 251–252
  Seishun Bridge, 239–240
  Shiosai Bridge, 25, 231–232, 233
  Tobu Bridge, 25, 28, 141, 235–236
  Tokimeki Bridge, 209–210
  Tonbo No Hashi Bridge, 191
  Umenoki-Todoro Park Bridge, 188–189
  Yumetsuri Bridge, 189, 190–191
Japan Engineering Consultants Co. Ltd., 210
Johnson Creek Bridge, Oregon, USA, 37, 141, 153, 237–239

Kajima Co. Ltd., 236
Kawaguchi, M., 236, 237
Kelheim Bridge, Germany, 173, 227, 228
  Kent Messenger Millennium Bridge, Maidstone, UK, 33, 36, 76, 83, 84, 90, 100, 102, 103, 104, 122, 198–201
Kikko Bridge, Japan, 196–198
Klockner Institute, Prague, Czech Republic, 122
Komatsubara et al., 232
Koushita Bridge, Japan, 72
Kreuzinger, H., 34

Kumagai *et al.*, 212
Kyoryo Consultants Co. Ltd., 213

Lake Geneva, bridge over, Switzerland, 1
Lake Hodges Bridge, San Diego, California, USA, 36, 194–195
LARSA program, 97
Leonhard and Adrä, 241
Leonhardt, F., 1, 10, 69
Leonhardt and Anda, 202
Leonhardt and Zellner, 7, 9
liana ropes, bridge with, 1, 2
Liebenberg, A.C., 9
Lifschutz Davidson, 244
lighting, 33
Lignon-Löex Bridge, Geneva, Switzerland, 179–180
Lin and Burns, 1, 10, 69, 171
loading test, Brno-Komin Bridge
   by heavy vehicles, photograph, 119
   by mechanical exciter, photograph, 121
   by running people, photograph, 121
   elevation and deformation, diagram, 120
loading test, DS-L ribbon stress structures
   sectional model, diagram, 122
   static lift, drag and torsion moment, diagram, 123–124
loading test, Elbe Bridge, 50, 51
loading test, Grants Pass Bridge, 122, 124
loading test, Johnson Creek Bridge, 153–154
   static model, deformation of deck at mid-span, table, 154
   static model, diagram, 153
   static model, load situated mid-span, 153
   static model, ultimate load, 154
loading test, Prague-Troja Bridge, 14, 119
   by 38 vehicles, photograph, 120
   calculated and measured natural frequencies, table, 122
   deflection at midspans, diagram, 120
   exciter-induced modes, diagram, 121
loading test, Sacramento River Bridge, Redding
   by 24 cars, photograph, 120
   by horse riding, photograph, 122
   deformation, diagram, 121
   sectional model, diagram, 124
   static lift, drag and torsion moment, diagram, 124
loading test, Vranov Lake, Czech Republic, 152
loading tests, static and dynamic, 119–126, 151–154; dynamic tests, 119, 120–126; wind tunnel tests, 122–126, 152; static tests, 119–120, 153–154
Lockmeadow Bridge, Maidstone, Kent, UK, 249–250

Macalloy bar stays, 242
Maeda Engineering Corporation, 189, 237
Maidstone Bridges, UK *see* Lockmeadow Bridge; Kent Messenger Millennium Bridge
Maier, W.M., 179
Malecon Bridge, Madrid, Spain, 173
Mathivat, J., 9, 39, 160
Max Bögl a Josef Krsl, 206
Max-Eyth-See Bridge, Stuttgart, Germany, 25, 134, 215–217
MC90 rheological functions, 67

McKenzie River Bridge, Oregon, USA, 28, 130, 131, 133, 135, 137, 138, 139, 223–224, 225
   construction sequences, diagram, 135
McLouglin Boulevard Bridge, Oregon, USA, 37, 210–212
Mejorada Bridge, Pari River, Peru, 2
Menai Straits Bridge, Wales, UK, 7
Menn, C., 9, 156
Mitsui Construction Co Ltd., 25
Morino-Wakuwaku Bridge, Japan, 212–213, 214
Mörsch creep function, 69
Mowat Construction Company, 212, 223, 247, 249
Muller, J., 17
Museum Bridge *see* Deutsches Museum, Munich, Germany

National Standards, 31
natural arch, Utah, USA, 14
Natural Consultant Co. Ltd., 231
natural modes and frequencies, 34, 35, 38, 40, 52–53, 100, 102, 103, 113, 115, 122, 149, 150, 151
Navratil, J., 63, 65
Neckar River Bridge, Germany, 155, 240–241
Nepal, example of a bridge, 7
Nihonkai Consultant Co. Ltd., 234
Nissetsu Consultant Co. Ltd., 191
Nordbahnhof Bridge, Stuttgart, Germany, 27, 131, 132
Nozomi Bridge, Japan, 140
Nymburk Bridge, Czech Republic, 74

OBEC, 194, 212, 223, 224, 247, 249
observation platforms, 124, 138, 152, 194, 222
Okino *et al.*, 252
Olomouc, Czech Republic, bridge over R35 expressway, 36, 76, 204–206
Olse River Bridge, Bohumin, Czech Republic, 36, 195–196
Oriental Construction Co. Ltd., 25, 28, 72, 140, 141, 213, 236
Osormort Bridge, Spain, 158

parametric study, static and dynamic analysis, cable-supported structures, 167
parametric study, static and dynamic analysis, suspension structures, 146–149
   2D structure, 146–147; diagrams, 147, 148
   3D structure, 147–149; diagrams, 148, 149
Pearce and Jobson, 1, 152
pedestrians, 1, 31, 33, 34, 39, 87, 99, 100, 121, 122, 149, 150, 157, 177, 182, 186, 187, 192, 202, 204, 212, 214, 217, 224, 227, 230, 237, 240, 244, 246, 247
   see also bicycles; horse-riding
Peramola Bridge, Spain, 214, 215
Peru
   Mejorada Bridge, Pari River, 2
   Varrugas Bridge, 10
Phorzheim 1 Bridge, Germany, 217, 218
Phorzheim 111 Bridge, Germany, 202–203
Picasso, P., 1
piers and abutments, 79–81, 82–84, 101, 105, 179, 180, 181, 184, 185, 186, 188, 191, 193, 195, 196, 204
   abutments, diagram, 81

piers and abutments (*continued*)
   bending moments in a structure with haunches, diagram, 80
   erection of the stress ribbon at abutments, diagram, 81
   erection of the stress ribbon at intermediate piers, diagram, 82
   intermediate piers, 81
   see also stress ribbon structures, transferring stress ribbon force to the soil
Pirner, M., 112, 120, 124, 152, 230
Podolny and Muller, 9, 22, 160
Podolny and Scalzi, 7, 160
Portuguese National Pavilion for EXPO '98, Lisbon, 3
Post-Tensioning Institute, 26
Prague-Troja Bridge, Czech Republic, 14, 36, 83, 88, 89, 119, 120, 121, 122, 182–183, 184, 185–186, 187
prestressed band, 72–79, 111, 199, 200
   bending moments in a structure with flexible supporting member, diagram, 78
   bending moments in a structure with saddle, diagram, 78
   bearing and prestressing tendons, diagram, 77
   deformation and bending moments, diagram, 73, 74
   modelling of a supporting member and saddles, diagram, 79
   stress ribbons at supports, diagram, 77
   typical sections, diagram, 75
prestressing, effects of, 55–61
   curved beam-equivalent load, 58
   equivalent forces, 56
   equivalent load, normal forces, and bending moments in curved beam, 60
   equivalent load, normal forces, and bending moments in straight beam, 60
   modelling the curved beam, 59
   modelling prestressing, 57
   radial forces acting on stiff and slender arch, 61
   types of prestressing, 56
Priestly *et al.*, 40
Punt da Suransuns Bridge, Switzerland, 203–204, 205
pylon, Delta Ponds Bridge, Oregon, USA, 147, 148, 149
pylon, erection of the, suspension structures, 135–137, 229, 230
   McKenzie River Bridge, casting of the tower, 135
   McKenzie River Bridge, lifting of the tower, 135
   Vranov Lake Bridge, 219–220; construction sequences, of the erection of the tower, diagram,13; partial lifting of the tower, 136, 137
   Willamette River Bridge, 223; erection of the tower, 136

Radbuza River Bridge, Plzen, Czech Republic, 107–113
railing, 31, 33
Rayor and Strasky, 192
Redding Bridge, California, USA, 5, 6, 75, 85, 96, 98, 99, 120, 121, 122, 124, 187–188
Redfield, C., 188, 191
Redfield and Strasky, 187, 191
Reula Bridge, Spain, 214
Reyes Construction Inc., 230
RM2000 program, 98
Roberts, T.M., 34, 35
rock anchors, 81, 84–86, 220
Roebling, J., 7

roofs, suspended, 1, 2, 3, 4, 5
Rosenstein 11 Bridge, Stuttgart, Germany, 201–202
Rosewood Golf Club Bridge, Japan, 172, 251–252
Ruck a Chucky Bridge, California, USA, 171
   diagram of balancing vertical forces and balancing transverse forces, 172

Saarinen, A., 3
Saddledome structure, 3, 4
Schlaich, J., 7, 202, 228
Schlaich, Bergermann & Partners, 15, 25, 27, 131, 132, 134, 173, 202, 203, 215, 217, 227, 228, 229
Schlaich and Seidel, 173, 227
Schlaich *et al.*, 3
SCIA program, 63
Scotland
   Almond River Bridge, 10
   Hadrian Bridge, 104, 105
Scott, R., 7, 39
Segadaes Tavares & Partners, 3
Segre River, Spain, bridges over, 213–215
Seible, F., 241, 242
Seishun Bridge, Japan, 239–240
seismic design, 40
self-anchored structures, 13, 15–16, 17, 106, 107, 155, 158, 160, 206, 208, 209, 217, 220, 224, 229, 232, 234, 235, 237, 238, 239, 240
   self-anchored suspension structures, 127, 133, 134, 140–141, 142, 146: erection and service, diagram, 128
Shasta Constructors Inc., 188
Shimizu Corporation, 172, 252
Shin Nippon Giken Co. Ltd., 191
Shiosai Bridge, Japan, 25, 231–232, 233
Shizuoka Construction Technology Center, 232
shrinkage of concrete *see* creep and shrinkage of concrete
Siza, A., 3
SKANSKA 209, 246
Smerda and Kristek, 63
socket types, diagram, 26
Spain
   Basella Bridge, 214
   Glorias Catalanas Bridge, Barcelona, 172, 251
   Malecon Bridge, Madrid, 173
   Osormort Bridge, 158
   Peramola Bridge, 214, 215
   Reula Bridge, 214
   Segre River, bridges over, 213–215
star alignment of the stress ribbon, 104–106
   plan and developed elevation with pier, diagram, 105
   plan and developed elevation with tension ring, diagram, 106
static and dynamic analysis, cable-stayed structures, 164–169
   bending of the cable, influence of local strengthening and influence of support by a spring, diagram, 167
   bending moments and shear forces in stay cables, diagram, 166
   Ernst modulus, diagram, 165
   modelling the cable-stayed structure, deck modelled by 3D bars and deck modelled by shell elements, diagram, 164
   normal stresses in stay cables, diagram, 164
   redistribution of bending moments, diagram, 168

redistribution of bending moments in a two-span beam, diagram, 168
static function of the stay cable, diagram, 164
stresses in stay cables, allowable range of stresses, dead load, max tension and min tension, 165
static and dynamic analysis, stress ribbon structures, 90–101
  analysis of a two-span structure, diagram, 98
  cast-in-place stress ribbon structure cast on the falsework, diagram, 91
  forces in the bearing tendons during the erection, diagram, 96
  Grants Pass Bridge, Oregon, USA, calculation model of the pier, diagram, 101
  Grants Pass Bridge, Oregon, USA, calculation model of the structure, diagram, 101
  Grants Pass Bridge, Oregon, USA, natural modes and frequencies, diagram, 102
  loading, deformation and stresses of precast stress ribbon structure, diagram, 93
  loading and deformation of the stress ribbon, diagram, 91
  Maidstone Bridge, UK, bending moments in the deck, diagram, 103
  Maidstone Bridge, UK, calculation model, diagram, 102
  Maidstone Bridge, UK, natural modes and frequencies, diagram, 103
  precast stress ribbon structure, diagram, 92
  Redding Bridge, California, USA, analysis as a cable, diagram, 98
  Redding Bridge, California, USA, bending moments and stresses at a support haunch, diagram, 99
  stage of service, static function, diagram, 97
  static function, diagram, 95; stage of erection, diagram, 95
  stress ribbon stiffened by a dead load, diagram, 91
  stress ribbon structure, deformation and bending moments and modelling of the deck, diagram, 99
  typical natural modes, diagram, 100
static and dynamic analysis, suspension structures, 141–154
  analysis of suspension structures, 145–146
  examples of analysis, 149–151
  initial stage of earth-anchored suspension structure, 142–145
  loading tests, 151–154
  parametric study, 146–149
Stevenson, R., 10
stiffening of structures, 17–23, 35, 91, 156, 157, 158
  beam stiffening of the cable, diagram, 15
  cable stiffening, diagram, 14
  deflection and moments in cable-stayed structure due to live load, diagram, 22
  deflection and moments in suspension structure due to live load, diagram, 21
  deflection, moments and normal forces in stress ribbon structure due to live load, diagram, 20
  deflection, moments and normal forces in stress ribbon structure due to temperature drop, diagram, 20
  deformations of the suspension structure, diagram, 23
  relative deformations of suspension and cable-stayed structures, diagram, 22
  stiffness of the stress ribbon structure, diagram, 18; table of maximum deformation of structures, 19
  structures stiffened by external tendons, 113–119
  studied structures, diagram, 19
  tension stiffening, diagram, 23

Strasky, J., 1, 29, 161, 183, 188, 192, 194, 195, 212, 217, 222, 223, 224, 247, 249
Strasky, Husty and Partners, 63, 196, 201, 206, 209, 246, 255
Strasky and Anatech, 230
Strasky and Kompfner, 230
Strasky and Pirner, 52, 182
Strasky and Rayor, 222
Strasky *et al.*, 63
stress ribbon bridge, use of term, 1
stress ribbon structures, 71–126, 139, 140, 145, 153
  curved stress ribbon bridge, 175–178
  erection of the deck, 86–90; battered micropiles, diagram, 87; construction sequences A, diagram, 87; construction sequences B, diagram, 90; drill shafts, diagram, 86; drill shafts and ground anchors, diagram, 87; erection of a segment, diagram, 88
  examples, 179–213; Bircherweid Bridge, 179; Blue Valley Ranch Bridge, 191–192; DS-L Bridges, 182–186 (Brno-Komin Bridge, 182, 183, 184; Prague-Troja Bridge, 182–183, 184, 185–186, 187); Freiburg Bridge, 180–181; Grants Pass Bridge, 192–194; Kikko Bridge, 196–198; Lake Hodges Bridge, 194–195; Lignon-Löex Bridge, 179–180; Maidstone Bridge (Kent Messenger Millennium Bridge) 198–201; McLoughlin Boulevard Bridge, 210–212; Morino-Wakuwaku Bridge, 212–213, 214; Olse River Bridge, 195–196; Olomouc Bridge (over R35 expressway), 204–206; Phorzheim 111 Bridge, 202–203; Punt da Suransuns Bridge, 203–204, 205; Redding Bridge, 187–188; Rosenstein 11 Bridge, 201–202; Svratka River Bridge, 206–209; Tokimeki Bridge, 209–210; Tonbo No Hashi Bridge, 191; Umenoki-Todoro Park Bridge, 188–189; Yumetsuri Bridge, 189, 190–191
  piers and abutments, 79–81, 82–84
  prestressed band, 72–79
  special arrangements, 101–106; cranked alignment of the stress ribbon, 104, 105; star alignment of the stress ribbon, 104–106; stress ribbon supported by inclined struts, 101, 103; stress ribbon suspended on stay cables, 104
  static and dynamic analysis, 90–101; analysis of the structure as a cable, 94–96 (erection stage, 94–95; service stage, 96); analysis of the structure as a geometrically non-linear structure, 96–99; designing structural members, 100–101; dynamic analysis, 99–100; examples of the analysis, 100; static function, 90–94
  static and dynamic loading tests, 119–126; dynamic tests, 119, 120–126; static tests, 119–120
  stress ribbon supported by arch, 106–113; dynamic model, 112–113; model test, 107–108; static model, 108–112; structural arrangement, 106–107
  structural arrangement, 71–72; geometry, 71–72; structural members, 72
  structures stiffened by external tendons, 113–119; diagram of cross section, 116; diagram of elevation and plan, 116; diagram of cross section and elevation, 114; diagram of deformation of stress ribbon, load on one half of length, load of whole length, temperature changes, 114, 117; diagram of natural modes and frequencies, 115; diagram of structural arrangement of stress ribbon, cross sections, 118; photographs of cross section, elevation and view, 118–119; table of number bearing and prestressing strands, 117, 118

stress ribbon structures (*continued*)
  transferring stress ribbon force to the soil, 81, 84–86; Maidstone Bridge, UK, abutment, elevation and cross-section, diagram, 84; rock anchors, diagram, 85, 86
Stripps Crossing at UCSD, USA, 241–242
Strömsund, Sweden, 8
structural members, 23–30, 72
  anchoring of cables: diagram, 26, 27; photograph (towers and struts), 28
  composite and prestressed concrete, diagram, 28
  deck, diagram, 24
  grouting of stay cables, 29
  grouting of suspension cables, diagram, 29
  static function, diagram, 72
  tension members, cables developed by steel industry and cables developed by pre-stressed concrete industry, diagram, 26
  tower, diagram, 24
structural systems, 13–23
  beam stiffening of the cable, diagram, 15
  cable and arch structure, diagram, 13
  cable stiffening, diagram, 14
  cable-supported structures, diagram, 15; long-span cable-supported bridges, diagram, 17
Studnickova, 122
Sumitomo Mitsui Construction Co, Ltd., 140, 189, 191, 196, 198, 210, 231, 232, 233, 234, 235, 240
Sunniberg Bridge, Switzerland, 156
  diagrams: calculation model and cross-section of deck; cross-section, longitudinal section, transverse bending moments and longitudinal bending moments, 157
surfacing, 32
suspension structures, 127–154, 175
  erection of the structures, 134–141
  examples, 213–240; Ayumi Bridge, 234–235; Deutsches Museum, Munich, Germany, pedestrian bridge built inside, 227–229; Ganmon Bridge, 232–234; Halgavor Bridge, 224, 226–227; Harbor Drive Bridge, 229–231; Inachus Bridge, 236–237, 238; Ishikawa Zoo Bridge, 231, 232; Johnson Creek Bridge, 237–239; Kelheim Bridge, 227, 228; Max-Eyth-See Bridge, 215–217; McKenzie River Bridge, 223–224, 225; Phorzheim 1 Bridge, 217, 218; Segre River, bridges over, 213–221; Seishun Bridge, 239–240; Shiosai Bridge, 231–232, 233; Tobu Bridge, 235–236; Vranov Lake Bridge, 217–222; Willamette River Bridge, 222–223;
  static and dynamic analysis, 141–154
  structural arrangement, 127–134
Svratka River Bridge, Brno, Czech Republic, 36, 206–209
Sweden, Strömsund, 8
Switzerland
  Bircherweid bridge over motorway N3, near Pfäfikon, 4, 179
  Diepoldsau Bridge, 160–161, 169
  Fribourg Bridge, 8
  Lake Geneva, bridge over, 1
  Lignon-Löex Bridge, Geneva, 179–180
  Punt da Suransuns Bridge, 203–204, 205
  Sunniberg Bridge, 156

Tacoma Bridge, USA, 7, 39

Tanaka *et al.*, 209
TDA program, 63, 67
TDV, 98
Technical University of Brno, Czech Republic, 63, 66
temperature, 20, 106, 114, 117, 132, 133, 167, 169
tension members, 3, 26, 27, 28, 35
time dependent analysis, 63–66, 169
Tobu Bridge, Japan, 25, 28, 141, 235–236
Tokimeki Bridge, Japan, 209–210
Tomishenko and Goodier, 50
Tonbo No Hashi Bridge, Japan, 191
towers, 22, 24–25, 27, 28, 132, 134, 135–137, 145, 151, 157, 159, 214, 216, 218, 221, 224, 225, 243, 244, 247, 251, 252, 254
Toyo Ito & Associates, 236
Troyano, L.F., 1
Troyano and Mantreola, 173
Troyano *et al.*, 214
Tsunomoto and Ohnuma, 235
Turkey, Bosporus bridge, 1, 2
TY Lin International, 171, 172, 195, 230

UK
  Halgavor Bridge, 224, 226–227
  Hungerford Bridge, London, 25, 27, 28, 160, 242–244
  Kent Messenger Millennium Bridge, Maidstone, 33, 36, 76, 83, 84, 90, 100, 102, 103, 104, 122, 198–201
  Lockmeadow Bridge, Maidstone, Kent, 249–250
  see also Scotland; Wales
Umenoki-Todoro Park Bridge, Japan, 188–189
USA
  arch, natural, Utah, 14
  Blue Valley Ranch Bridge, Colorado, 99, 191–192
  Brooklyn Bridge, New York, 7, 8, 17
  Delta Pond Bridge, Oregon, 37, 163, 247–249
  Dulles International Airport terminal, Washington DC, 1, 2, 34–39, 39–40, 71
  Gateway Bridge (1–5), Oregon, 37, 162–163, 246–247
  Grants Pass Bridge, Oregon, 5, 32, 71, 76, 82, 83, 88, 89, 100, 101, 102, 122, 124, 192–194, 239
  Johnson Creek Bridge, Oregon, 37, 141, 153, 237–239
  Lake Hodges Bridge, San Diego, California, 36, 194–195
  McKenzie River Bridge, Oregon, 28, 130, 131, 133, 135, 137, 138, 139, 223–224, 225
  McLouglin Boulevard Bridge, Oregon, 37, 210–212
  Redding Bridge, California, 5, 6, 75, 85, 96, 98, 99, 120, 121, 122, 124, 187–188
  Ruck a Chucky Bridge, California, 171, 172
  Stripps Crossing at UCSD, 241–242
  Tacoma Bridge, 7, 39
  Virginia Railroad Bridge, 15
  Willamette River Bridge, Oregon, 23, 24, 37, 131, 132, 133, 136, 138, 150, 151, 152, 222–223

Varrugas Bridge, Peru 10
Verescagin rule, 44
vibrations, physiological effect of, 34–39, 99–100, 147–148, 177, 186, 188, 192, 224, 226–227
  acceleration of analysed footbridges, 38

lock-in effect, 35–39
pacing and jumping frequencies: Hz, 34
psychological classifications, 34
results of analysis, table, 36–38
typical natural modes, 35
vertical and horizontal load, 35
vertical vibration, 34–35
vibration due to wind, 39
Virginia Railroad Bridge, USA, 15
Vltava River Bridge, Ceske Budejovice, Czech Republic, 36, 38
Völkel *et al.*, 10, 240
vortex shedding, 39
Vranov Lake Bridge, Czech Republic, 8, 9, 10, 37, 127, 132, 133, 136, 139, 140, 149, 150, 151, 152, 217–222

Wales, Menai Straits Bridge, 7
Walther, R., 4, 10, 34, 39
Walther *et al.*, 10, 34, 39, 161
Weisz, B., 180
Wenaweser, B., 180

Werrekuss Bridge, Germany, 15
Weser River Bridge, Germany, 131
West Wind Labs, 194
Wildish Company, 223
Willamette River Bridge, Oregon, USA, 23, 24, 37, 131, 132, 133, 136, 138, 150, 151, 152, 222–223
wind, 34, 39–40, 99, 112, 113, 150, 194, 226
  flutter, 39
  vibration due to, 39
wind tunnel tests, 122–126, 152, 234, 244
Winkler's springs, 48
Wittfoht, H., 1, 127
Wolfensberger, R., 179, 180
World War II, 8, 242
WSP, 244

Yumetsuri Bridge, Japan, 189, 190–191

Židlochovice Bridge, Czech Republic, 157, 158, 161, 162
Zoo, Köln, 1